# Recombinant DNA and Biotechnology

# Recombinant DNA and Biotechnology

Contributors

**Vitali Alexeev, Alyson Pidich et al.**

**AURIS**
Reference

**www.aurisreference.com**

# Recombinant DNA and Biotechnology

Contributors: Vitali Alexeev, Alyson Pidich et al.

**Published by Auris Reference Limited**
www.aurisreference.com

United Kingdom

**Recombinant DNA and Biotechnology**

ISBN: 978-1-78154-964-3

British Library Cataloguing in Publication Data
A CIP record for this book is available from the British Library

Printed in the United Kingdom

Exclusively distributed by CBS Publishers & Distributors Pvt. Ltd.

Sales & Distribution Rights only for India, Pakistan, Bangladesh, Sri Lanka, Nepal and Bhutan. This book is not to be sold outside these territories.

# Contents

# List of Abbreviations

| | |
|---|---|
| AAV | Adeno-Associated Virus |
| AIDS | Acquired Immunodeficiency Syndrome |
| BCS | Bovine Chorionic Somatomammotropin |
| CDR | Complementarity Determining Regions |
| CPD | Cyclobutane Pyrimidine Dimer |
| CS | Chorionic Somatomammotropin |
| CS | Cockayne Syndrome |
| DC | Dendritic Cells |
| DSB | Double Strand Break |
| GST | Glutathione S-Transferase |
| HCS | Human Chorionic Somatomammotropin |
| HR | Homologous Recombination |
| IGS | Inter-Genic Spacer |
| IRES | Internal Ribosomal Entry Site |
| ITS | Internal Transcribed Spacer |
| MHC | Major Histocompatability Complex |
| NER | Nucleotide Excision Repair |
| NES | Nuclear Export Signals |
| PCR | Polymerase Chain Reaction |
| RPE | Retinal Pigment Epithelium |
| SLIC | Sequence and Ligation Independent Cloning |
| SSA | Single Strand Annealing |
| TAA | Tumor-Associated Antigen |
| TCR | T Cell Receptor |
| USER | Uracil-Specific Excision Reagent |
| XLMR | X-Linked Mental Retardation |
| XP | Xeroderma Pigmentosum |

# List of Contributors

**Vitali Alexeev**
Department of Dermatology and Cutaneous Biology, Jefferson Medical College, Thomas Jefferson University, Philadelphia, Pennsylvania, USA

**Alyson Pidich**
Temple University School of Medicine, Philadelphia, Pennsylvania, USA

**Daria Marley Kemp**
Department of Dermatology and Cutaneous Biology, Jefferson Medical College, Thomas Jefferson University, Philadelphia, Pennsylvania, USA

**Olga Igoucheva**
Department of Dermatology and Cutaneous Biology, Jefferson Medical College, Thomas Jefferson University, Philadelphia, Pennsylvania, USA

**Jorge Angel Ascacio-Martínez**
Department of Biochemistry and Molecular Medicine, School of Medicine, Autonomous University of Nuevo León, Monterrey Nuevo León, Av. Madero Pte. s/n Col. Mitras Centro, Monterrey, N.L., México

**Hugo Alberto Barrera-Saldaña**
Department of Biochemistry and Molecular Medicine, School of Medicine, Autonomous University of Nuevo León, Monterrey Nuevo León, Av. Madero Pte. s/n Col. Mitras Centro, Monterrey, N.L., México

**Kei Adachi**
Department of Microbiology and Molecular Genetics, USA
University of Pittsburgh School of Medicine, USA

**Hiroyuki Nakai**
Department of Microbiology and Molecular Genetics, USA
University of Pittsburgh School of Medicine, USA

**Carolina Quayle**
Dept. of Microbiology, Institute of Biomedical Sciences, University of Sao Paulo, Brazil

**Carlos Frederico Martins Menck**
Dept. of Microbiology, Institute of Biomedical Sciences, University of Sao Paulo, Brazil

**Keronninn Moreno Lima-Bessa**
Dept. of Cellular Biology and Genetics, Institute of Biosciences Federal University of Rio Grande do Norte, Brazil

**Bhupal Ban**
Department of Biochemistry and Molecular Biology, Tulane University School of Medicine, New Orleans, Louisiana, USA

**Diane A. Blake**
Department of Biochemistry and Molecular Biology, Tulane University School of Medicine, New Orleans, Louisiana, USA

**Zhenyu Shi**
School of Chemistry, University of Melbourne, Parkville, Victoria, Australia, Bio21 Molecular Science and Biotechnology Institute, Parkville, Victoria, Australia

**Anthony G. Wedd**
School of Chemistry, University of Melbourne, Parkville, Victoria, Australia, Bio21 Molecular Science and Biotechnology Institute, Parkville, Victoria, Australia

**Sally L. Gras**
School of Chemistry, University of Melbourne, Parkville, Victoria, Australia, Bio21 Molecular Science and Biotechnology Institute, Parkville, Victoria, Australia

**Changrim Lee**
Department of Biochemistry, College of Life Science & Biotechnology, Yonsei University, Seoul, Republic of Korea

**Seokbong Hong**
Department of Biochemistry, College of Life Science & Biotechnology, Yonsei University, Seoul, Republic of Korea

**Min Hye Lee**
Department of Biochemistry, College of Life Science & Biotechnology, Yonsei University, Seoul, Republic of Korea

**Hyeon-Sook Koo**
Department of Biochemistry, College of Life Science & Biotechnology, Yonsei University, Seoul, Republic of Korea

**Lisa Mathiasen**
FIRC (Foundation for Italian Cancer Research) Institute of Molecular Oncology (IFOM), via Adamello 16, 20139, Milan, Italy

**Chiara Bruckmann**
FIRC (Foundation for Italian Cancer Research) Institute of Molecular Oncology (IFOM), via Adamello 16, 20139, Milan, Italy

**Sebastiano Pasqualato**
Crystallography Unit, Department of Experimental Oncology, European Institute of Oncology, Via Adamello 16, Milan, 20139, Italy

**Francesco Blasi**
FIRC (Foundation for Italian Cancer Research) Institute of Molecular Oncology (IFOM), via Adamello 16, 20139, Milan, Italy

**Adam D. Brown**
Department of Cellular and Structural Biology, University of Texas Health Science Center at San Antonio, San Antonio, Texas, United States of America
Greehey Children's Cancer Research Institute, University of Texas Health Science Center at San Antonio, San Antonio, Texas, United States of America

**Brian W. Sager**
Department of Cellular and Structural Biology, University of Texas Health Science Center at San Antonio, San Antonio, Texas, United States of America
Greehey Children's Cancer Research Institute, University of Texas Health Science Center at San Antonio, San Antonio, Texas, United States of America

**Aparna Gorthi**
Department of Cellular and Structural Biology, University of Texas Health Science Center at San Antonio, San Antonio, Texas, United States of America
Greehey Children's Cancer Research Institute, University of Texas Health Science Center at San Antonio, San Antonio, Texas, United States of America

**Sonal S. Tonapi**
Department of Cellular and Structural Biology, University of Texas Health Science Center at San Antonio, San Antonio, Texas, United States of America
Greehey Children's Cancer Research Institute, University of Texas Health Science Center at San Antonio, San Antonio, Texas, United States of America

**Eric J. Brown**
Abramson Family Cancer Research Institute, Department of Cancer Biology, University of Pennsylvania School of Medicine, Philadelphia, Pennsylvania, United States of America

**Alexander J. R. Bishop**
Department of Cellular and Structural Biology, University of Texas Health Science Center at San Antonio, San Antonio, Texas, United States of America

Greehey Children's Cancer Research Institute, University of Texas Health Science Center at San Antonio, San Antonio, Texas, United States of America
Cancer Therapy and Research Center, University of Texas Health Science Center, San Antonio, Texas, United States of America

**Joseph Edward Ironside**
Institute of Biological, Environmental and Rural Sciences, Aberystwyth University, Aberystwyth, United Kingdom

# Preface

Recombinant DNA (rDNA) molecules are DNA molecules formed by laboratory methods of genetic recombination (such as molecular cloning) to bring together genetic material from multiple sources, creating sequences that would not otherwise be found in the genome. Recombinant DNA is possible because DNA molecules from all organisms share the same chemical structure. They differ only in the nucleotide sequence within that identical overall structure. There are three different methods by which Recombinant DNA is made. They are Transformation, Phage Introduction, and Non-Bacterial Transformation. This book, Recombinant DNA and Biotechnology, emphasizes major techniques of product development and manufacturing in biotechnology are discussed, with a focus on recombinant DNA technology. For recombinant-DNA technology, the issues of gene isolation, gene cloning, protein expression, scale-up (manufacturing), and quality assurance are addressed. First chapter focuses on recombinant DNA technology in emerging modalities for melanoma immunotherapy. Second chapter gives an approach on genetic engineering and biotechnology of growth hormones. In third chapter, we provide an overview of how wtAAV and rAAV alter the fate of the host cells through DDR, and how DDR processes the viral genomic DNA by exerting DNA repair machinery to establish the lytic and latent life cycles of wtAAV and transduction of rAAV. Fourth chapter mainly focuses on the XP syndrome, deficiencies in NER can also lead to other genetic diseases, such as trichothiodystrophy (TTD), Cockayne syndrome (CS), cerebro–oculo– facial–skeletal syndrome (COFS) and UV-sensitive syndrome (UVsS), all of which have photosensitivity as a common feature. Fifth chapter overviews on recombinant antibodies and nonantibody scaffolds for immunoassays. Immunoassays are frequently applied to the analysis of both low molecular ligands and macromolecular drugs, and are also applied in such important areas as the quantitation of biomarkers that indicate disease progression and immunogenicity of therapeutic drug candidates. A new building-brick-style parallel DNA assembly framework for simple and flexible batch construction is presented in sixth chapter. In seventh chapter, we examined the roles of two PHF8 homologs, JMJD-1.1 and JMJD-1.2, in the model organism *C. elegans* in response to DNA damage. In eighth chapter, we report the production of large amounts of soluble and pure recombinant human PBX1:PREP1 complex in an active form capable of binding DNA. In ninth chapter, we provide evidence that retinal pigment epithelium (RPE) cells lacking ATR have decreased density with abnormal morphology, a decreased frequency of HR and an increased level of chromosomal damage. Last chapter focuses on diversity and recombination of dispersed ribosomal DNA and protein coding genes in microsporidia. recombinant antibodies and non-antibody scaffolds for immunoassays.

# Chapter 1

# RECOMBINANT DNA TECHNOLOGY IN EMERGING MODALITIES FOR MELANOMA IMMUNOTHERAPY

Vitali Alexeev[1], Alyson Pidich[2], Daria Marley Kemp[1], and Olga Igoucheva[1]

[1]Department of Dermatology and Cutaneous Biology, Jefferson Medical College, Thomas Jefferson University, Philadelphia, Pennsylvania, USA

[2]Temple University School of Medicine, Philadelphia, Pennsylvania, USA

## INTRODUCTION

The history of immunotherapy of cancer dates back to 1890s when New York surgeon William Coley used Streptococcus and Serratia bacterial extracts to treat cancer. Up to the mid-1930s 'Coley's mixed toxins," were used to treat various tumors. Better understanding of the human immune system led to the identification of a number of tumor-associated antigens (TAAs) in the 1980s [1] and development of various immunotherapeutic approaches. Of particular relevance to melanoma immunotherapy was the identification of various antigens expressed specifically in melanocytes and, respectively, in the majority of melanomas. These melanoma-associated antigens include tyrosinase (Tyr), a key enzyme in melanin biosynthesis, tyrosinase-related proteins 1 and 2 (TRP1, TRP2), gp100 (aka pmel17), Melan-a, and MART1. These and several other melanoma-associated antigens formed the basis for the immunologic targeting of the tumor. Up to date, multiple peptide, dendritic cell, adjuvant, lymphocyte, antibody, DNA and virus-based strategies were tested in pre-clinical and clinical studies with varying degrees of success. In recent years, identification of the specific antigenic MHC class I epitopes, advancements in genetic engineering, gene delivery, and cell-based therapeutic approaches allowed development of the novel melanoma-targeting immunotherapeutics.

# GENETIC ENGINEERING OF ANTIGEN-SPECIFIC T CELLS

## Recombinant T Cell Receptors

Identification of the tumor-reactive T cells among a population of the tumor-infiltrating lymphocytes led to the development of the T cell-based therapies, particularly to the strategy known as adoptive T cells transfer. This strategy is based on the isolation of the tumor-infiltrating lymphocytes following analysis of their ability to target tumor cells and clonal expansion of tumor-reactive T cells via stimulation of cell proliferation with anti-CD3 and antiCD28 antibodies in the presence of IL-2. Upon obtaining a large quantity ($> 10^8$ cells), these cells are infused back to a tumor-bearing patient along with the lymphodepleting chemotherapy to temporary knock down circulating immunocytes and repetitive administration of the IL-2 (Fig. 1).

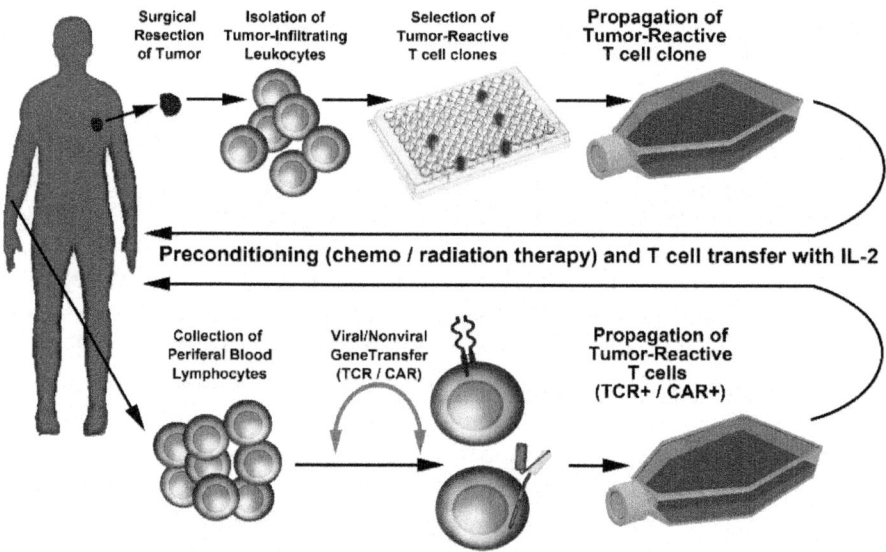

**Figure 1.** Clinical application of the T cell-mediated tumor immunotherapy. Diagram on the top depicts application of the Tumor-Infiltrating Lymphocytes (TILs). Diagram on the bottom illustrates application of the genetically engineered (TCR and CAR-modified) T cells.

Presently, 87 clinical trials using TIL are completed or on-going. These clinical trials are aimed at treatment of multiple cancers including: Malignant Melanoma, Nasopharyngeal Carcinoma, Hepatocellular Carcinoma, Breast Carcinoma, Leukemia, Lymphoma, Multiple Myeloma, Plasma Cell Neoplasm,

Kidney Cancer, Metastatic Colorectal Cancer, Metastatic Gastric Cancer, Metastatic Pancreatic Cancer, Metastatic Hepatocellular Carcinoma, Cervical Cancer, Oropharyngeal Cancer, Vaginal Cancer, Anal Cancer, Penile Cancer, Non-Small Cell Lung Cancer, Brain and Central Nervous System Tumors. Several completed clinical trials on malignant melanoma clearly demonstrated that infusing TILs along with IL-2 and pre-conditioning with reduced-intensity circulating lymphocyte-depleting chemotherapy mediates tumor-targeting immune response in up to 50% of patients [2]. The highest response rate up to 70 % with up to 30% complete remission lasting for up to 3 years was reported when radiation sensitization was used in conjunction with the transfer of the tumor-reactive TILs.

Despite the success of the pioneering studies at the Surgery Branch of the US National Cancer Institute and the consequent clinical trials, this approach, although holding much promise in treating melanoma, is facing several challenges that limit broad application of the TIL-based immunotherapy. As TILs are isolated from resected tumors, the first and foremost requirement is the eligibility for surgery, which should be conducted, preferably, in the facility equipped for the isolation of TILs, identification and expansion of the tumor-reactive T cells. *Ex vivo* stimulation and propagation of TILs to large quantities required for the effective immunotherapy is time-consuming, labor-retaining, and expensive. Although recent clinical studies showed that infusion of the minimally cultured TILs without pre-selection for tumor reactivity provide a rather high response rate [3], the search for a better melanoma-targeting strategy is on-going.

Nevertheless, isolation of the individual melanoma-reactive T cell clones allowed the development of another immunotherapeutic approach – generation of the T cells expressing recombinant antigen specific T cell receptors (TCRs). TCRs are members of the immunoglobulin family proteins. Each TCR consists of 2 different membrane-anchored chains that are joined by the disulfide bridges to form heterodimer. About 95% of the T cells are characterized by the expression of the α and β chains, whereas the remaining 5% express γ and δ chains. Respectively, T cells expressing these receptors are often referred to as α/β and γ/δ T cells. Each chain is comprised of the variable and constant regions. The variable domain of both α- and β-chains have three hypervariable regions also known as complementarity determining regions (CDR), however, the β-chain has an additional area of hypervariability that is not involved in antigen binding. TCR α and γ chains are generated within T cells by VJ recombination, whereas β and δ chains by the V(D)J recombination. Currently, the majority of the TILs selected for the ability to target tumors are α/β T cells expressing respective TCR chains that determine T cell specificity to an antigenic peptides presented by the major histocompatability complex (MHC) proteins. Therefore, it was proposed that sequences encoding tumor antigen recognizing TCR chains can be obtained from tumor-reactive T cells and then used for the gene transfer into patient-derived lymphocytes, thereby creating large quantities of tumor-reactive T cells. The first TCRs specific to melanocytic antigens MART-1 and gp-100 were cloned in 1990s. Pioneering clinical studies using human peripheral blood lymphocytes transduced with these TCRs demonstrated melanoma regression in lymphodepleted patients [4] (Fig. 1). Although these and other initial clinical studies demonstrated a feasibility of the recombinant T cells-based approach, they also revealed multiple challenges. For example, the ability of recombinant TCR chains to interact and pair with the endogenous chains could lead to the generation of 4 different TCRs in a single cell (Fig. 2). Chain misparing decreases the expression of the function, tumor-reactive TCRs and therefore reduces T cell-mediated tumor targeting. To overcome misparing, several strategies were proposed. Recent pre-clinical and clinical studies demonstrated that replacement of the human TCR constant region with murine counterpart reduced misparing and allowed enhanced expression of the functional TCRs and improved T cell functional activity [5]. It was also reported that targeted mutagenesis and generation of the additional cystein residues in recombinant α and β chains permitted stronger pairing of these chains, higher expression of functional TCRs and improved T cell function [6,7]. Recent studies also showed that targeting of the endogenous chains by siRNA allows higher expression of the functional recombinant TCR. Of particular interest is the proposed approach to encode siRNA along with the TCR chains to concurrently express recombinant and inhibit translation of the endogenous chains [8]. Protein engineering was also employed to improve pairing of the recombinant chains. Thus, substitution of specific amino acids within constant regions of the antigen-specific TCRs supported paring and enhanced functional activity of these receptors [9]. It remains to be determined which of these recombinant DNA-based methods will provide better targeting of melanoma (Fig. 2). Nevertheless, recent studies using chimeric murine-human hybrid highly avid tyrosinase-specific TCR demonstrated a favorable clinical outcome [10].

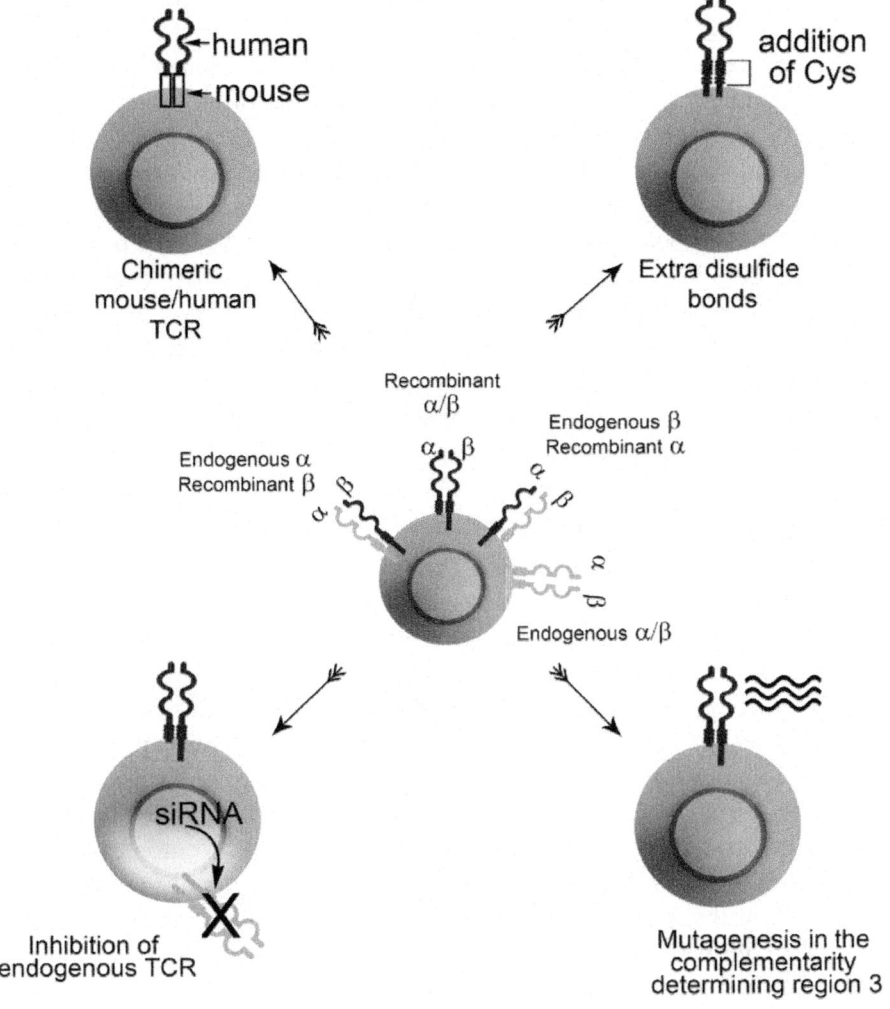

**Figure 2.** Strategies aimed at the improvement of the recombinant TCR pairing. Expression of the recombinant TCR may lead to the generation of 4 different TCRs within a cell (center). Different strategies designed to improve tumor-specific recombinant TCR pairing and activity include: generation of hybrid molecules containing the constant region from murine TCR, addition of disulfide bonds, alteration of the amino acid sequence within the TCR chains, and siRNA-mediated inhibition of the endogenous TCR gene expression (see text for details).

It is apparent that both $\alpha$ and $\beta$ chains of the antigen-specific TCR should be expressed in each individual T cell. To date, internal ribosomal entry site (IRES) elements [11], double promoters [12], or co-infection with several viral vectors [13] were used to express several heterologous proteins in cells. However, these methods have their imperfections. For instance, in IRES-mediated co-expression, the upstream protein is usually more strongly transcribed than the downstream protein. Expression of the proteins from two different or biscistronic promoters or the use of multiples viruses also do not provide equal concurrent expression of multiple transgenes. A more promising approach involves the use of the self-processing viral peptide bridges such as 2A or 2A-like peptides described in Picornaviridae [14]. In picornavirus, these sequences share a highly conservative 18 amino acids motif mediating cleavage between C-terminal glycine and N-terminal proline of the 2B peptide. At present 2a and 2A-like sequences are commonly refer to as *cis*-acting hydrolase elements that allows ribosome skipping and cellular expression of multiple, discrete proteins in essentially equimolar quantities derived from a single ORF. To ensure concurrent expression of both α and β chains of the transgenic TCR an A2 sequence is most commonly used for quantitative co-expression of these heterologous proteins.

Transfer of the recombinant TCR genes into the T cells is another somewhat limiting factor for the broad application of the genetically engineered T cells for melanoma immuno-targeting. Currently, for human applications, a gene transfer platform that can mediate stable transfer of the TCR genes is retroviral system [15]. Lentiviral vectors and transposons were also tested [16, 17]. Use of retroviruses provided several advantages including a rather high infectivity and rapid integration of the transgene into host genome. With multiple vector backbones, virus packaging cell lines, and well-established GMP protocols, a retroviral system offers relative simplicity of viral vector construction and production of viruses. Since retroviruses can infect only dividing cells, stimulation of the T cell proliferation must be done prior to the gene transfer. Also, these viruses have limited capacity for the packaging. For instance, high virus titers cannot be obtained with larger retroviral vectors. Although an average size of a viral vector encoding typical α/β TCR is around 7 kb, this limits possible alternative approaches such as inclusion of various regulatory elements or another transgenes that may enhance T cell activation. Use of the viral system also presents certain safety concerns relevant to the random integration of the transgenes into the host genome that may result in the activation of oncogenes or inactivation of tumor suppressors. This may lead to the various adverse invents including development of a lymphoproliferative disease resembling leukemia due, in part, to the integration of the retroviral gene transfer vehicle near an oncogene [18, 19]. Thus far, the development of lymphoma-like symptoms has not been reported in patients treated with recombinant T cells. It is also essential to note that production of the TCR-encoding cGMP virus substantially increases the cost of the treatment with recombinant T cells. On the contrary to the retrovirus-based gene transfer, lentiviruses can infect non-dividing cells and therefore can be used for the gene transfer into quiescent T cells. Although "safe" lentiviral systems are developed to minimize the chance of producing replication-competent virus (eg. ViaraSafe from Cell Biolabs), transduction of patient-derived T cells for the adoptive transfer will always present some degree of risk.

Besides viral approaches, non-viral gene transfer may also be used for the expression of the TCRs in T cells. Recently, a Sleeping Beauty Transposon System was tested for the transduction of the T cells [17]. Sleeping Beauty Transposon System consists of two components - the transposon, composed of inverted terminal repeat sequences (IRs) with desired genetic material in between, and a SB transposase enzyme. Most recently, a number of IRs and hyperactive transposases with increasing enzymatic activities were developed to mediate transposition of transposon-encoding proteins into the genomic DNA [20]. Although transposition of SB transposons appears to be unregulated, it has certain advantages over viral based approaches. For instance, expression of transgenes, TCRs in particular, could be regulated by specific promoters that provide either T cell specific expression (eg. CD3 promoter; [21, 22]), or high level of expression (eg. elongation factor 1 promoter; [23,24]. Promoters may also be selected for further specific applications (discussed below). On the contrary to the viral gene transfer, non-viral systems also permit significantly simpler production of the cGMP-grade material (plasmid DNA) and lesser safety testing. Up to date, the Sleeping Beauty transposon-mediated approach was shown to mediate a long-term stable integration of the T-cell receptor genes targeting melanoma-derived antigen, MART-1, in laboratory settings (Fig. 3b). This system provided 50% efficiency of the TCR integration into the genome of the T cells and sustained functional reactivity of lymphocytes to the antigen-positive melanoma [25].

Other non-viral strategies could be useful in genetic engineering of the T cells. For example, integrase-mediated insertion of the genetic material may provide stable, site-directed integration of the transgenes (TCRs) into T cell genome. This strategy involves integrase from the *Streptomyces* phage ΦC31 that catalyzes unidirectional recombination between attP motifs in phage and attB sites in bacterial genomes. Usually attP and attB sites are cleaved and joined to each other, generating two hybrid sequences (attL and attR) that flank the integrated phage genome. However, ΦC31 integrase can also recognize several endogenous sequences in eukaryotic chromosomes as attP sites and integrate attB-bearing transgenes into them (Fig. 3c). Such pseudo attP sites were found in every mammalian genome with more than 100 ΦC31 integration sites identified in human cells. Thus far, only three preferred sites located in human Xq22.1, 8p22, 19q13.31 loci are commonly used by this enzyme [26, 27]. Therefore, ΦC31-integrase-based system is somewhat similar to the SB transposone system (Fig. 3b, c). Yet, it provides better specificity of the transgene integration. We recently tested whether ΦC31 can efficiently integrate transgenes into the T cells. Our initial data using GFP-encoding reporter plasmid with short (34bp) attB site demonstrated that nucleofection reaction provides rather efficient transduction of the transgene and ΦC31 integrase-encoding plasmids into T cells (Fig. 4a) and stable, integrase-dependent insertion of the reporter into both CD4+ and CD8+ T cells (Fig. 4b). Transduction of the T cells with tyrosinase-specific TCR (described in 10) ligated into the attB-harboring mammalian expression vector also resulted in the sustained expression of this melanoma-specific TCR and the ability of the T cells to target antigen-positive melanoma cells *in vitro* (Fig. 4c)

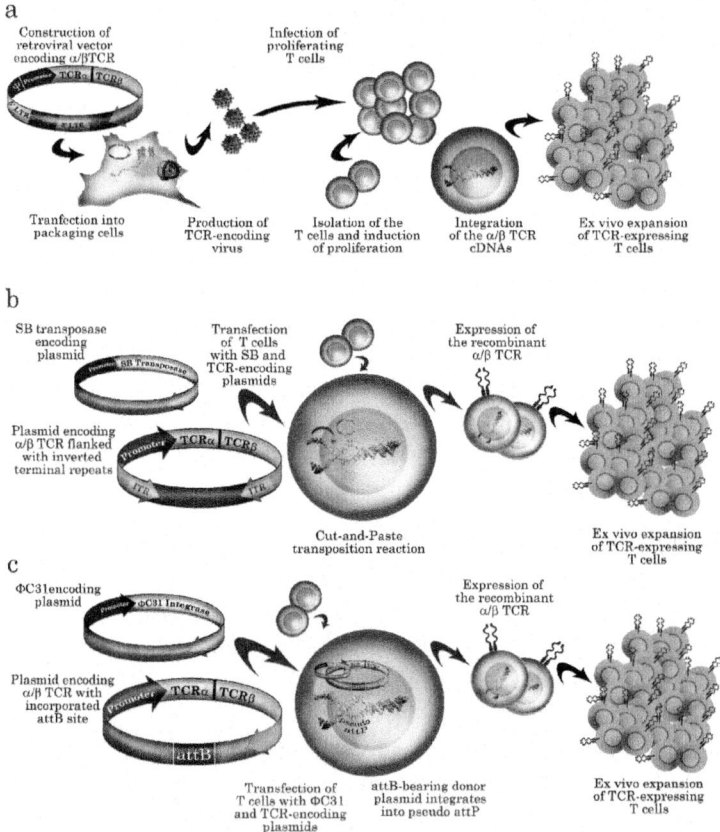

**Figure 3.** Schematic diagram depicting genetic engineering of the tumor-targeting T cells expressing recombinant TCR. Diagrams depict generation of the recombinant T

cells via (a) retrovirus-mediated gene transfer, (b) Sleeping Beauty transposon-mediated gene transposition, and (c) ΦC31 integrase-mediated gene insertion (see text for details).

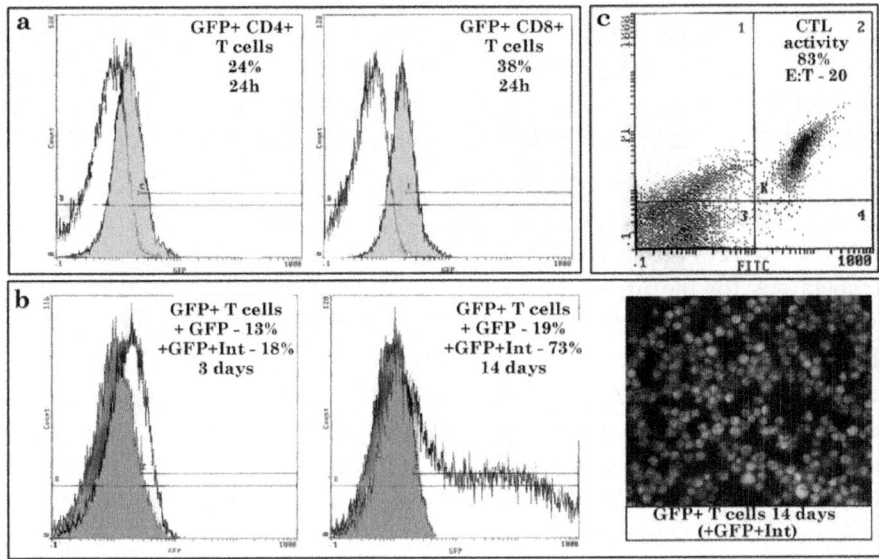

**Figure 4.** ΦC31 integrase-mediated genetic modification of the T cells. (a) Example showing typical nucleofection reaction of the reporter gene (GFP) into freshly isolated T cells 24 hours post Nucleofection (representative FACS profiles). (b) Analysis of the GFP expression after Nucleofection of the T cells in the presence of absence of the ΦC31 integrase-encoding plasmid (representative profiles and the direct fluorescent microscopy) 3 and 14 days after Nucleofection. (c) FACS-based analysis of the CTL activity of the recombinant T cells generated to express tyrosinase-specific TCR against human melanoma in vitro (representative profile). Times and percentages are indicated in the panels; Shaded Profiles – GFP-positive cells (see text for details).

Collectively, viral and non-viral strategies for the genetic engineering of the T cells expressing melanoma-specific TCRs are suitable for the *ex vivo* production of large quantities (more than $10^8$ cells) of the tumor-specific T cells that can be used for the adoptive T cell transfer. However, clinical utility of the non-viral approaches remains to be elucidated. In spite of T cell transduction strategy, it is clear that the ability to generate melanoma-specific recombinant T cell receptors allowed significant advancement in the development of the clinically-applicable TCR-based approach for melanoma immunotherapy. Its primary advantages are in the use of a natural and a rather well-understood mechanism of the T cell function and the ability to select/generate multiple melanoma-reactive TCRs that can be used alone or in combination. Currently, several melanoma-targeting TCRs specific to tyrosinase, MART-1, and gp100 were cloned. One can envision generation of TCR-encoding cDNA banks that could be utilized for the generation of different melanoma-reacting T cells from the pool of patient-derived T cells to target several TAAs. However, this strategy has several disadvantages including restriction of specific TCRs to one HLA type, dependence from the expression and presentation of an antigen, limited intracellular signaling from the recombinant α/β TCRs, mispairing of TCR chains, and the inability to target non-protein tumor antigens.

At present time, about 20 clinical studies involving melanoma-specific T cells expressing recombinant TCR were conducted in US alone (some of them reviewed in [28]). The result of some of the completed trials opened new perspectives for the improvement of the TCR-based strategies. For instance, adoptive transfer of the T cells genetically engineered to express highly avid MART-1-specific TCR has achieved objective clinical responses in a 13% of treated patients [29]. Analysis of CTL-resistant tumor cell revealed that these resistant clones exhibited hyperactivation of the NF-κB survival pathway and overexpression of the antiapoptotic Bcl-2, Bcl-x, Bcl-$_{xL}$, and Mcl-1 genes [30]. These studies suggest that sensitivity of melanoma to the recombinant T cells could be increased by the pharmacological inhibition of the NF-κB pathway and/or Bcl-2 family members. Multiple investigative studies are on-going to further improve recombinant TCR-based approach.

## Chimeric Antigen Receptors

Independently, an alternative approach involving recombinant DNA technology was developed to generate tumor-targeting T cells. It utilizes fusion of the variable chain of the tumor-antigen-specific antibody, TCR constant region, and intracellular signaling domains. Initially, these structures were called T-bodies [31]. They are comprise of the single-chain antibody (sFv), TCR transmembrane domain and the intracellular signaling domain of the TCR-ζ. One of the first tumor associated antigens targeted by T cells expressing T-bodies was erbB2 (HER2/neu) receptor that is over-expressed in multiple cancers [32]. Later, a more general term – chimeric antigen receptors or CARs emerged. As compared to the TCRs, CARs allow overcoming dependency on HLA type, antigen presentation, and restricted intracellular signaling of the recombinant α/β TCRs. Initial studies with T-bodies (and recombinant TCRs) demonstrated a rather short lifespan of the engineered T cells and the inability of the recombinant receptors to fully support persistence of the T cell. To address this issue, several studies were conducted to identify the most potent CAR structures by testing several signaling molecules involved in T cell activation (Fig. 5). It was demonstrated that fusion of TCR-ζ with the intracellular domain of CD28 can augment cytokine production by CAR-expressing T cells upon encountering antigen and enhance antitumor efficacy [33]. Inclusion of CD134 (OX-40) into CAR structure also led to the elevated tumoricidal activity of the recombinant T cells [34]. Comparative analysis of the different CARs comprised of TCR-ζ signal transduction domain, CD28 and/or CD137 (4-1BB) intracellular domains demonstrated that addition of the CD137 supports T cell function to a greater extent as compared to other constructs [35]. Collectively, addition of these signaling domains to the CAR structure allowed overcoming (to certain extent) inefficient effector function and anergic status of the T cells.

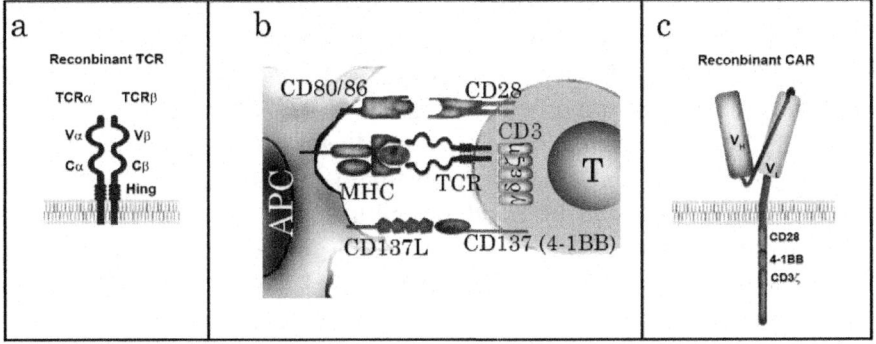

**Figure 5.** Recombinant TCR and CAR structures. (a) Diagram depicting recombinant TCR structure. (b) Diagram illustrating molecular interactions involved in TCR-mediated pro-proliferation and pro-survival intracellular signaling including engagement of the CD28, CD3, and CD137 (4-1BB). (c) Diagram depicting recombinant CAR structure with CD3ξ, CD137 (4-1BB), and CD28 signaling domains (see text for details).

As recognition of target cells by CAR depends on the antibody, CARs can recognize not only polypeptides but also non-protein molecules such as tumor-associated glycolipids and carbohydrates. However, antibody-mediated binding require surface expression of an antigen and strict selection of TAAs to avoid autoimmune side effects. Also, use of the mouse monoclonal antibody sequences in CAR design may lead to the unwanted immune recognition of the CAR-expressing T cells and limit long-term clinical use [36, 37]. Nevertheless, existence of a large number of the tumor antigen-specific antibodies and robust anti-tumor response by CAR-expressing T cells suggest great clinical utility of these recombinant molecules. Currently, in US alone there are 18 clinical trials aimed at treatment of various malignancies with CAR-engineered leukocytes, with 16 trials in the recruitment phase. Eight of them are aimed at targeting different B cell malignancies with anti-CD19-CAR. Three trials are intended to test HER2-specific CAR-modified T cells for the treatment of sarcoma, glioblastoma, and advanced Her2-positive lung malignancy.

With regard to melanoma, several CAR designs were tested for the ability to target this malignancy. Thus, recent studies demonstrated that treatment of melanoma xenografts in nude mice using engineered T cells expressing tandem CAR (CD28/TCRζ) specific to ganglioside GD3 with IL2 supplementation led to complete remissions of the established tumors in 50 % of treated animals [38]. As GD3 is often over-expressed in melanoma, this approach could be potent in eliminating melanoma in human patients.

Another attractive target for the CAR-mediated T cell therapy for melanoma is a high molecular weight melanoma-associated antigen (HMW-MAA) encoded by CSPG4 gene. This is a cell-surface proteoglycan expressed on more than 90% of the tumors. Recent studies on targeting of this antigen using CAR that is comprised of the anti- HMW-MAA antibody chain and intracellular signaling domains of the CD28, CD137, and CD3ζ demonstrated that T cell genetically modified to express this CAR were cytolytic to the HMW-MAA–positive melanoma cells, produced cytokine and proliferate *in vitro* [39]. The potential clinical utility of the CAR-mediated HMW-MAA targeting was emphasized by another recent animal study [40]. Analysis of a few human melanoma biopsies revealed the presence of less than 2% of specific tumor cells co-expressing CD20 and HMW-MAA. Implantation of tumors containing these CD20+ HMW-MAA+ cells into immuno-deficient mice resulted in a rapid growth of tumors. Targeting of these pre-established lesions with T cells expressing either CD20 or HMW-MAA-directed CAR showed elimination of lesions in nearly 90% of treated animals. CD20-specific engineered T cells were unable to eradicated melanoma lesions artificially expressing CD20 suggesting that native expression of the antigen is required for effective targeting. These studies provided additional evidence that direction of the T cells toward HMW-MAA via genetic engineering can permit effective elimination of tumor lesions.

As progression of most tumors including melanoma depends on the microenvironment, T-cell mediated targeting of the microenvironmental components could also be a viable strategy for melanoma immunotherapy. Particularly, tumor survival was shown to be dependent on the *de novo* formation of the intratumoral blood vessels characterised by high levels of the vascular endothelial growth factor receptor 2 (VEGFR2/KDR). Also, a number of studies associated high levels of VEGFR2 expression with various tumor stroma cells including subsets of macrophages, immature monocytes, immature dendritic cells and immuno-suppressive CD4+CD25+ regulatory T cells (Treg) [41-46]. Therefore, it was suggested that targeting of VEGFR2 –positive cells in tumor stroma may provide clinical benefits and tumor regression. In support of this notion, recent studies demonstrated that the direction of the T cells toward VEGFR-2 via CAR provide an effective means to eliminate pre-established experimental melanoma. Thus, using an animal model, it was shown that after systemic transplantation, anti–VEGFR-2 CAR and IL-12–co-transduced T cells infiltrated the tumors, expanded and persisted within tumor mass leading to tumor regression [47]. The anti-tumor effect was dependent on targeting of IL-12–responsive host cells via activation of anti–VEGFR-2 CAR-T cells and release of IL-12. Based on this data, one clinical trial aimed at the assessment of safety and effectiveness of cell therapy was initiated to treat recurrent and relapsed cancer by using anti-VEGFR2 CAR-modified T cells.

Presently, there is an accumulating body of evidence suggesting clinical utility of the T cell genetically engineered to express melanoma antigen-specific CARs. It is likely that in the near future CAR-mediated targeting of different melanoma antigens will evolve into general practice of cancer immunotherapy.

## DNA VACCINATION

Another immunotherapeutic approach directly relevant to recombinant DNA is genetic or DNA vaccination. The original idea of DNA vaccination emanated from the observations that intramuscular injection of DNA encoding influenza A virus protein resulted in the robust activation of the immune responses that protected the host from viral challenge [48]. Generally, DNA-mediated activation of immune response involves multiple processes. First, plasmid DNA should be delivered intracellularly and expressed in the host cells. Next, in most of cases the antigen has to be secreted from the cells and picked up by the dendritic cells (DC), processed and presented in the context of the MHC class II to the CD4+ T helper (Th) cells. Alternatively, if the antigen is expressed directly in the DCs, it could be processed intracellularly and presented via MHC class I molecules, leading to the activation of the CD8+ T cells and induction of the cytotoxic immune responses. Initial studies on DNA vaccination were carried out using an intramuscular route of vaccine administration (Fig. 6). This allowed high levels of antigen expression and secretion from the elongated muscle cells into perimysium, the resident site of the intramuscular DCs. Later, DNA vaccination through the skin was suggested to be superior over the intramuscular route. Skin has evolved as a barrier to prevent the entry of pathogens, with efficient immune surveillance complex including Langerhans cells, dendritic cells, lymphocytes, and other leukocytes. Skin is also rich in lymphatic vasculature network that provides an efficient route for DC and T cell trafficking. Depending on the physical methods of into-skin DNA delivery, DNA-based vaccines can be targeted to specific locations in the skin [49].

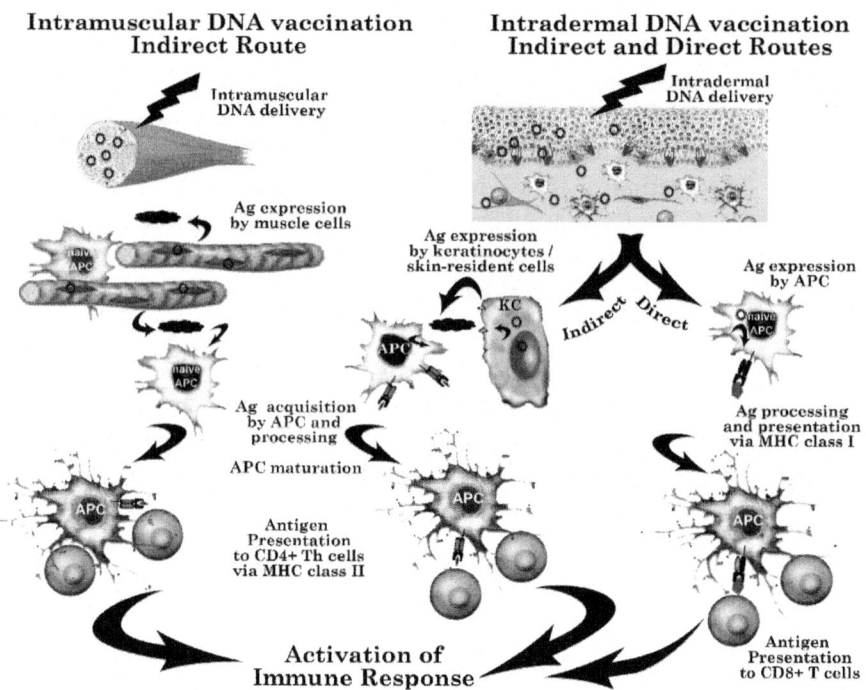

**Figure 6.** Intramuscular and Intradermal DNA vaccination. Intramuscular and intradermal sites are used for DNA vaccination. The former allows high level of antigen (Ag) expression in muscle cells and MHC class II Ag processing and presentation whereas the latter permits expression of the Ag in the Antigen-Presenting Cells (APC) and direct presentation of the antigenic peptides to the CD8+ cytotoxic T cells (see text for details).

The DNA vaccination approach has several advantages over other types of vaccinations: (i) multiple expression vectors coding for different antigen and co-stimulatory molecules can be concurrently delivered into the skin (or the muscle); (ii) the use of cell-type-specific promoters can provide specificity of protein expression; (iii) protein expression from designed plasmids can be controlled by inducible promoters, the use of ubiquitous chromatin opening elements (UCOE), or chemically (e.g. sodium butyrate). Also of note is the relative simplicity and inexpensiveness of the cGMP grade DNA vaccine production and pre-clinical testing. These attractive characteristics of DNA vaccines have prompted extensive research within the past 10 years.

Multiple studies on pre-clinical animal models of melanoma and other cancers have been conducted. Studies on the canine model of aggressive and metastatic melanoma (stages II-IV) demonstrated that xenogeneic vaccination of dogs with DNA vaccine coding for human tyrosinase led to an excellent clinical response in the majority of vaccinated dogs. A long-term survival of dogs with advanced stage IV disease with bulky lung metastases (on average 400 days) was observed [50]. Vaccinated dogs with stage II/III disease also had long-term survivals (on average 500 days) with no evidence of melanoma on necropsy. Overall, median survival time for all treated dogs was 389 days. Another canine model study [51] showed that xenogeneic DNA vaccination induces melanoma-specific antibody response, which coincides with observed clinical responses. As a result, in 2010 Merial, an animal health company has gained full-licensure from the U.S. Department of Agriculture (USDA) for ONCEPT™ Canine Melanoma DNA Vaccine. Up to date, ONCEPT is the first and only USDA-approved therapeutic vaccine for the treatment of cancer in either animals or humans. (The first DNA vaccine was licensed by the USDA in 2005 for prevention of West Nile virus infection in horses).

However, presently only a few human clinical trials on DNA vaccination were conducted. One of such study, aimed at the evaluation of the immune response in patients with hormone-refractory prostate cancer showed that DNA vaccination with a prostate-specific antigen (PSA) encoding plasmid given with GM-CSF and IL-2 is safe in doses of up to 900 μg, and that the vaccination can induce cellular and humoral immune responses [52].

Similar to the reference above canine studies, DNA vaccines were shown to be effective in mouse melanoma models when mice were vaccinated with heterologous DNA encoding human melanoma-associated antigen gp-100 [53]. This vaccination regimen was augmented by the GM-CSF and was most effective in the prophylactic setting. It was also effective in suppressing pre-established melanoma. However, vaccinations with autologous mouse melanoma antigens were less successful. Nevertheless, the relative simplicity of modifying recombinant DNA allowed testing of various genetic alterations aimed at breaking the immunologic tolerance and enhancing immune responses to DNA vaccines. For example, concurrent vaccination with DNA encoding several melanoma-specific epitopes can be used. This approach was tested in several studies with different degree of success. As a result, vaccination of mice with $gp100_{25-33}$ and $TRP-2_{181-188}$ encoding minigene was effective in preventing melanoma development [54]. As many of the melanoma MHC class I epitopes were characterized for melanoma including those derived from tyrosinase, TRP1, TRP2, gp-100, MART-1, and MC-1R (some of them shared between mouse and human MHC molecules [55], one can envision generation of an ultimate genetic immunogen capable of targeting several melanoma-associated antigens.

Recombinant DNA technology has also allowed introduction of immuno-augmentation sequences into the DNA vaccine composition. Identified universal pan HLA DR helper binding epitope (PADRE; KXVAAWYLKA) was shown to enhance immunogenicity of both peptide and DNA vaccines [56, 57]. Other studies demonstrated augmentation of melanoma-specific immune responses via direct fusion of the DNA vaccine with the VP22 protein of the herpes simplex virus-1 [58].

Besides introducing immuno-enhancing alteration to the DNA vaccine, other strategies could be employed to enhance DNA vaccination efficacy including addition of the immuno-enhancing molecules to vaccine composition, alteration of the microenvironment at vaccine administration site, and use of the prime-boost immunization regimens. Recent studies demonstrated that antibody-mediated inhibition of the cytotoxic T lymphocyte antigen 4 (CTLA-4) enhances melanoma-specific immune response. This strategy was recently tested in treatment of stage III-IV melanoma and the drug (Ipilimumam) was approved by the FDA as first anti-melanoma immunotherapeutic [59, 60]. CTLA-4 presents its immuno-inhibitory function during activation of the T cells by the antigen-presenting cells. It also inhibits TCR-mediated intracellular signaling in activated T cells and down-modulating T cell mediated immunity. Therefore, it is possible that inhibition of CTLA-4 in conjunction with DNA vaccination may provide significant enhancement of the vaccine-mediated immune response induction. Although providing CTLA-4 inhibiting antibodies like Ipilimumab along with DNA vaccination is not feasible, other options could be explored. For example, recently characterized genetically engineered lipocalin (LCN2) exhibits a strong cross-species antagonistic activity to CTLA-4 [61]. It is likely that this molecule could be included into DNA vaccine composition to enhance DC-mediated activation of the T cells. Other immuno-modulatory strategies may include addition of CD40 ligand, which was shown to stimulate expression of maturation markers CD80, CD86 and IL-12 in APC [62, 63] and its ability to activate $CD8^+$ T cells and increase cell-mediated immunity [64, 65]. Addition of different cytokines and growth factors including GM-CSF, IL-2, IL12 for stimulation/support of the T cells was also tested in several studies (as exemplified in preceding paragraphs) and could be further explored. Alteration of microenvironment via application chemokines to recruit specific sets of the leukocytes to the vaccine administration site may also provide a favorable milieu for the launch of the effective DNA-vaccine induced immune response [66, 67]. These and many other strategies can be proposed; however, the clinical utility of the DNA vaccination combination with other approaches remains to be determined. Nevertheless, presently in the US alone, 10 clinical studies utilizing xenogenic (mouse) or human DNA vaccines coding for melanoma associated antigens have been completed.

In these trials, tyrosinase, gp75, gp100, and TRP2 were used as antigens. Although most of these studies are already completed, currently no study results are posted nor are follow-up reports available on patient survival and characterization of immune response. Nevertheless, DNA vaccination remains to be a promising modality that could provide cost-effective and generic immunotherapy for patients with melanoma and other cancers.

## OTHER STRATEGIES INVOLVING RECOMBINANT DNA TECHNOLOGY

At the present time, almost every immuno-therapeutic approach utilizes recombinant DNA in one way or another. Understanding of the immuno-regulatory functions of DCs and the molecular mechanisms involved in the capture, processing and presentation of antigens by DCs allowed the development of the DC-based vaccines. Initially, in the mid 1990's, several pre-clinical and clinical studies were conducted using autologous DCs pulsed with melanoma-associated antigens. These studies demonstrated that antigen-loaded DC can trigger active melanoma specific immune responses [68, 69]. However, it became apparent that enforced expression of the antigens in DC rather than loading of these cells with peptides allows for presentation of the tumor-derived antigens via MHC class I complex and priming of the CD8+ T cells to elicit cytotoxic immune response. Moreover, to provide DC specific expression of the antigens, long and short CD11c promoters were characterized and used in several studies [70, 71]. These promoters allow effective and cell type specific expression of the antigens in DCs, as well as more efficient priming and activation of the T cells *in vitro* and *in vivo*. Considering a necessity of the direct interaction of the DC with T lymphocytes, application of T cell recruiting chemokines was also explored recently. These pioneering studies demonstrate that forced expression of the secondary lymphoid chemokine, CCL21, in antigen loaded DCs enhances their ability to recruit and activate T cells [72, 73]. One clinical phase I clinical trial using CCL21 transduced DCs pulsed with MART-1 and gp100 was completed in 2012. Altogether, a total of 64 clinical trials aimed at targeting of melanoma using dendritic cells are listed. Thirty nine of them are completed with no reports yet available. The majority of these trials in some way utilize recombinant DNA technology.

## CONCLUSION

During last decade, various melanoma-specific immunotherapeutics that utilize recombinant DNA have been developed and tested in pre-clinical and clinical studies with varying degrees of clinical success. Many of these approaches, including recombinant TCRs and CARs, have already demonstrated promising clinical results, thus providing us with the hope that in the near future melanoma immunotherapy will become curable for melanoma patients.

## ACKNOWLEDGEMENTS

Presented here data regarding clinical trials was obtained through registry and results database of publicly and privately supported clinical studies (clinicaltrials.gov). Experimental data presented on Fig. 4 was obtained in Dr. Alexeev's laboratory. Plasmid DNA encoding tyrosinase-specific TCR was kindly provided by Dr. S.A. Rosenberg (NCI, NIH). ΦC31 integrase-encoding plasmid was obtained from Dr. M.P. Calos (Stanford University, CA).

## REFERENCES

1.    Knuth, A., Wolfel, T., Klehmann, E., Boon, T. and Meyer zum Buschenfelde, K.H. (1989) Cytolytic T-cell clones against an autologous human melanoma: specificity study and definition of three antigens by immunoselection. *Proceedings of the National Academy of Sciences of the United States of America*, 86, 2804-2808.

2.    Dudley, M.E., Yang, J.C., Sherry, R., Hughes, M.S., Royal, R., Kammula, U., Robbins, P.F., Huang, J., Citrin, D.E., Leitman, S.F. *et al.* (2008) Adoptive cell therapy for patients with metastatic melanoma: evaluation of intensive myeloablative chemoradiation preparative regimens. *J Clin Oncol*, 26, 5233-5239.

3.    Dudley, M.E., Gross, C.A., Langhan, M.M., Garcia, M.R., Sherry, R.M., Yang, J.C., Phan, G.Q., Kammula, U.S., Hughes, M.S., Citrin, D.E. *et al.*CD8+ enriched "young" tumor infiltrating lymphocytes can mediate regression of metastatic melanoma. *Clinical cancer research : an official journal of the American Association for Cancer Research*, 16, 6122-6131.

4.    Goff, S.L., Johnson, L.A., Black, M.A., Xu, H., Zheng, Z., Cohen, C.J., Morgan, R.A., Rosenberg, S.A. and Feldman, S.A. Enhanced receptor expression and in vitro effector function of a murine-human hybrid MART-1-reactive T cell receptor following a rapid expansion. *Cancer immunology, immunotherapy : CII*, 59, 1551-1560.

5.    Cohen, C.J., Zhao, Y., Zheng, Z., Rosenberg, S.A. and Morgan, R.A. (2006) Enhanced antitumor activity of murine-human hybrid T-cell receptor (TCR) in human lymphocytes is associated with improved pairing and TCR/CD3 stability. *Cancer research*, 66, 8878-8886.

6.    Cohen, C.J., Li, Y.F., El-Gamil, M., Robbins, P.F., Rosenberg, S.A. and Morgan, R.A. (2007) Enhanced antitumor activity of T cells engineered

to express T-cell receptors with a second disulfide bond. *Cancer research*, 67, 3898-3903.

7. Kuball, J., Dossett, M.L., Wolfl, M., Ho, W.Y., Voss, R.H., Fowler, C. and Greenberg, P.D. (2007) Facilitating matched pairing and expression of TCR chains introduced into human T cells. *Blood*, 109, 2331-2338.

8. Okamoto, S., Mineno, J., Ikeda, H., Fujiwara, H., Yasukawa, M., Shiku, H. and Kato, I. (2009) Improved expression and reactivity of transduced tumor-specific TCRs in human lymphocytes by specific silencing of endogenous TCR. *Cancer research*, 69, 9003-9011.

9. Govers, C., Sebestyen, Z., Coccoris, M., Willemsen, R.A. and Debets, R. T cell receptor gene therapy: strategies for optimizing transgenic TCR pairing. *Trends Mol Med*, 16, 77-87.

10. Frankel, T.L., Burns, W.R., Peng, P.D., Yu, Z., Chinnasamy, D., Wargo, J.A., Zheng, Z., Restifo, N.P., Rosenberg, S.A. and Morgan, R.A. Both CD4 and CD8 T cells mediate equally effective in vivo tumor treatment when engineered with a highly avid TCR targeting tyrosinase. *J Immunol*, 184, 5988-5998.

11. Douin, V., Bornes, S., Creancier, L., Rochaix, P., Favre, G., Prats, A.C. and Couderc, B. (2004) Use and comparison of different internal ribosomal entry sites (IRES) in tricistronic retroviral vectors. *BMC biotechnology*, 4, 16.

12. Baron, U., Freundlieb, S., Gossen, M. and Bujard, H. (1995) Co-regulation of two gene activities by tetracycline via a bidirectional promoter.*Nucleic acids research*, 23, 3605-3606.

13. Mastakov, M.Y., Baer, K., Kotin, R.M. and During, M.J. (2002) Recombinant adeno-associated virus serotypes 2- and 5-mediated gene transfer in the mammalian brain: quantitative analysis of heparin co-infusion. *Molecular therapy : the journal of the American Society of Gene Therapy*, 5, 371-380.

14. Ryan, M.D., King, A.M. and Thomas, G.P. (1991) Cleavage of foot-and-mouth disease virus polyprotein is mediated by residues located within a 19 amino acid sequence. *J Gen Virol*, 72 ( Pt 11), 2727-2732.

15. Baum, C., Schambach, A., Bohne, J. and Galla, M. (2006) Retrovirus vectors: toward the plentivirus? *Molecular therapy : the journal of the American Society of Gene Therapy*, 13, 1050-1063.

16. Frecha, C., Levy, C., Cosset, F.L. and Verhoeyen, E. (2010) Advances in the field of lentivector-based transduction of T and B lymphocytes for gene therapy. *Molecular therapy : the journal of the American Society of Gene Therapy*, 18, 1748-1757.

17.   Hackett, P.B., Largaespada, D.A. and Cooper, L.J. (2010) A transposon and transposase system for human application. *Molecular therapy : the journal of the American Society of Gene Therapy*, 18, 674-683.

18.   Hacein-Bey-Abina, S., von Kalle, C., Schmidt, M., Le Deist, F., Wulffraat, N., McIntyre, E., Radford, I., Villeval, J.L., Fraser, C.C., Cavazzana-Calvo, M. *et al.* (2003) A serious adverse event after successful gene therapy for X-linked severe combined immunodeficiency. *The New England journal of medicine*, 348, 255-256.

19.   Nam, C.H. and Rabbitts, T.H. (2006) The role of LMO2 in development and in T cell leukemia after chromosomal translocation or retroviral insertion. *Molecular therapy : the journal of the American Society of Gene Therapy*, 13, 15-25.

20.   Mates, L. (2011) Rodent transgenesis mediated by a novel hyperactive Sleeping Beauty transposon system. *Methods Mol Biol*, 738, 87-99.

21.   Chatterjee, M., Hedrich, C.M., Rauen, T., Ioannidis, C., Terhorst, C. and Tsokos, G.C. (2012) CD3-T Cell Receptor Co-stimulation through SLAMF3 and SLAMF6 Receptors Enhances RORgammat Recruitment to the IL17A Promoter in Human T Lymphocytes. *The Journal of biological chemistry*, 287, 38168-38177.

22.   Clevers, H., Lonberg, N., Dunlap, S., Lacy, E. and Terhorst, C. (1989) An enhancer located in a CpG-island 3' to the TCR/CD3-epsilon gene confers T lymphocyte-specificity to its promoter. *The EMBO journal*, 8, 2527-2535.

23.   Hahm, S.H., Yi, Y., Lee, D.K., Noh, M.J., Yun, L., Hwang, S. and Lee, K.H. (2004) Construction of retroviral vectors with enhanced efficiency of transgene expression. *Journal of virological methods*, 121, 127-136.

24.   Morita, S., Kojima, T. and Kitamura, T. (2000) Plat-E: an efficient and stable system for transient packaging of retroviruses. *Gene therapy*, 7, 1063-1066.

25.   Peng, P.D., Cohen, C.J., Yang, S., Hsu, C., Jones, S., Zhao, Y., Zheng, Z., Rosenberg, S.A. and Morgan, R.A. (2009) Efficient nonviral Sleeping Beauty transposon-based TCR gene transfer to peripheral blood lymphocytes confers antigen-specific antitumor reactivity. *Gene therapy*, 16, 1042-1049.

26.   Sclimenti, C.R., Thyagarajan, B. and Calos, M.P. (2001) Directed evolution of a recombinase for improved genomic integration at a native human sequence. *Nucleic acids research*, 29, 5044-5051.

27.   Groth, A.C., Olivares, E.C., Thyagarajan, B. and Calos, M.P. (2000) A phage integrase directs efficient site-specific integration in human cells.

*Proceedings of the National Academy of Sciences of the United States of America*, 97, 5995-6000.

28. Park, T.S., Rosenberg, S.A. and Morgan, R.A. (2011) Treating cancer with genetically engineered T cells. *Trends in biotechnology*, 29, 550-557.

29. Morgan, R.A., Dudley, M.E., Wunderlich, J.R., Hughes, M.S., Yang, J.C., Sherry, R.M., Royal, R.E., Topalian, S.L., Kammula, U.S., Restifo, N.P. *et al.* (2006) Cancer regression in patients after transfer of genetically engineered lymphocytes. *Science*, 314, 126-129.

30. Jazirehi, A.R., Baritaki, S., Koya, R.C., Bonavida, B. and Economou, J.S. (2011) Molecular mechanism of MART-1+/A*0201+ human melanoma resistance to specific CTL-killing despite functional tumor-CTL interaction. *Cancer research*, 71, 1406-1417.

31. Eshhar, Z. (1997) Tumor-specific T-bodies: towards clinical application. *Cancer immunology, immunotherapy : CII*, 45, 131-136.

32. Moritz, D., Wels, W., Mattern, J. and Groner, B. (1994) Cytotoxic T lymphocytes with a grafted recognition specificity for ERBB2-expressing tumor cells. *Proceedings of the National Academy of Sciences of the United States of America*, 91, 4318-4322.

33. Koehler, H., Kofler, D., Hombach, A. and Abken, H. (2007) CD28 costimulation overcomes transforming growth factor-beta-mediated repression of proliferation of redirected human CD4+ and CD8+ T cells in an antitumor cell attack. *Cancer research*, 67, 2265-2273.

34. Finney, H.M., Akbar, A.N. and Lawson, A.D. (2004) Activation of resting human primary T cells with chimeric receptors: costimulation from CD28, inducible costimulator, CD134, and CD137 in series with signals from the TCR zeta chain. *J Immunol*, 172, 104-113.

35. Milone, M.C., Fish, J.D., Carpenito, C., Carroll, R.G., Binder, G.K., Teachey, D., Samanta, M., Lakhal, M., Gloss, B., Danet-Desnoyers, G. *et al.*(2009) Chimeric receptors containing CD137 signal transduction domains mediate enhanced survival of T cells and increased antileukemic efficacy in vivo. *Molecular therapy : the journal of the American Society of Gene Therapy*, 17, 1453-1464.

36. Kershaw, M.H., Westwood, J.A., Parker, L.L., Wang, G., Eshhar, Z., Mavroukakis, S.A., White, D.E., Wunderlich, J.R., Canevari, S., Rogers-Freezer, L. *et al.* (2006) A phase I study on adoptive immunotherapy using gene-modified T cells for ovarian cancer. *Clinical cancer research : an official journal of the American Association for Cancer Research*, 12, 6106-6115.

37. Lamers, C.H., Willemsen, R., van Elzakker, P., van Steenbergen-Langeveld, S., Broertjes, M., Oosterwijk-Wakka, J., Oosterwijk, E., Sleijfer, S., Debets, R. and Gratama, J.W. (2011) Immune responses to transgene and retroviral vector in patients treated with ex vivo-engineered T cells.*Blood*, 117, 72-82.

38. Lo, A.S., Ma, Q., Liu, D.L. and Junghans, R.P. (2010) Anti-GD3 chimeric sFv-CD28/T-cell receptor zeta designer T cells for treatment of metastatic melanoma and other neuroectodermal tumors. *Clinical cancer research : an official journal of the American Association for Cancer Research*, 16, 2769-2780.

39. Burns, W.R., Zhao, Y., Frankel, T.L., Hinrichs, C.S., Zheng, Z., Xu, H., Feldman, S.A., Ferrone, S., Rosenberg, S.A. and Morgan, R.A. (2010) A high molecular weight melanoma-associated antigen-specific chimeric antigen receptor redirects lymphocytes to target human melanomas. *Cancer research*, 70, 3027-3033.

40. Schmidt, P., Kopecky, C., Hombach, A., Zigrino, P., Mauch, C. and Abken, H. (2011) Eradication of melanomas by targeted elimination of a minor subset of tumor cells. *Proceedings of the National Academy of Sciences of the United States of America*, 108, 2474-2479.

41. Duignan, I.J., Corcoran, E., Pennello, A., Plym, M.J., Amatulli, M., Claros, N., Iacolina, M., Youssoufian, H., Witte, L., Samakoglu, S. *et al.*Pleiotropic stromal effects of vascular endothelial growth factor receptor 2 antibody therapy in renal cell carcinoma models. *Neoplasia (New York, N.Y*, 13, 49-59.

42. Larrivee, B., Pollet, I. and Karsan, A. (2005) Activation of vascular endothelial growth factor receptor-2 in bone marrow leads to accumulation of myeloid cells: role of granulocyte-macrophage colony-stimulating factor. *J Immunol*, 175, 3015-3024.

43. Murdoch, C., Muthana, M., Coffelt, S.B. and Lewis, C.E. (2008) The role of myeloid cells in the promotion of tumour angiogenesis. *Nature reviews*, 8, 618-631.

44. Suzuki, H., Onishi, H., Wada, J., Yamasaki, A., Tanaka, H., Nakano, K., Morisaki, T. and Katano, M. VEGFR2 is selectively expressed by FOXP3high CD4+ Treg. *European journal of immunology*, 40, 197-203.

45. Udagawa, T., Puder, M., Wood, M., Schaefer, B.C. and D'Amato, R.J. (2006) Analysis of tumor-associated stromal cells using SCID GFP transgenic mice: contribution of local and bone marrow-derived host cells. *Faseb J*, 20, 95-102.

46. Zhang, N., Fang, Z., Contag, P.R., Purchio, A.F. and West, D.B. (2004) Tracking angiogenesis induced by skin wounding and contact hypersensitivity using a Vegfr2-luciferase transgenic mouse. *Blood*, 103, 617-626.

47. Chinnasamy, D., Yu, Z., Kerkar, S.P., Zhang, L., Morgan, R.A., Restifo, N.P. and Rosenberg, S.A. Local delivery of interleukin-12 using T cells targeting VEGF receptor-2 eradicates multiple vascularized tumors in mice. *Clinical cancer research : an official journal of the American Association for Cancer Research*, 18, 1672-1683.

48. Ulmer, J.B., Donnelly, J.J., Parker, S.E., Rhodes, G.H., Felgner, P.L., Dwarki, V.J., Gromkowski, S.H., Deck, R.R., DeWitt, C.M., Friedman, A. *et al.*(1993) Heterologous protection against influenza by injection of DNA encoding a viral protein. *Science*, 259, 1745-1749.

49. Alexeev, V. and Uitto, J. (2005) *DNA pharmaceuticals for skin diseases*. Wiley-VCH, Weinheim.

50. Bergman, P.J., McKnight, J., Novosad, A., Charney, S., Farrelly, J., Craft, D., Wulderk, M., Jeffers, Y., Sadelain, M., Hohenhaus, A.E. *et al.* (2003) Long-term survival of dogs with advanced malignant melanoma after DNA vaccination with xenogeneic human tyrosinase: a phase I trial. *Clinical cancer research : an official journal of the American Association for Cancer Research*, 9, 1284-1290.

51. Liao, J.C., Gregor, P., Wolchok, J.D., Orlandi, F., Craft, D., Leung, C., Houghton, A.N. and Bergman, P.J. (2006) Vaccination with human tyrosinase DNA induces antibody responses in dogs with advanced melanoma. *Cancer Immun*, 6, 8.

52. Pavlenko, M., Roos, A.K., Lundqvist, A., Palmborg, A., Miller, A.M., Ozenci, V., Bergman, B., Egevad, L., Hellstrom, M., Kiessling, R. *et al.* (2004) A phase I trial of DNA vaccination with a plasmid expressing prostate-specific antigen in patients with hormone-refractory prostate cancer. *British journal of cancer*, 91, 688-694.

53. Rakhmilevich, A.L., Imboden, M., Hao, Z., Macklin, M.D., Roberts, T., Wright, K.M., Albertini, M.R., Yang, N.S. and Sondel, P.M. (2001) Effective particle-mediated vaccination against mouse melanoma by coadministration of plasmid DNA encoding Gp100 and granulocyte-macrophage colony-stimulating factor. *Clin Cancer Res*, 7, 952-961.

54. Xiang, R., Lode, H.N., Chao, T.H., Ruehlmann, J.M., Dolman, C.S., Rodriguez, F., Whitton, J.L., Overwijk, W.W., Restifo, N.P. and Reisfeld, R.A. (2000) An autologous oral DNA vaccine protects against murine

melanoma. *Proceedings of the National Academy of Sciences of the United States of America*, 97, 5492-5497.

55.  Parkhurst, M.R., Fitzgerald, E.B., Southwood, S., Sette, A., Rosenberg, S.A. and Kawakami, Y. (1998) Identification of a shared HLA-A*0201-restricted T-cell epitope from the melanoma antigen tyrosinase-related protein 2 (TRP2). *Cancer research*, 58, 4895-4901.

56.  Bettahi, I., Dasgupta, G., Renaudet, O., Chentoufi, A.A., Zhang, X., Carpenter, D., Yoon, S., Dumy, P. and BenMohamed, L. (2009) Antitumor activity of a self-adjuvanting glyco-lipopeptide vaccine bearing B cell, CD4+ and CD8+ T cell epitopes. *Cancer immunology, immunotherapy : CII*, 58, 187-200.

57.  del Guercio, M.F., Alexander, J., Kubo, R.T., Arrhenius, T., Maewal, A., Appella, E., Hoffman, S.L., Jones, T., Valmori, D., Sakaguchi, K. *et al.*(1997) Potent immunogenic short linear peptide constructs composed of B cell epitopes and Pan DR T helper epitopes (PADRE) for antibody responses in vivo. *Vaccine*, 15, 441-448.

58.  Engelhorn, M.E., Guevara-Patino, J.A., Merghoub, T., Liu, C., Ferrone, C.R., Rizzuto, G.A., Cymerman, D.H., Posnett, D.N., Houghton, A.N. and Wolchok, J.D. (2008) Mechanisms of immunization against cancer using chimeric antigens. *Molecular therapy : the journal of the American Society of Gene Therapy*, 16, 773-781.

59.  Hodi, F.S., O'Day, S.J., McDermott, D.F., Weber, R.W., Sosman, J.A., Haanen, J.B., Gonzalez, R., Robert, C., Schadendorf, D., Hassel, J.C. *et al.*Improved survival with ipilimumab in patients with metastatic melanoma. *The New England journal of medicine*, 363, 711-723.

60.  Schartz, N.E., Farges, C., Madelaine, I., Bruzzoni, H., Calvo, F., Hoos, A. and Lebbe, C. Complete regression of a previously untreated melanoma brain metastasis with ipilimumab. *Melanoma research*, 20, 247-250.

61.  Schonfeld, D., Matschiner, G., Chatwell, L., Trentmann, S., Gille, H., Hulsmeyer, M., Brown, N., Kaye, P.M., Schlehuber, S., Hohlbaum, A.M. *et al.*(2009) An engineered lipocalin specific for CTLA-4 reveals a combining site with structural and conformational features similar to antibodies.*Proceedings of the National Academy of Sciences of the United States of America*, 106, 8198-8203.

62.  Feder-Mengus, C., Schultz-Thater, E., Oertli, D., Marti, W.R., Heberer, M., Spagnoli, G.C. and Zajac, P. (2005) Nonreplicating recombinant vaccinia virus expressing CD40 ligand enhances APC capacity to stimulate specific CD4+ and CD8+ T cell responses. *Human gene therapy*, 16, 348-360.

63. Sato, T., Terai, M., Yasuda, R., Watanabe, R., Berd, D., Mastrangelo, M.J. and Hasumi, K. (2004) Combination of monocyte-derived dendritic cells and activated T cells which express CD40 ligand: a new approach to cancer immunotherapy. *Cancer immunology, immunotherapy : CII*, 53, 53-61.

64. Peter, I., Nawrath, M., Kamarashev, J., Odermatt, B., Mezzacasa, A. and Hemmi, S. (2002) Immunotherapy for murine K1735 melanoma: combinatorial use of recombinant adenovirus expressing CD40L and other immunomodulators. *Cancer gene therapy*, 9, 597-605.

65. Rousseau, R.F., Biagi, E., Dutour, A., Yvon, E.S., Brown, M.P., Lin, T., Mei, Z., Grilley, B., Popek, E., Heslop, H.E. *et al.* (2006) Immunotherapy of high-risk acute leukemia with a recipient (autologous) vaccine expressing transgenic human CD40L and IL-2 after chemotherapy and allogeneic stem cell transplantation. *Blood*, 107, 1332-1341.

66. Guo, J.H., Fan, M.W., Sun, J.H. and Jia, R. (2009) Fusion of antigen to chemokine CCL20 or CXCL13 strategy to enhance DNA vaccine potency.*International immunopharmacology*, 9, 925-930.

67. Novak, L., Igoucheva, O., Cho, S. and Alexeev, V. (2007) Characterization of the CCL21-mediated melanoma-specific immune responses and in situ melanoma eradication. *Molecular cancer therapeutics*, 6, 1755-1764.

68. Nestle, F.O., Alijagic, S., Gilliet, M., Sun, Y., Grabbe, S., Dummer, R., Burg, G. and Schadendorf, D. (1998) Vaccination of melanoma patients with peptide- or tumor lysate-pulsed dendritic cells. *Nature medicine*, 4, 328-332.

69. 69.Thurner, B., Haendle, I., Roder, C., Dieckmann, D., Keikavoussi, P., Jonuleit, H., Bender, A., Maczek, C., Schreiner, D., von den Driesch, P. *et al.*(1999) Vaccination with mage-3A1 peptide-pulsed mature, monocyte-derived dendritic cells expands specific cytotoxic T cells and induces regression of some metastases in advanced stage IV melanoma. *The Journal of experimental medicine*, 190, 1669-1678.

70. Brocker, T., Riedinger, M. and Karjalainen, K. (1997) Targeted expression of major histocompatibility complex (MHC) class II molecules demonstrates that dendritic cells can induce negative but not positive selection of thymocytes in vivo. *The Journal of experimental medicine*, 185, 541-550.

71. Ni, J., Nolte, B., Arnold, A., Fournier, P. and Schirrmacher, V. (2009) Targeting anti-tumor DNA vaccines to dendritic cells via a short CD11c promoter sequence. *Vaccine*, 27, 5480-5487.

72. Mule, J.J. (2009) Dendritic cell-based vaccines for pancreatic cancer and melanoma. *Annals of the New York Academy of Sciences*, 1174, 33-40.

73. Terando, A., Roessler, B. and Mule, J.J. (2004) Chemokine gene modification of human dendritic cell-based tumor vaccines using a recombinant adenoviral vector. *Cancer gene therapy*, 11, 165-173.

# Chapter 2

# GENETIC ENGINEERING AND BIOTECHNOLOGY OF GROWTH HORMONES

Jorge Angel Ascacio-Martínez and Hugo Alberto Barrera-Saldaña

Department of Biochemistry and Molecular Medicine, School of Medicine, Autonomous University of Nuevo León, Monterrey Nuevo León, Av. Madero Pte. s/n Col. Mitras Centro, Monterrey, N.L., México

## INTRODUCTION

In its modern conception, biotechnology is the use of genetic engineering techniques to manipulate microorganisms, plants, and animals in order to produce commercial products and processes that benefit man. These techniques, which are the backbone of the biotechnological revolution that began in the mid 1970s, have permitted the isolation and manipulation of specific genes and the development of transgenic microorganisms that produce mainly eukaryotic proteins of therapeutic use, such as vaccines, enzymes, and hormones.

Biotechnology is present in diverse areas such as food production, degradation of industrial waste, mining, and medicine. Recent achievements include drug production in transgenic animals and plants, as well as the commercial exploitation of gene sequences generated by the human genome project and similar projects of plants and animals of commercial interest that are and will be in process.

Human growth hormone was, after insulin, the second product of this new technology. This product was developed and commercialized initially by Genentech, and was used clinically for treating growth problems and dwarfism (1). Furthermore, growth hormones from different animal species have also been produced in transgenic organisms and these have been used in different examples in the aquatic animal and livestock sectors.

## THE GROWTH HORMONE (GH) FAMILY

GHs belong to a family of proteins with structural similarity and certain common functions that include prolactin (Prl), somatolactin (SL), chorionic somatomammotropin (CS), proliferin (PLF) and proteins related to Prl (PLP) (2). This family represents one of the most physiologically diverse protein groups that have evolved by gene duplication. The two most studied members of this family have been GH and Prl, which have been described from primitive fish to mammals; however, other members of the family are not so amply distributed or studied.

## Structure of Growth Hormones

GHs (see Figure 1), in general, have a molecular weight of around 22,000 Daltons (22 kDa or simply 22k) and do not require post-translational modifications. They are synthesized in somatotrophs in the hypophysis, intervening as an important endocrine factor in postnatal somatic growth and lactation.

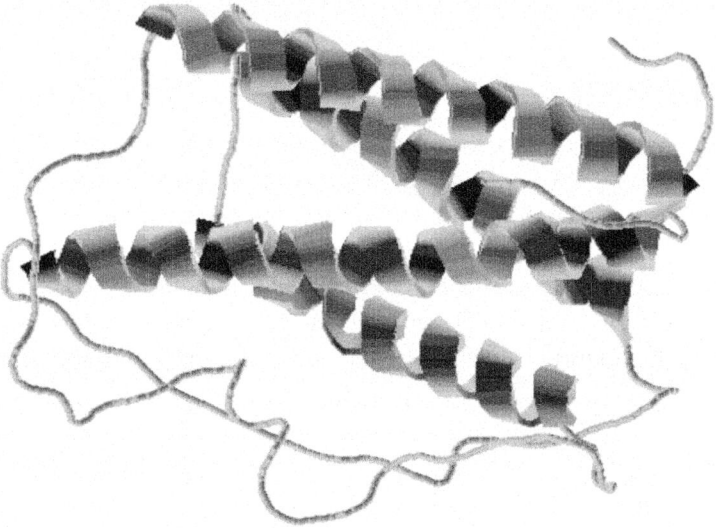

**Figure 1**. Growth hormones' consensus tridimensional structure. The GHs have in general 190 aminoacidic residues, four alpha helixes, and two sulphide bonds.

## Hormones of the Human Growth Hormone Family

### HGH22k

HGH22k (or HGHN) is the main product of the GH gene (hGH-N) active in the hypophysis and it is responsible for postnatal growth as well as being an important modulator of carbohydrate, lipid, nitrogen and mineral metabolism. It is the best known hormone and the only one of the HGH family that has been commercialized.

As mentioned, besides being the cure for hypophyseal dwarfism, HGH22k postulated benefits are as an anabolic in athletics and for the treatment of trauma because of its postulated regenerative properties (3).

## HGH20k

In addition to the mRNA of HGH22k, an alternative processing pathway of the primary transcript of the hGH-N gene generates a second mRNA that is responsible for the production of the 20k isoform of HGH or HGH20k. Its smaller size is due to elimination of the first 45 nucleotides of the third exon of the mRNA and of the amino acids that correspond to positions 32-46 of the hormone, producing a protein with 176 amino acid residues (4).

This isoform comprises approximately 10% of all the GH produced in the hypophysis and although it has not been shown to be the etiological agent of any known disease, it is known that its levels are significantly higher in patients with active acromegaly and in those with anorexia nervosa (5).

The administration of exogenous HGH20k suppresses endogenous secretion of HGH22k in healthy subjects, which suggests that the regulation of secretion of both hormones is physiologically similar (6). In vitro findings suggest that both hormones can equally stimulate bone remodeling and allow anabolic effects on skeletal tissue when they are administered in vivo to laboratory animals (7).

## HGHV

Several isoforms also derive from the GH gene expressed in the placenta (hGH-V)(Table 1). The most abundant mRNA from this gene in the placenta at terminus also codifies for a 22 kDa isoform. A less abundant isoform (HGHV2) originate from a species of mRNA that retains the fourth intron and due to this, it codifies for a 26 kDa protein that anchors to the membrane and which could have a local action (8). A 25 kDa protein is also derived by glycosylation of residue 140 of asparagine from the 22 kDa isoform (9, 10). Finally, two new transcripts of this gene have recently been identified: one already known that as in the case of the HGH20k also produces a 20 kDa protein, and another novel splicing variation that results in a mRNA known as hGHV3, that traduces into a 24 kDa isoform (11).

**Table 1.** HGH-V isoforms generated by alternative splicing and processing

| Isoform | Size | Length | Characteristic |
|---|---|---|---|
| • HGH-V22k | 22kDa | 191aa | Main isoform. |
| • HGH-V25k | 25kDa | 191aa | Glycosylated version of HGH-V22. |
| • HGH-V2 | 26kDa | 230aa | Retains the fourth intron. |
| • HGH-V20K | 20kDa* | 176aa | Deletion of aa residues 32 to 46. |
| •HGH-V3 | 24kDa* | 219aa | Alternate processing at level of exon 4. |

*Only the mRNAs that codify each have been identified.

During pregnancy, while hypophyseal HGHN progressively disappears from the maternal circulation until undetectable values are reached at weeks 24 to 25, HGHV progressively increases until birth, suggesting that it has a key role during human gestation (12). It has also been found that in cases of intrauterine growth restriction, circulating levels of HGHV measured between week 31 and birth are lower than those reported in normal pregnancy (13, 14, 15).

Finally, although low levels of this hormone have been associated with intrauterine growth retardation, cases of hGH-V gene deletion have also been reported, but without an apparent pathology (16).

## Human Chorionic Somatomammotropin (HCS)

HCS is detected in maternal serum from the fourth week of gestation, increasing throughout the pregnancy in a linear fashion, and reaching high production levels of a couple of grams per day at the end of gestation. These actions result in both elevation of glucose and amino acids in the maternal circulation. These are in turn used by the fetus for his/her development. It also generates free fatty acids (by lipolytic effect), which are used as an energy source by the fetus (17, 18). Little is known about the HCS physiological role, and still is not known its action mechanism. Producing rHCS by biotechnology will help to advance these investigations.

## In Vitro Bioassays for GHs and CSHs

As stated above except for HGH22k, the functions of the rest of hormones of the human GH family have been not completely defined. Their biological activities are being studied, classifying them into at least two general categories:

a.    Somatogenic activities. These involve linear bone growth and alterations in carbohydrate metabolism; effects that are in part mediated by local and hepatic generation of insulinlike growth factor-I (IGF-I). The somatogenic activity of HGHV has been studied by stimulating body weight increase in hypophysectomized rats, reporting a linear increase comparable to that produced by HGH22k (19).

b.    Lactogenic activities. These include stimulation of lactation and reproductive functions (20). The lactogenic activities of this hormone have been studied using a cell model (by mitogenic response to Nb2 cells) and a response that is parallel to HGH22k has been reported, although it is significantly less (19).

## The Human GH Locus

Besides the two hGH genes (normal and variant), three HCSs complement the multigenic HGH family from the human genome and these are arranged in the following order: HGHN, HCS-1, HCS-2, HGH-V y HCS-3 (21, 22) (Figure 2). While HCS-1 appears to be a pseudogene, HCS-2 and HCS-3 are very active in the placenta and interestingly; mature versions of the hormones that they codify are identical (23).

In the last few years, in our laboratory, all the hGH and HCS genes have been cloned and expressed in cell culture, and the factors that affect their levels of expression have been particularly studied (24).

In the same way, and using polymerase chain reaction (PCR) with consensus primers, several new genes and complementary DNAs (cDNAs) to the mRNA of numerous GHs have been isolated in our laboratory, mainly from mammals (unpublished results).

# GROWTH HORMONE OF ANIMAL ORIGIN

## Bovine Growth Hormone (BGH)

Bovine growth hormone (BGH) or bovine somatotropin improves the efficiency of milk production (per unit of food consumed) (25), and the production (body weight) and composition (muscle: fat ratio) of meat (26). In the case of milk cows, this permits a reduction in the number of animals needed for milk production and a subsequent savings in maintenance, feeding, water, drugs, etc. It also reduces the production of manure, and nitrogen from urine and methane (27).

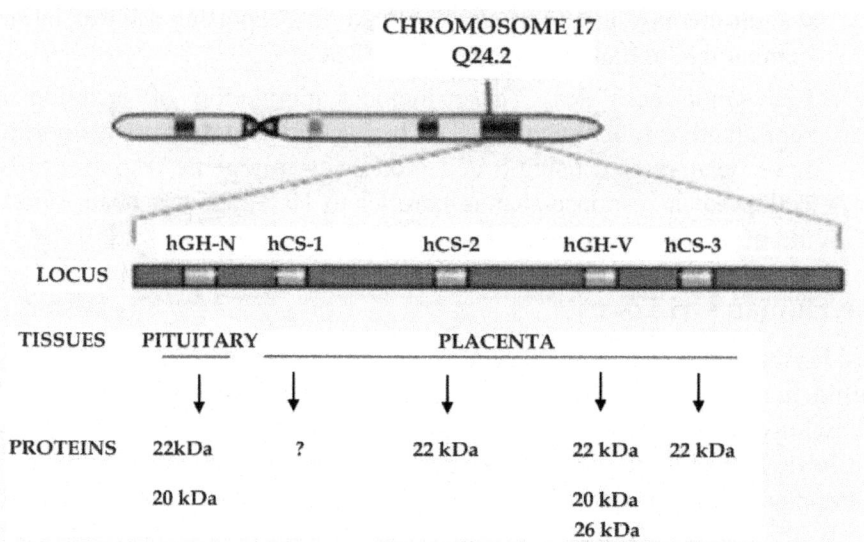

**Figure 2.** HGH-HCS multigenic complex. Located on Chromosome 17, every gene is indicated; the tissue where they are expressed and the proteic isoforms that are produced are shown.

Milk from cows treated with rBGH, does not differ from that of untreated cows (28, 29). The characteristics that have been evaluated include the freezing point, pH, thermal properties, susceptibility to oxygenation, and sensory characteristics, including taste; in fact all organoleptic properties are conserved. Also, differences have not been found in the properties necessary for producing cheese, including initial growth of the culture, coagulation, acidification, production and composition (29).

rBGH is administered subcutaneously and is dispensed as a long-acting suspension that is applied in a determined period of time. The taste of bovine meat and milk treated with rBGH is not altered, but the fat content is less.

## Caprine Growth Hormone (CHGH)

For small ruminants there are studies in lactating goats in which the administration of rBGH increased milk production 23% and stimulated mammary gland growth more than in those that were frequently milked, with it being similar to prolactin (30). However, the production of recombinant CHGH, which is identical to ovine and thus can be used in both animals, had not been reported, until we achieve its expression on the methylotrophic yeast Pichia pastoris. (See section 7.2).

## Equine Growth Hormone (ECGH)

With regard to horses, GH is used in the prevention of muscle wasting, in the repair of tendons and fractured bones, as well as for the treatment of anovulation in mares. Besides this, it is also used for repairing muscle tissue, to tonify and invigorate race horses, and for improving physical conditions in older horses by restoring nitrogen balance. It can also stimulate growth and early maturity in young horses, increase milk production in lactating mares and promote wound healing, especially of bone and cartilage (31, 32), as occurred in the case shown in Figure 3.

**Figure 3.** Uses of equine GH. The race horse "Might and Power" (right) became the winner of the Melbourne Cup in 1997. But in 1999, a tendon from one of its hooves was severely damaged. The horse was treated with ECGH, recovered and in 2000 was able to return to horse racing (32).

## Canine Growth Hormone (CFGH)

With regard to the dog (Canis familiaris), each day there is more evidence of the role that its GH (CFGH) plays in bone fracture treatment, in which the hormone helps reduce the bone restoration period (33). It is no less important in the treatment of obesity in dogs, thanks to the metabolism activation produced by the hormone in removing fatty acids, and in general, in counteracting symptoms related to the presence or absence of the same GH. Also, since this hormone is identical to pig GH (PGH) (33), its virtues are valid for the application of CFGH in the porcine industry, where it generates leaner meat (34), which is of greater value.

## Feline Growth Hormone (FCGH)

Although there is very little literature on cat GH (FCGH), the benefits identified in other GHs apply to this feline species, since these animals present the symptoms mentioned before for dogs, which are caused by the absence or low concentration of FCGH (dwarfism and alopecia, among others). Also, as referred to in the literature, biological tests of adipogenic activity in culture cells use cat serum (which contains FCGH) instead of bovine serum, because FCGH lacks adipogenicity (17).

Therefore, recombinant production of this GH would be useful in the mentioned tests. It is important to point out the usefulness that recombinant FCGH would have in future research on the metabolic study and role of this hormone in this and other feline species, including of course, large wild cats in captivity.

# BIOLOGICAL POTENTIAL OF GHS

## Growth Hormones of Human Origin

Although HGH22k is widely commercialized and more functions now have been recognized to it (Table 2), the same does not occur with the other proteins and isoforms from this family; essentially the 20 kDa isoform of HGH, HGHV, also the isoform of 20 kDa of the latter (HGHV20k), and lastly, HCS. Partly because of this, many of their functions and mechanisms of action are still unknown.

**Table 2.** New functions atributed to HGH22k

| Immunization and healing | Mental function |
|---|---|
| • Resistance to common diseases<br>• Ability to heal<br>• Healing of old lesions<br>• Healing of other lesions<br>• Ulcer treatment | • Emotional stability<br>• Memory<br>• General aspect and attitude<br>• Mental energy and clarity |
| **Skin and hair** | **Muscle strength and tone** |
| • Skin elasticity<br>• Skin thickness<br>• Skin texture<br>• Growth of new hair<br>• Disappearance of wrinkles<br>• Skin hydration | • Increase in energy in general<br>• Increase muscle strength<br>• Promotion of muscle mass gain |

| Sexual factors | Circulatory system |
|---|---|
| <ul><li>Duration of an erection</li><li>Increase in libido</li><li>Potential/frequency of sexual activity</li><li>Regulation and control of the menstrual cycle</li><li>Positive effects in the reproductive system</li><li>Increase in breast-milk volume</li></ul> | <ul><li>Improvement in circulation</li><li>Stabilization of blood pressure</li><li>Improvement in cardiac function</li></ul> |
| Bone | Fats |
| <ul><li>As treatment for bone fractures</li><li>Osteoporosis treatment</li><li>Increases flexibility of the back and joints</li></ul> | <ul><li>Increases "good" cholesterol (HDL) levels</li><li>Reduces fat</li></ul> |

(Taken from Elian y cols., 1999), (3).

It is believed that some of the hormone's less abundant natural variants, such as HGH20k, could retain desirable properties of the principal hormone and lack some of its other undesirable effects, such as its diabetogenic effect, which occurs with prolonged use (35).

## Growth Hormones of Animal Origin

The biotechnological potential of GHs could be enormous, since besides its use in species of the same origin, it has been demonstrated that the GHs of mammals have activity in phylogenetically lower animals. For example, BGH and porcine GH (PGH) have been used experimentally for the treatment of hypophyseal dwarfism in dogs (36) and cats (37). Regarding farm animals, porcine, bovine, caprine and ovine livestock have been treated with exogenous GH to improve production, since it increases food conversion efficiency, growth rate, weight gain, and milk and meat production. What is surprising is the finding that BGH stimulates salmon growth, and even more interesting that bovine chorionic somatomammotropin (BCS) works even better (38).

# EXPRESSION SYSTEMS FOR GROWTH HORMONES

## The History of Human Recombinant GH

As previously mentioned, among the first cDNAs cloned and expressed in the bacteria Escherichia coli is precisely HGH (1). This expression system has been used since 1985 for the production of recombinant HGH by Genentech (protropin), which was later followed by Lily (humatrope), Biotech (biotropin), Novo Nordisk (norditropin), Serono (serostim), and others.

## Different Biotechnological Hosts

Since the recombinant protein is frequently recovered from E. coli with undesirable modifications (extra methionine, incorrect folding, aggregated forms, etc.) and contaminated with highly pyrogenic substances, toilsome purification schemes are needed to obtain it with the desired purity, structure and biological activity. For this, subsequent efforts have focused on the search for better expression systems, with production being attempted with *Saccharomyces cerevisiae* (39), *Bacillus subtilis* (40), mammal cell cultures (41), as well as transgenic animals (42). Unfortunately, these expression systems do not offer a production level greater than that of E. coli and therefore in most cases they are not profitable.

In our laboratory, we succeeded in producing HGH22k in E. coli by fusing it with maltose binding protein (rHGH-MBP) in 1994 (unpublished results). However, due to the fact that to recover the hormone, whether from the periplasm or the cytoplasm, complicated strategies were needed, together with the limitations of the bacterial systems for folding and processing foreign proteins correctly, we proposed searching for an expression system that allows synthesizing the protein, purifying it with greater ease while retaining functionality. Thus, the evaluation of different expression systems was started in our laboratories, considering the methylotrophic yeast *Pichia pastoris* as the best (43).

## *Pichia pastoris* as a Biotechnological Host for GHs

Yeasts offer the best of both prokaryotes and eukaryotes, since, in addition to performing some of the post-translational modifications that are common in superior organisms, they are easily grown in flasks and bioreactors, like bacteria, using simple and inexpensive culture media (44).

*P. pastoris* is a methylotrophic yeast (capable of growing in methanol as its only carbon source) that performs post-translational modifications, produces recombinant protein levels of one or two orders of magnitude above that of *Saccharomyces cerevisiae* (45), is capable of secreting heterologic proteins into the culture media (where the levels of native protein are very low), and in contrast with the latter, can be cultivated at cell densities of more than 100 g/L of dry weight (46).

# RECOMBINANT GROWTH HORMONES

In our laboratories, we identified as a scientific objective and a technological advantage, the construction and evaluation of GH protein producing *P. pastoris* strains. This as a first step in evaluating its potential in medicine as well as in animal health and productivity; searching to develop both infrastructure and experience in producing, purifying, and testing its biological activity.

Also, as previously mentioned, mammalian GHs have activity in phylogenetically inferior animals, nevertheless potentially adverse reactions to heterologic GHs can be triggered, which is why having a GH specific-species would avoid these undesirable side effects.

Regarding human hormones, we proposed constructing productive strains for the HGH22k, the HGH20k, the HGHV, and the HCS. With regard to animal GHs, we channeled our efforts into building strains to produce GHs from bovines (BGH), caprines (CHGH), ovines (OGH), equines (ECGH), canines (CFGH), porcines (PGH) and felines (FCGH); all based on the *Pichia pastoris* yeast expression system.

For this, the following experimental strategy was proposed:

- Obtain, clone, and manipulate cDNA from these hormones.
- Construct and insert into the genome of P. pastoris the hormones' expression cassettes.
- Develop the fermentation processes for each new strain.
- Implement the purification schemes of the recombinant hormones.
- Evaluate in vitro the bioactivity of the semipurified recombinant hormones.

As a result of this experimental work, we achieved the followings:

- Using different methodological approaches (RT-PCR, mutagenic PCR, subcloning, etc.) we cloned the cDNAs of the hormones of interest.
- Through genetic engineering manipulations, we converted the cloned cDNAs into expression cassettes capable of functioning in *Pichia pastoris*.
- The respective expression cassettes were integrated into the *Pichia pastoris* genome by homologous recombination.
- Through an inducible (with methanol) expression system, we were able to overproduce and recover from the culture media each of the respective recombinant hormones (rGHs/rHCSs).
- The data from the physicochemical and biological characterizations showed that the methodology described herein generates heterologous proteins that are identical to their natural counterparts and biologically active.

# TECHNOLOGICAL PLATFORM FOR THE PRODUCTION OF RECOMBINANT GHS

## Overall Strategy

As depicted in figures 4, the following are the two stages of the overall strategy in which the work was divided:

- Construction of *P. pastoris* strains carrying the hormones' expression cassettes producing rGHs/HCSs.
- Production and characterization of the recombinant hormones.

## Construction of Propagating GH cDNA Plasmids (pBS-XGHs)

Oligonucleotides for GHs cDNAs amplification by PCR were designed based on consensus nucleotide sequences of GHs (mature region) of related mammals. Extra restriction sites were added on their flanks (XhoI and AvrII) to facilitate insertion of the amplicon into the expression vector. With these, each of the hormones' cDNAs was amplified from plasmids previously constructed in our laboratory carrying the respective nucleotide sequences. Each amplicon was cloned into propagating plasmid such as the pBS II KS plasmid (+) and subsequently subcloned into the yeast expression vector pPIC9 at its multicloning site, between the restriction sites XhoI and AvrII (after previous purification of the corresponding fragment and vector), thereby giving rise to each of the pPIC9-XGH expression plasmids.

In CHGH's case, which differs from BGH in a single aa residue, a different strategy was implemented. Site-directed mutagenesis was used relying on a primer to convert codon 130 of BGH cDNA into one corresponding to CHGH. A 345 bp region containing the mutated GH cDNA was thus amplified, which was cloned in pBS and later transferred into pPIC9- BGH to converted it into pPIC9-CHGH (49).

## Construction of Expression Plasmids (pPIC9-XGHs) for Each Hormone

Preparative digestions of pBS-XGH and pPIC9 with the enzymes XhoI and AvrII were performed for all GHs except for CHGH. For CHGH, ApaI and XmaI (natural site) enzymes, which release a 133 bp fragment containing the mutagenized codon for CHGH, were used. This was purified and linked into the previously digested pPIC9-BGH vector in the same sites, replacing the fragment to originate the pPIC9-CHGH expressor vector The ligation reactions between pPIC9 and each cDNA were used to transform competent Ca++

cells of XL1-Blue Escherichia coli. PCR was used to verify that the resulting tranformants indeed carried each pPIC9-XGH, where "x" corresponds to each of the sequences of the hormone in question. The candidate clones produced by PCR with AOX1 primers for an amplicon of 1050 bp, since the expression cassette for each hormone is flanked by long regions of the AOX1 gene. While strains that were not integrated into the "cassette" gave rise to an amplicon of only 500 bp.

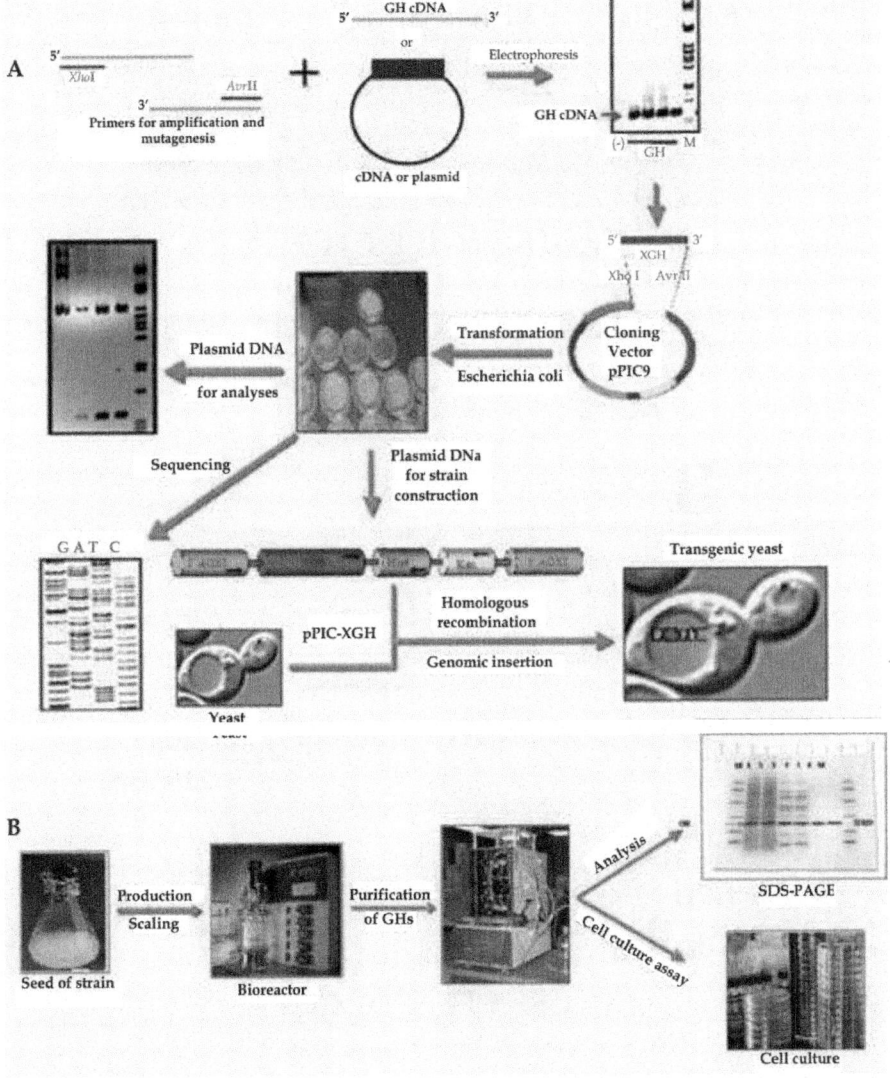

**Figure 4.** General strategy for strain construction and recombinant hormones production. (A) Genetic engineering phase. The steps followed to construct and characterize

new strains of GHs and HCSs producing Pichia pastoris are shown. Protocols followed were based in different techniques (47, 48). (B) Biotechnology phase. The steps followed for the production and scaling, semipurification and bioassay of each of the recombinant hormones are shown.

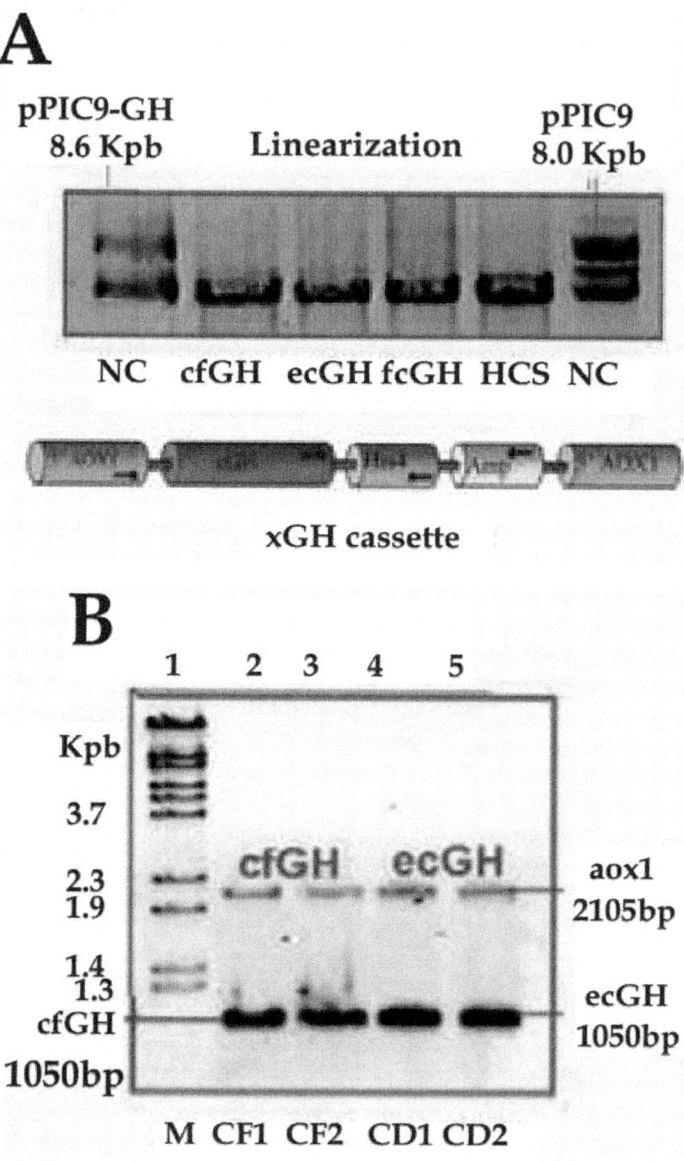

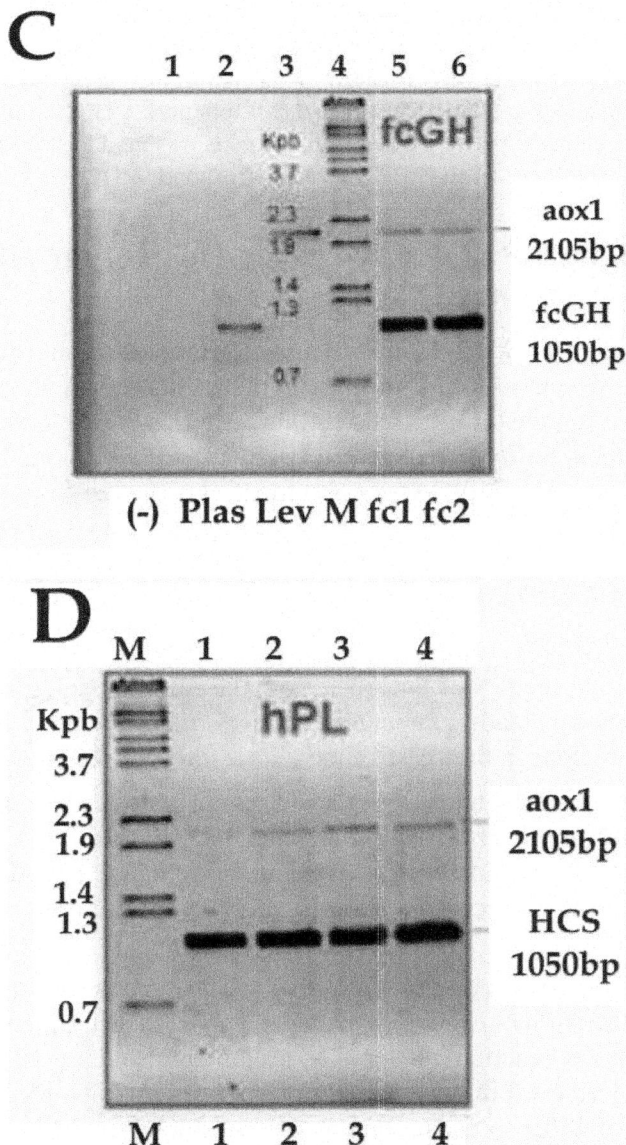

**Figure 5.** Detection of the expression "cassettes" of cfGH, ecGH, fcGH and hCS in P. pastoris genome. Analysis by PCR with AOX1 primer yeast strains transfected. In each case, the 1050 bp corresponds to the expression cassette of the recombinant hormone in question, while the 2105 bp to the AOX1 gene of the yeast itself. A) The diagram shows the linearized "cassette" of XGH and the gel products (which transfected into *Pichia pastoris* integrate the "cassette" into the genome) with SacI enzyme: cfGH = dog GH, ecGH = horse GH, fcGH = cat GH and HCS = human CS; in lane 1 NC-

GH = uncut plasmid pPIC9 and in lane 6 NC = uncut pPIC9 plasmid. B) CF (1 and 2) = dog GH lanes 2 and 3 respectively, and CD (1 and 2) = horse GH lanes 4 and 5, respectively. C) (-) = negative PCR lane 1, Plas= amplification positive control lane 2, Lev = *Pichia pastoris* genomic DNA lane 3, M= pb marker lane 4 and fc (1 and 2) = cat GH lanes 5 and 6 respectively. D) Lanes 1 to 4 correspond to strains with the HCS "cassette". M = marker-bp λ BsteII. The gels correspond to 1% agarose.

# CONSTRUCTION OF GHS PRODUCING *PICHIA PASTORIS* STRAINS

The *Pichia pastoris* GS115 strain has a mutation in the histidinol dehydrogenase (his4) gene, which prevents it from synthesizing histidine. The class of plasmids used to transform it contains this gene (his4). The transformants are selected for their ability to restore growth in a medium lacking histidine. The plasmid vectors of the pPIC series and those constructed to express the GHs are of this class.

## Insertion of GHs Expression Cassettes into the Genome of *P. pastoris*

Each pPIC-XGH vector was linearized with the enzyme SacI, transformed into the yeast previously made competent for transformation and left exposed to the homologous regions in the yeast genome necessary for recombination.

After incubating the DNA with competent cells, transformation reactions were plated to recover clones needing no histidine to grow (HIS+ transformants). Then transformants were analyzed on their genomic DNAs by PCR using AOX1 primers to verify the presence of the transgenic hormone expressing cassette.

Verification of integration into the yeast genome of the expression cassette of the hormones was achieved in agarose gel by confirming that the amplification reaction rendered a prominent band of 1050 bp, which corresponds to the expression cassette of the hormone involved in each case and another of 2105 bp corresponding to the endogenous gene AOX1 of the yeast genome (Figure 5). In addition, each hormone "cassette" was subjected to nucleotide sequencing to verify that all they corresponded to the expected growth hormones.

## Analysis of New *Pichia pastoris* Strains´ Phenotypes

*Pichia pastoris* strains were grown and the biomass was adjusted to low cell density (0.5 u at 600 nm). These were transferred to induction medium with 0.5% methanol and grown for 100 hours with the addition of methanol every 24 hours to compensate for its evaporation. Biomass growth was analyzed under methanol as the sole carbon source. The Mut+ phenotype strains metabolize methanol more rapidly, achieving significantly higher cell densities than their Muts counterparts that metabolize more slowly, appreciating a slight increase in biomass under the same fermentation conditions.

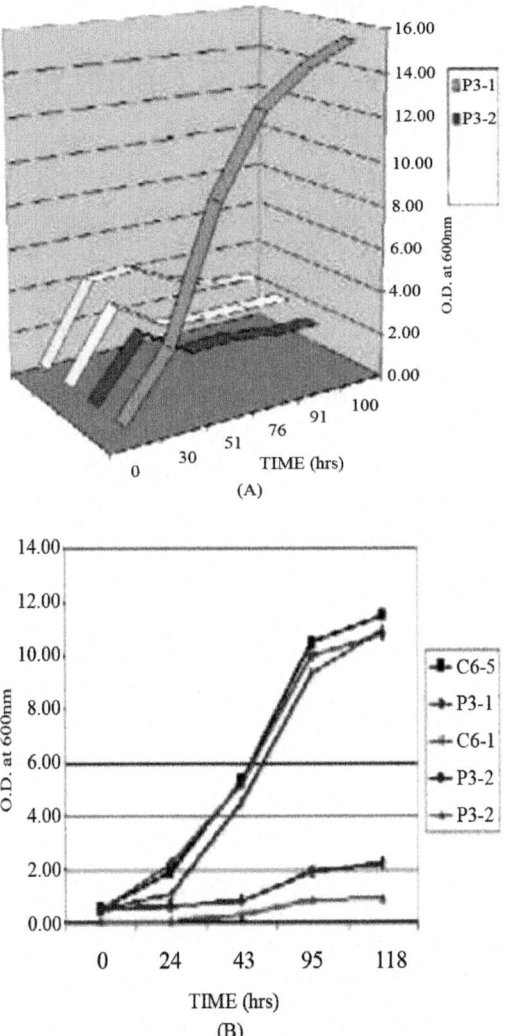

**Figure 6.** Mut phenotype characterization in Pichia pastoris strains. Growth kinetics in minimal medium using methanol as sole carbon source. Induction was started at the density of 0.5 U and ended after about 100 hrs. (A) Plot of the samples P3-1 and 2 = dog GH 1 and 2 strains, CS3 CS-2 = 2 and strain human strains pPIC9 = "mock" with the pPIC9 plasmid. (B) Graph of the C6-5T samples = horse GH, P3-1T =dog GH, C6-4T = horse GH, GH P3-Q2 = P3-dog and dog-2B = GH.

An analysis of the growth of the strains after 100-hour fermentation with 0.5% methanol identified the Mut phenotype of each strain.

After fermentation, the strains found to be Mut+ reached about 15 optical units at 600 nm, while those that were Muts did not exceed 2 units (Fig. 6). In the strains that had been built previously, their Mut phenotype was inferred when these were fermented in the bioreactor.

## PRODUCTION AND ANALYSIS OF RECOMBINANT HORMONES IN THE FLASK

To test the fermentation of strains, a biomass was generated in a flask. This was inoculated with a colony of each strain in 25 ml of biomass producing culture medium (BMGY) (50). This was incubated at 30°C at 250 rpm for 24 to 48 hours for the first stage of growth until a biomass with an OD of 600 nm of 10 was reached.

For the second stage, which is the induction of the recombinant hormone production, yeasts were harvested by centrifugation and the packed cells were washed with 30 mL sterile water, then these were pelleted and resuspended in fresh cassette induction medium (with methanol) (BMMY) (50). The induction was maintained by adding methanol every 24 hours to a final concentration of 1% to compensate for loss by evaporation. The experiment lasted 96 hrs. Figure 7 shows the process that was followed.

When analyzing the polyacrylamide gels of proteins from the culture media of each strain, we observed that all constructions produced and directed the secretion into the medium of the recombinant hormone in question to a greater or lesser extent. For the particular case of CFGH it was observed that a strain of Muts phenotype displayed better production of recombinant protein than its counterpart Mut+. Strain of HGH-V proved to be the least productive (Fig. 8).

Figure 9 shows the gel with the results of the production of all recombinant strains generated in *Pichia pastoris*. They all produced different amounts of their respective hormone at the level of 22 kDa, except for the HGH20k, which, migrated below the rest of the recombinant hormones.

The percentage of each recombinant hormone in the culture medium was estimated by densitometry of each gel. For this we used the Gel-Doc software by BIO-RAD (Hercules, CA. EUA) and the ImageJ program (51). The results of estimation of the percentage of each hormone in relation to background proteins from *Pichia pastoris* were: HCS = 65%, CFGH = 60%, HGH22k = 30%, ECGH = 30%, BGH = 25%, FCGH = 25%, CHGH = 25%, HGH20k = 12% and HGHV = 8%.

Production kinetics was carried out for CHGH strain in a flask with a volume of 50 ml of rich medium. Samples were taken at 24, 48, 72, 96 and 120 hours of induction with methanol with restitution every 24 hours of 1% methanol. Bradford protein determination showed that the production of total protein secreted into the culture medium was 20µg/mL by densitometry and 60% represented CHGH giving us 12µg/mL of production of the recombinant hormone.

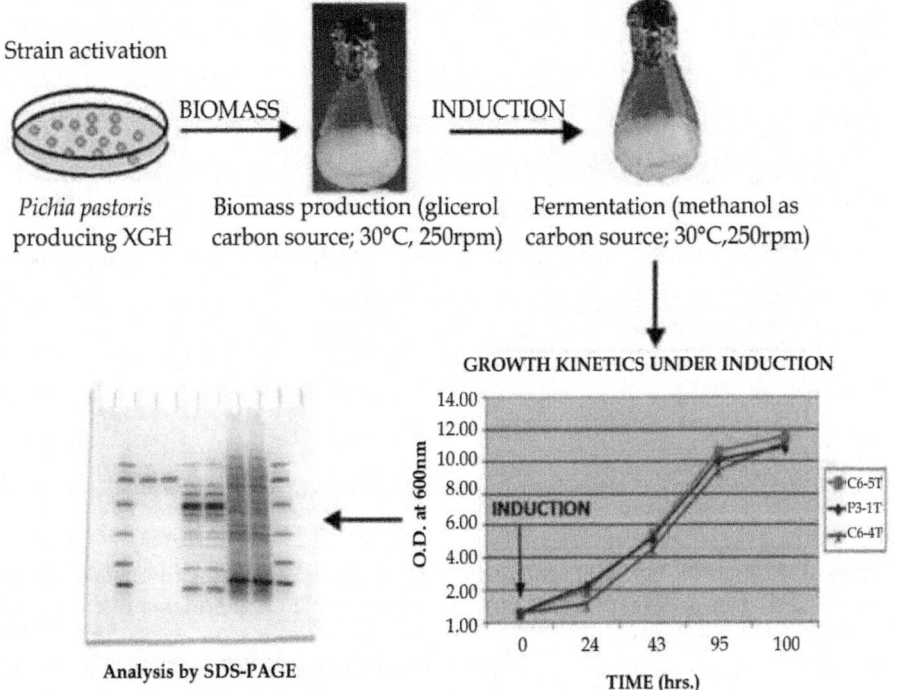

**Figure 7.** Outline of the fermentation process. General procedure for the biotechnological production of recombinant hormones by fermentation of each strain. Strains were plated to activate them, incubated in liquid medium to generate a biomass flask, and the induced transgene expression by adding methanol. The culture medium was analyzed by SDS-PAGE in search of the hormones that are migrating around 22 kDa

## PRODUCTION SCALING IN THE BIOREACTOR

When passing to a bioreactor and increasing the scale, it is possible to obtain protein concentrations 20 to 200 times greater than in flasks. In the fermentor *Pichia pastoris* reaches high cell densities greater than 100 g/L of dry weight (46).

The fermentor was Bioflo 3000 (1 liter) of New Brunswick Scientific (NBSC) (NJ. EUA). The type of fermentation conducted was in fed-batch. The parameters monitored were scheduled addition of substrates to the fermentor, pH, percentage of dissolved oxygen, agitation, temperature, and aeration. The process involved three basic steps: 1) obtaining high densities of biomass, 2) induction of the cassette expression of each hormone with methanol and 3) harvest of biomass and culture medium containing the recombinant protein. Figure 10 shows the steps followed for the recombinant production of each hormone.

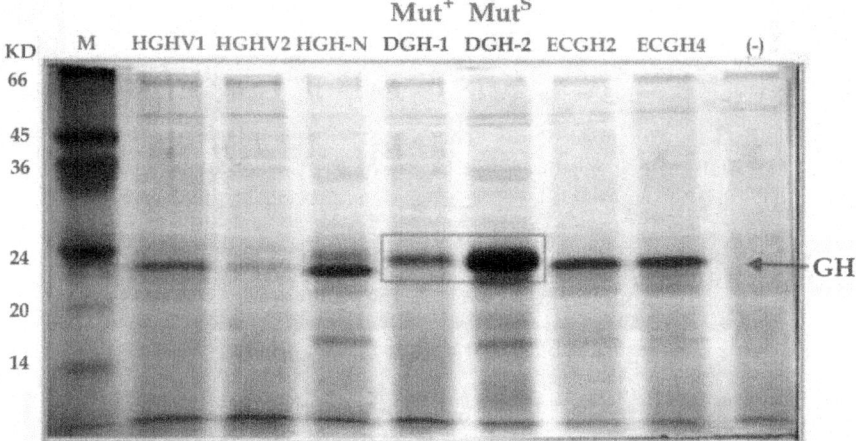

**Figure 8.** Production in *Pichia pastoris* of recombinant GHs (CFGH, HGH, HGH-V and ECGH) at flask level. The bands correspond to the GHs proteins resolved at the level of 22 kDa that come from the culture media induced with methanol. The lanes are: M = molecular weight marker, HGHV1 = HGH variant strain 1; HGHV2 = HGH variant strain 2; HGH = normal pituitary HGH of 22 kDa; DGH-1 = Dog GH strain 1; DGH-2 = dog or *Canis familiaris* GH strain 2; ECGH-2 = horse GH strain 2, ECGH-4 = horse GH strain 4 and the last lane identified as (-) = negative control of PCR. Mut+= methanol utilization plus; Muts = "methanol utilization slow". Note the prominent band of the DGH-2 corresponding to the Muts phenotype, compared to the lower intensity of Mut+. The samples correspond to 500 μL concentrates of the original media. Gel corresponds to one of 15% polyacrylamide-SDS stained with Coomassie blue.

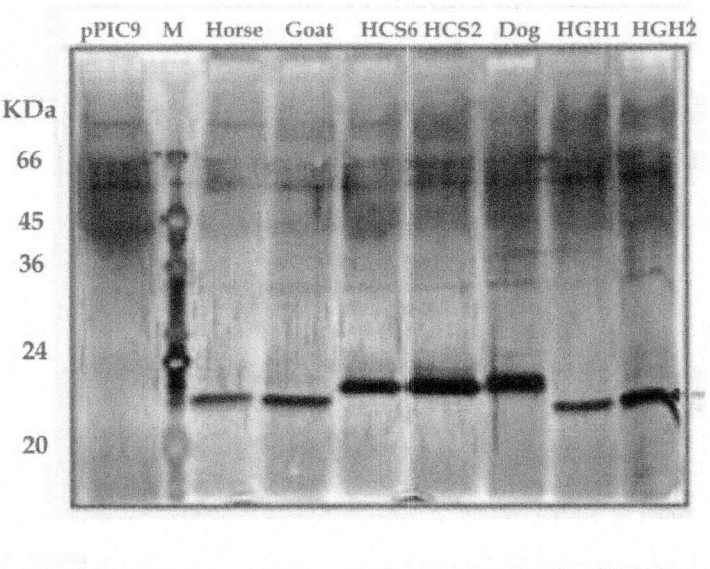

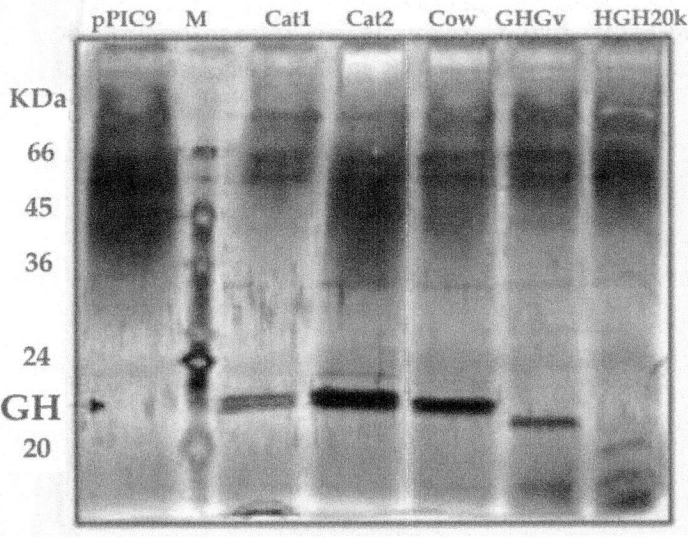

**Figure 9.** Flask production in all strains yielding recombinant hormones. In all cases a prominent band is seen (except for HGH20k) at the 22 kDa level for each hormone. They are seen in their respective lane in each case indicated by their name. The lanes of the left side gel show the proteins from the culture media with recombinant hormones: horse = ECGH, goat = CHGH, HCS (6 and 2) = Human chorionic somatomammotropin clones 6 and 2, dog = CFGH and HGH (1 and 2) = cloned human GH 1 and 2. In the right gel lanes: cat (1 and 2) = GH 1 and 2 from cat; cow = BGH; HGHv = HGH placental variant and HGH20k = isoform of 20 kDa of hGH. The pPIC9 lane refers to the "mock" strain of *Pichia pastoris*. SDS-PAGE 15% gels are silver stained.

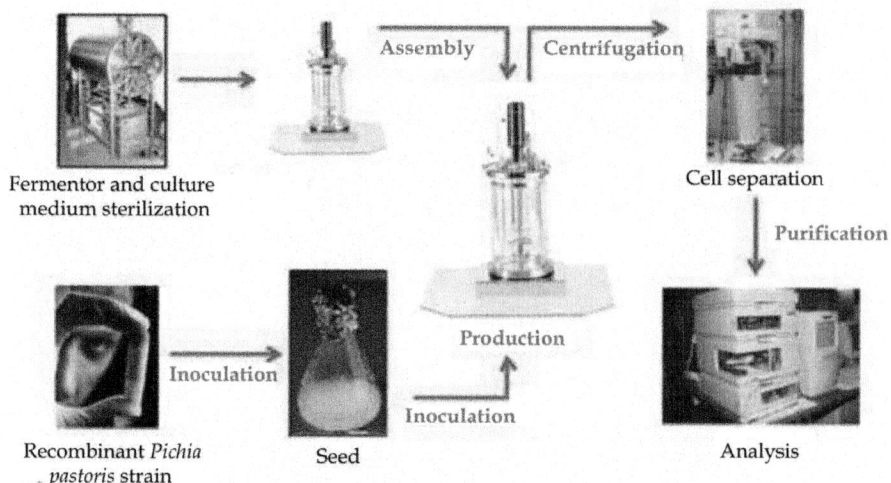

**Figure 10.** Recombinant hormone production bioreactor. This illustrates the stages of the biotechnological process of production, from the preparation of the fermentor and the medium, to the analysis of proteins in the fermented culture medium.

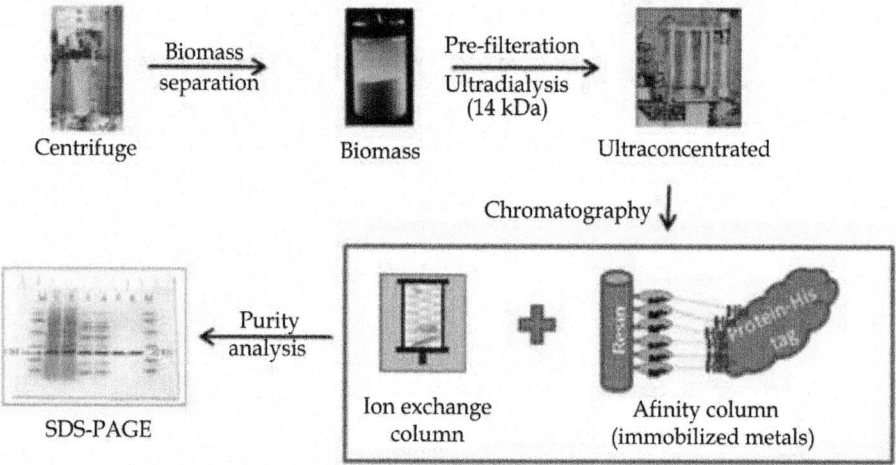

**Figure 11.** "DownStream" Process for recombinant hormones. The medium was separated from the biomass by centrifugation. This was pre-filtered with a 0.45µm membrane, and ultra-dialyzed in a cutoff membrane of 10 kDa. It was then passed through an anion exchange column (FF-QS) and/or in an affinity column by immobilized metal (IMAC). Purity was analyzed in PAGE-SDS, quantified, and finally lyophilized and stored for later use.

# SEMIPURIFICATION OF RECOMBINANT HORMONES

Figure 11 shows the process of purification or "downstream" that was followed for each recombinant hormone produced.

Each culture medium containing the recombinant hormone was ultra-dialyzed and ultraconcentrated. The pore membrane used was 14 kDa. At the same time it was ultradyalized. The ultraconcentrate obtained was lyophilized to preserve the samples and all the powder was recovered in 50 mL plastic tubes and weighed. Total protein was quantified in each lyophilized culture medium with the Bradford method (52). Samples were stored at -20°C until use.

Each sample was prepared for loading into a chromatographic column. Five mg of the total protein from each sample was adjusted to the conditions of the loading buffer. A column was loaded with Q-Sepharose fast flow anion exchange resin; column filling with resin was carried out by gravity flow. After passing the sample, washing was done with 10 mL of loading buffer. To recover the proteins from the column, 15 mL of elution buffer (loading buffer plus NaCl) was passed with an ionic strength increased sequentially with NaCl. Total protein was measured by the Bradford method (52) for each of the fractions recovered to see what percentage of the total they represented.

The collected fractions were visualized on discontinuous polyacrylamide gel concentrations of 4-15% under denaturing conditions (SDS-PAGE), and stained with Coomassie and silver techniques (53). The fractions were subjected to lyophilization, the powder was recovered and weighted, total protein was measured by the Bradford method. In addition, using PAGE-SDS the percentage of purity of the monomer was determined with ImageJ software. The samples were stored at -20°C until used for further analysis or to determine their biological activity.

# TESTING THE BIOLOGICAL ACTIVITY OF RECOMBINANT HORMONES

## Lactogenic Activity Bioassay

The biological activity of hormones produced in *Pichia pastoris* recombinant was determined by their ability to promote proliferation of the Nb2 cell line, which comes from rat lymphoma (54, 55). GHs were also tested for their somatogenic activity in the adipocyte differentiation model based on the glyceraldehyde 3-phosphate dehydrogenase (GPDH) assay.

The cell-free culture medium was dialyzed and each hormone was quantified by gel densitometry. Dilutions of each recombinant hormones in the culture medium dialyzed were tested. Cell proliferation was determined by tetrazolium salt assay (MTT) (56) and was expressed as the average of three repetitions, in comparison with the positive control recombinant rat prolactin (rRPRL) and the negative control (culture medium without hormone). The concentrations tested were 0.001, 0.01, 0.1, 1, 10 and 50 nM of the following hormones, CFGH, ECGH, FCGH, HGH, HCS and RPRL. Upon completion of the testing time, we proceeded to measure the effect of hormones on cell proliferation by MTT assay (56). The activity was evaluated by color generation based on the reduction of tetrazolium salt (methyl 3- [4,5-dimethylthiazol-2-yl] -2,5-diphenyl tetrazolium) from yellow to purple forming crystals by Nb2 cell metabolism. An increase of living cells is reflected by increased metabolic activity. This increase directly correlates with the formation of absorbance monitored formazan crystals; i.e., the greater the number of cells, the greater the increase in cell metabolism with greater formazan formation and greater biological activity.

The bioassay was carried out in triplicate in a humid atmosphere with 5% $CO_2$-95% air at 37°C. After the incubation period of 3 days, we added 10µL of MTT (to a final concentration of 0.5 mg/mL) to each well. Samples were incubated for 4 hrs under the atmospheric conditions mentioned above. After this time, we added 100 uL of formazan solubilizing solution (10% SDS in 0.01 M HCl) to each of the wells. These were left incubating overnight under the same atmospheric conditions. It was verified that the precipitate of formazan purple crystals had completely dissolved and absorbance was measured at 590 nm using an ELISA plate reader.

## Somatogenic Bioassay

To test the biological activity of goat GH, we assayed fibroblasts from cell line 3T3-F442A (pre-adipocytes) and with the glyceraldehyde 3-phosphate dehydrogenase (GPDH) assay as described in (18, 57). Cells were exposed to different concentrations of CHGH in the medium. The positive control was 10% fetal bovine serum (FBS) (v/v) the negative control was 0.25% FBS (v/v). The cells were incubated for 12 days, were harvested and proteins extracted; the supernatants were frozen in aliquots for subsequent tests. Specific activity of GPDH was measured in cell extracts by NADH oxidation measured in a spectrophotometer at 340 nm; 40 µg of protein was used. CHGH demonstrated biological activity in the essay. The same was doing with the others hormones showing biological activity.

# CONTRIBUTIONS

These are the first reports in the literature of the production by biotechnology of the recombinant GHs described here using the Pichia pastoris methylotrophic system. The new strains of Pichia pastoris constructed with GH cassettes such as canine-porcine, equine, feline, and caprine, complemented our strains previously constructed for GHs of 22 y 20 kDa, HCS and bovine GH (unpublished results), making our collections of GH clones the largest one for this protein family in the world as far as we know.

All hormones were efficiently produced, processed, and secreted into the culture media. Each hormone constituted the main proteic band among proteins secreted by Pichia pastoris analyzed by SDS-PAGE. The phenotypes Muts resulted the best for producing GHs, as was the case for CFGH (58) and for HCS (Ascacio-Martínez y Barrera-Saldaña, unpublished results).

All our constructed strains had correct processing of their heterologous secretion Saccharomyces cerevisiae alpha mating factor signal peptide in the maturation of the recombinant GHs. They were secreted into the culture media in their native and bioactive form (49, 59, 60).

Using a bioreactor increased the production of the recombinant proteins 10 to 20 times compared to an Erlenmeyer flask. Ionic chromatography was a good option in all cases for semipurification of rGHs and HCS in this system. All hormones showed biological activity in the Nb2 essay, showing that human GH had more activity than animal GHs. The same happened in the pre-adypocite system (3T3), concluding that *Pichia pastoris* produces, processes, and secretes rGHs in the bioactive form.

Biotechnological platforms were developed that made possible to move from the construction of the producer clones to the bioassay of the semipurified recombinant protein. With the technology here developed we have acquired the capacity to advance in scaling the production of veterinary and livestock rGHs for their field tests.

# PERSPECTIVES

We have developed an efficient expression system and laboratory fermentor-scale production biotechnological platform with which to partner with the productive sector to produce virtually any GH of human and animal origin with acceptable quantity, quality and activity to start field evaluations. Doing that, would allow us investigate their full potential in animal biotechnology, to then offer them as an option in veterinary treatments and to stimulate cattle production and the health of competition animals. In addition to the obvious veterinary or livestock application, their availability will also allow the discovery of unexpected biological activities for animal wellness.

## ACKNOWLEDGEMENTS

The authors want to thank J.M. Reyes, J.P. Palma, C.N. Sanchez, L.L. Escamilla, H.L. Gallardo, E.L. Cab, R.G. Padilla and M. Guerrero for their support, experiments during their thesis and valuable contributions to the information here reviewed. Authors thank Sergio Lozano for his critical reading of the manuscript.[

## REFERENCES

1.    Goeddel D.V., Heyneker H.L., Hozumi T., Arentzen R., Itakura K., Yansura D.G., Ross M.J., Miozarri G., Crea R., Seeburg P. Direct expression in Escherichia coli of a DNA sequence coding for human growth hormone. Nature 281(5732): 544-548, (1979).

2.    Niall H.D., Hogan M.L., Sayer R., Rosenblum I.Y., Greenwood, F.C. Sequences of pituitary and placental lactogenic and growth hormones: evolution from a primordial peptide by gene duplication. Proc. Natl. Acad. Sci. USA, 68: 866-869, (1971).

3.    Elian, G., Jamieson, J., Gross, S. Staying Young: Growth Hormone and Other Natural Strategies to Reverse the Aging Process. Age Reversal Press. First Edition pp:120, (1999).

4.    De Noto F, Rutter J.W., Goodman H.M. Human growth hormone DNA sequence and mRNAstructure: possible alternative splicing. Nucleic Acid Res. 9: 3719-30, (1981).

5.    Tsushima T., Katoh Y., Miyachi Y., Chihara K., Teramoto A., Irie M., Hashimoto, Y. Serum concentrations of 20K human growth hormone in normal adults and patients with various endocrine disorders. Study Group of 20K hGH. Endocr. J. 47 Suppl: S 17-21, (2000).

6.    Hashimoto Y., Kamioka T., Hosaka M., Mabuchi K., Mizuchi A., Shimazaki Y., Tsuno M., Tanaka T. Exogenous 20 K growth hormone (GH) suppresses endogenous 22K GH secretion in normal men. J. Clin. Endocrinol. Metab. 85(2): 601-606, (2000).

7.    Wang D.S., Sato K., Demura H., Kato Y., Maruo N., Miyachi Y. Osteo-anabolic effects of human growth hormone with 22K- and 20K Daltons on human osteoblast-like cells.Endocr. J. 46(1): 125-132, (1999).

8.    Cooke N.E., Ray J., Emery J.G., Liebhaber S.A. Two distinct species of human growth hormone-variant mRNA in the human placenta predict

the expression of novel growth hormone proteins. J. Biol. Chem. 263: 9001-9006, (1988).

9.  Ray J., Jones B., Liebhaber S.A., y Cooke N.E. Glycosylated human growth hormone variant. Endocrinology 125: 566-568, (1989).

10. Frankenne F., Scippo M., Van Beeumen J., Igout A., Hennen G. Identification of placental human growth hormone as the growth hormone-V gene expression product. J. Clin. Endocrinol. Metab. 71: 15-18, (1990).

11. Boguszewski C.L., Svensson P.A., Jansson T., Clark R., Carlsson M.S. Carlsson B. Cloning of two novel growth hormone transcripts expressed in human placenta. J. Clin. Endocrinol. Metab. 83(8): 2878-2885, (1988).

12. Frankenne F., Closset J., Gomez F., Scippo M.L., Smal J., Hennen G. The physiology of growth hormones (GHs) in pregnant women and partial characterization of the placental GH variant. J. Clin. Endocrinol. Metab. 66(6): 1171-1180, (1988).

13. Mirlesse V., Frankenne F., Alsat E., Poncelet M., Hennen G., Evain-Brion D. Placental growth hormone levels in normal pregnancy and in pregnancies with intrauterine growht retardation. Pediatr Res. 34: 439-442, (1993).

14. Chowen J.A., Evain B.D., Pozo J., Alsat E., García Segura L.M., Argente J. Decreased expression of placental growth hormone in intrauterine growth retardation. Pedriatr. Res. 39 (4 Pt 1): 736-739, (1996).

15. Pardi G., Marcini A.M., Cetin I. Pathophysiology of intrauterine growth retardation: rol of the placenta. Acta Pediatr. Suppl. 423: 170-172, (1997).

16. Rygaard K., Revol A., Esquivel-Escobedo D., Beck B.L., Barrera-Saldaña H.A. Absence of human placental lactogen and placental growth hormona (HGH-V) during pregnancy: PCR análisis of the deletion. Human Genet. 102(1): 87-92, (1998).

17. Morikawa M., Green H., Lewis U.J. Activity of human growth hormone and related polypeptides on the adipose convertion of 3T3 cells. Molecular and Cellular Biology 4(2): 228-231, (1984).

18. Juarez-Aguilar, E., Castro-Munozledo, F., Guerra-Rodriguez, N.E., Resendez-Perez, D., Martinez-Rodriguez, H.G., Barrera-Saldana, H.A., Kuri-Harcuch, W. Functional domains of human growth hormone necessary for the adipogenic activity of hGH/hPL chimeric molecules. J. Cell Sci. 112(18):3127-3135, (1999).

19. MacLeod J.M., Worsley I., Ray J., Friesen H.G., Liebhaber S.A., Cooke N.E. Human growth hormone-variant is a biologically active somatogen and lactogen. Endocrinol. 128(3): 1298-1302, (1991).

20. Cooke N.E. Prolactin: normal synthesis and regulation. En DeGroot L.J. (ed.) Endocrinology.Saunders, Philadelphia., 1: 384-407, (1989).

21. Chen E.Y., Liao Y.C., Smith D.H., Barrera-Saldaña H.A., Gelinas R.E., Seeburg P.H. The human growth hormone locus: nucleotide sequence, biology, and evolution. Genomics 4(4): 479-97, (1989).

22. Barrera-Saldaña H.A. Growth hormone and placental lactogen: biology, medicine and biotechnology. Gene 211: 11-18, (1998).

23. Barrera-Saldaña H.A.,Seeburg P.H., Saunders G.F. Two structurally differentgenes produce the secreted human placental lactogen hormone. J. Biol. Chem. 258: 3787- 3793, (1983).

24. Canizales-Espinosa, M., Martínez-Rodríguez, H.G., Vila, V., Revol, A., Castillo-Ureta, H., Jiménez-Mateo, O., Egly, J.M., Castrillo, J.L. and Barrera-Saldaña, H.A. Differential strength of transfected human growth hormone and placental lactogen gene promoters. J. Endocr. Genet. 4 (1): 25-36, (2005).

25. Peel C.J. Bauman D.E. Somatotropin and lactation. J. Dairy Sci. 70: 474-486, (1987).

26. Etherton T.D., Kensinger R.S. Endocrine regulation of fetal and postnatal meat animal growth.J Anim Sci. 59(2): 511-528, (1984).

27. Bauman D.E. Regulation of nutrient partitioning: homeostasis, homeorhesis and exogenous somatotropin. Keynote lecture. En: Seventh International Conference on Production Disease in Farm Animals, F.A. Kallfelz pp. 306-323, (1989).

28. Sun M. Market sours on milk hormone. Science 17: 246(4932): 876-877, (1989).

29. Juskevich J.C., Guyer C.G. Bovine growth hormone: human food safety evaluation. Science 24: 249(4971): 875-884, (1990).

30. Boutinaud M., Rulquin H., Keisler D.H., Djiane J., Jammes H. Use of somatic cells from goat milk for dynamic studies of gene expression in the mammary gland. J. Anim. Sci. 80(5): 1258-69, (2002).

31. Stewart F., Tuffnell, P.P. Cloning the cDNAfor horse growth hormone and expression in Escherichia coli. J. Mol. Endocr. 6: 189-196, (1991).

32. http://www.jockeysite.com/stories/e_melbournecup1.htm

33. Ascacio-Martínez J.A., Barrera-Saldaña, H.A. A dog growth hormone cDNA codes for mature protein identical to pig growth hormone. Gene 143: 299-300, (1994).

34. Evock, C.M., Etherton, T.D., Chung C.S., Ivy, R.E. Pituitary porcine growth hormone (pGH) and a recombinant pGH analog stimulate pig

growth performance in a similar manner. J. Anim. Sci. 66(8): 1928-1941, (1988).

35. Daughaday W.H. The anterior pituitary. Williams textbook of Endocrinology. Ed. Philadelphia, WB Saunders. 568-613, (1985).

36. Eigenmann J.E. Diagnosis and treatment of dwarfism in a german shepherd dog. J. Am. Anim. Hosp. Assoc. 17: 798-804, (1981).

37. Muller G.H., Kirk R.W., Scott D.W. Small animal dermatology. Philadelphia. WB. Saunders. 4th Edition, 575-657, (1989).

38. Devlin R.H., Byatt J.C., Maclean E., Yesaki T.Y., Krivi G.G., Jaworski E.G., Clarke W.C. Bovine placental lactogen is a potent stimulator of growth and displays strong binding to hepatic receptor sites of coho salmon. General and Comparative Endocrinology, 95: 31-41, (1994).

39. Tokunaga T., Iwai S., Gomi H., Kodama K., Ohtsuka E., Ikehara M., Chisaka O., Matsubara K. Expression of synthetic human growth hormone gene in yeast. Yeast. 39: 117-120, (1985).

40. Franchi E., Maisano F., Testori S.A, Galli G., Toma S., Parente L., Ferra F.D., y Grandi G. A new human growth hormone production process using a recombinant Bacillus subtilisstrain. J. Biotechnology 18: 41-54, (1991).

41. Pavlakis G.N., Hizuka N., Gorden P., Seburg P.H., Hamer D.H. Expression of two human growth hormone genes in monkey cell infected by simian virus 40 recombinants. Proc. Natl. Acad. Sci. 78: 7398-7402, (1981).

42. Kerr D.E., Liang F., Bondioli K.R., Zhao H., Kreibich G., Wall R.J., Sun T.T. The bladder as a bioreactor: Urothelium production and secretion of growth hormone into urine. Nature Biotechnology 16: 75-78, (1997).

43. Escamilla-Treviño L.L., Viader Salvado J.M., Barrera Saldaña H., Guerrero Olazaran, M. Biosynthesis and secretion of recombinant human growth hormone. In: Pichia Pastoris. Biotechnology Letter 22: 109-114, (2000).

44. Romanos M.A., Scorer C.A., Clare. J.J. Foreign gene expression in yeast: A review. Yeast 8: 423-488, (1992).

45. Faber K.N., Harder W., Veenhuis, M. Review: Yeasts as factories for the production of foreign proteins. Yeast 11: 1131-1344, (1995).

46. Siegel R.S., Brierley R.A. Methylotrophic yeast Pichia pastoris produced in high-celldensity fermentation with high cell yields as vehicle for recombinant protein production. Biotechnol. Bioeng. 34: 403-404, (1989).

47. Ausubel, F.M., Brent, R., Kingston, R.E., Moore, D.D., Seidman, J.G., Smith, J.A., Struhl, K. Short Protocols in Molecular Biology, fourt ed., Wiley, Massachusetts, (1999).

48. Sambrook, J., Fristsch, E., Maniatis, T. Molecular Cloninig: A Laboratory Manual. Segunda Edición. Cols. Spring Harbor Laboratory Press. Cold Spring Harbor, (1989).

49. Reyes-Ruíz, J.M., Ascacio-Martínez, J.A., Barrera-Saldaña, H.A. Derivation of a growth hormone gene cassette for goat by mutagenesis of the corresponding bovine construct and its expression in Pichia pastoris. Biotechnology Letters. 28(13):1019- 25, (2006).

50. Invitrogen. Products for Gene Expression and Analysis. Instruction manual. Pichia Expression Kit. Protein Expression. A Manual of Methods for Expression of Recombinant Proteins in Pichia pastoris. Version L., (2000).

51. Rasband, W.S. ImageJ. U. S. National Institutes of Health, Bethesda, Maryland, USA, http://imagej.nih.gov/ij/, 1997-2011.

52. Bradford, M. (1976). A rapid and sensitive method for the quatitation of microgram quantities of protein utilizing the principle of protein-dye binding. Anal. Biochem., 72:248-254.

53. Merril, C.R. Gel Staining Techniques. Guide to Protein Purification: Methods in Enzymology (Methods in Enzymology Series, Vol 182). Murray P., Deutscher John N. y Abelson. pp. 477, (1990).

54. Tanaka, T., Shiu, R.P., Gout, P.W., Beer, C.T., Noble, R.L., Friesen, H.G. A new sensitive and specific bioassay for lactogenic hormones: measurement of prolactin and growth hormone in human serum. J. Clin. Endocrinol. Metab. 51(5):1058-1063, (1980).

55. Lawson, D.M., Sensui, N., Haisenleder, D.H., Gala, R.R. Rat lymphoma cell bioassay for prolactin: observations on its use and comparison with radioimmunoassay. Life Sci. 31(26):3063-3070, (1982).

56. Gerlier, D., Thomasset, N. Use of MTT colorimetric assay to measure cell activation. J. Immunol. Methods. 94(1-2):57-63, (1986).

57. Castro-Munozledo, F., Beltran-Langarica, A., Kuri-Harcuch, W., Commitment of 3T3- F442A cells to adipocyte differentiation takes place during the first 24-36 h after adipogenic stimulation: TNF-alpha inhibits commitment. Exp. Cell Res. 284, 161- 170, 2003.

58. Ascacio-Martínez, J.A., Barrera-Saldaña H.A. Production and secretion of biologically active recombinant canine growth hormone by Pichia pastoris. Gene. 340(2):261- 266, (2004).

59. Palma-Nicolás, J.P., Ascacio-Martínez, J.A., Revol-de-Mendoza A. y Barrera-Saldaña, H.A. Production of recombinant human placental variant growth hormone in Pichia pastoris. Biotechnology Letters. 27(21):1695-1700, (2005).

60. Treerattrakool S, Eurwilaichitr L, Udomkit A, Panyim S. Secretion of Pem-CMG, a peptide in the CHH/MIH/GIH family of Penaeus monodon, in Pichia pastoris is directed by secretion signal of the alpha-mating factor from Saccharomyces cerevisiae. J Biochem. Mol. Biol. 35(5):476-81, (2002).

# Chapter 3

# THE ROLE OF DNA REPAIR PATHWAYS IN ADENO-ASSOCIATED VIRUS INFECTION AND VIRAL GENOME REPLICATION / RECOMBINATION / INTEGRATION

Kei Adachi[1,2] and Hiroyuki Nakai[1,2]

[1]Department of Microbiology and Molecular Genetics, USA
[2]University of Pittsburgh School of Medicine, USA

## INTRODUCTION

Cellular DNA is constantly being damaged not only by extrinsic factors such as ionizing radiation and environmental carcinogens but also by intrinsic agents such as reactive oxygen species arising during normal cellular metabolism. Of the myriad of DNA lesions, inflicted by extrinsic and intrinsic genome damaging agents, DNA double strand break (DSB) is the most threatening. Replication fork arrest at DNA lesions could also be a threat since stalled replication forks, if fail to restart appropriately, induce DNA strand breaks. When cells encounter such strand breaks and other types of DNA damage, they mount a DNA damage response (DDR) (Harper & Elledge, 2007) that senses DNA damage and initiates a cascade of signal transduction pathways consequently culminating in cell cycle arrest, DNA repair and/or apoptosis when the DNA lesions become irreparable. Although cells are equipped with such DNA damage sensing and repair machinery primarily to handle damaged cellular DNA, triggers and receivers of DDR are not necessarily the cells' own genetic materials. DDR can also be provoked by essentially non-damaged DNA exogenously introduced into cells, most commonly viral genetic materials in nature and recombinant DNA (e.g., viral vectors for gene delivery) in laboratory.

During virus-host interaction, viruses manipulate DDR upon infection of cells in a way that benefits their life cycles, while host cells fight against them to eliminate the invaders. DDR is detrimental to viral life cycles in many instances; therefore, DDR is often viewed as an innate antiviral host defense mechanism. For example, adenoviruses express viral proteins that block the cellular non-homologous end joining (NHEJ) pathway, which, unless inactivated, concatemerizes viral genomes and prohibits viral genome packaging into virions (Evans & Hearing, 2005; Stracker et al., 2002). Viruses may also take advantage of DDR in their life cycle as seen in retroviruses, which exploit the NHEJ pathway to complete insertion of their genomic materials into host cellular DNA (Daniel et al., 1999; Li et al., 2001). Similar but distinct types of "intervention" by viruses on DDR have been found in many other viruses (Lilley et al., 2007; Weitzman et al., 2004, 2010). Thus, understanding DDR and DNA repair machinery is imperative for elucidating the biology of viruses and viral vectors, and conversely, studying virus biology provides new insights into fundamental biological processes elicited by DNA damage. In this context, interactions between viruses and host DDR and DNA repair machinery have recently gained attention and established a new area of basic research. Importantly, this field of study is relevant to gene therapy research in overcoming its limitations and drawbacks and improving the current molecular therapy approaches.

Adeno-associated virus (AAV) represents a good example for exploring this new research field, which studies the interactions between viruses and DNA repair machinery. AAV has become increasingly popular as a promising gene delivery vehicle. Wild type AAV (wtAAV) is replication defective and recombinant AAV (rAAV) is devoid of virally encoded genes. Despite their replication-defective nature and/or lack of expression of viral proteins, there are significant interactions between virus and host DNA repair machinery, which determine the fates of the virus and the host cells following infection. In this chapter, we provide an overview of how wtAAV and rAAV alter the fate of the host cells through DDR, and how DDR processes the viral genomic DNA by exerting DNA repair machinery to establish the lytic and latent life cycles of wtAAV and transduction of rAAV.

## ADENO-ASSOCIATED VIRUS (AAV)

Adeno-associated virus (AAV) is a non-enveloped replication-defective animal virus of approximately 20 nm in diameter (Figure 1a). It belongs to Dependovirus, a genus of the family Parvoviridae, which has a viral capsid in the simplest icosahedral shape composed of 60 units of viral structural proteins. Productive AAV replication requires co-infection of a helper virus such as adenoviruses and herpesviruses. A virion has an approximately 5-kb single-stranded DNA genome of either plus or minus polarity at an equal probability. AAV serotype 2 and many other serotypes are prevalent in human populations worldwide and up to 80% of adult humans have been infected with AAV in their childhood (Boutin et al., 2010; Calcedo et al., 2009; Erles et al., 1999). AAV is generally considered as a nonpathogenic virus, and clinical relevance of AAV infection in humans appears to be limited to male infertility (Erles et al., 2001), early miscarriage in pregnant women (Burguete et al., 1999; Pereira et al., 2010) and protection against cervical cancer (Su & Wu, 1996; Walz & Schlehofer, 1992) although some studies have shown negative results (Strickler et al., 1999). The current relevance of AAV in biological and medical research primarily stems from its benefits as a tool for gene delivery and genetic engineering of the cellular genome and as a refined agent for inducing DDR without damaging the cellular genome (Table 1).

## Wild Type Adeno-Associated Virus Serotype 2 (wtAAV2)

### *Structural Organization of wtAAV2*

AAV was first identified as a contaminant in adenovirus stocks in early 1960s (Atchison et al., 1965). Since infectious wild type AAV2 (wtAAV2) clones were generated from recombinant plasmids (Samulski et al., 1982), AAV2 has been most extensively studied for viral capsid structure, genome organization, virally encoded protein functions, AAV life cycle and infection pathways (reviewed in Berns & Parrish, 2007; Carter et al., 2009; Smith & Kotin, 2002).

**Table 1**. AAV as biological agents and tools

| AAV-derived agents and tools | Applications |
|---|---|
| rAAV of any serotypes | • Delivery of exogenous genetic materials to cells with no toxicity<br>• Targeted genetic manipulation of cells (*i.e.,* precise introduction of insertion, deletion or a small mutation at a defined location in the cellular genome in a predicted manner)<br>• Introduction of DDR without damaging the cellular genome<br>• Identification of DNA breakage sites in the cellular genome |
| wtAAV2 | • Introduction of DDR without damaging the cellular genome<br>• Tumor cell-specific killing |
| Rep68/78 | • Site-specific insertion of exogenous genetic materials at the AAVS1 site in the human chromosome 19q13.42 |

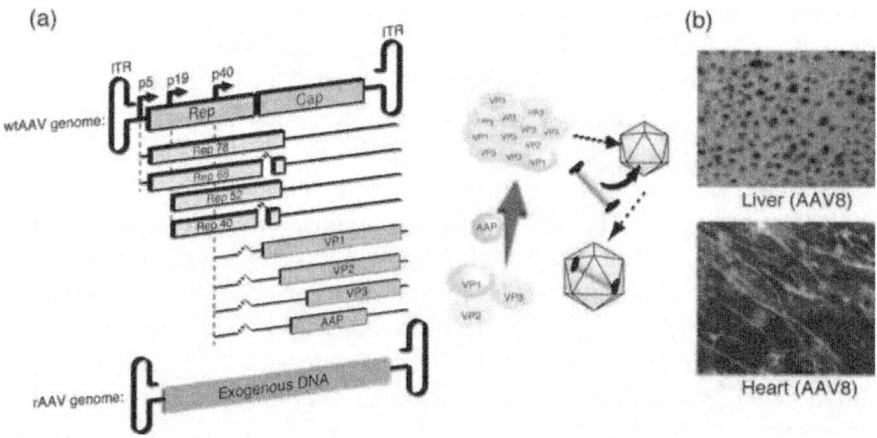

**Figure 1**. Wild-type AAV (wtAAV) and recombinant AAV (rAAV). (a) Structural organization of wtAAV and rAAV. wtAAV genome is a single-stranded DNA with ITRs at the both ends. The two viral genes (the rep and cap genes) encode five non-structural (Rep and AAP) and three structural (VP) proteins, and they are controlled by the three viral promoters (p5, p19, and p40). AAV virion particle consists of VP1, VP2, and VP3. AAP supports assembly of VP proteins. rAAV genome is devoid of the viral components except for the ITRs, and contains an exogenous DNA of interest. (b) Representative photomicrographs of XGal-stained sections of the murine liver and heart transduced AAV8-EF1α-nlslacZ and AAV8-CMV-lacZ vectors at a high dose (7.2.x $10^{12}$ viral particles per mouse), respectively. All the hepatocytes and cardiomyocytes express the lacZ marker gene product, which turns transduced cells blue by the XGal staining.

WtAAV2 has a single-stranded DNA genome of 4679 nucleotides (nt) in length (GenBank accession no. AF043303) that comprises the region encoding 2 viral genes (e.g., the rep and the cap genes), their promoters (p5, p19 and p40 promoters), a polyadenylation signal, and two 145-nt inverted terminal repeats (ITR) forming T-shaped DNA hairpins at each viral genome terminus (Figure 1a). WtAAV2 expresses a total of 3 structural proteins (VP1, VP2 and VP3 from the cap gene) and 5 non-structural proteins (Rep40, Rep52, Rep68, and Rep78 from the rep gene and AAP from an alternative open reading frame from the cap gene). VP proteins form the AAV viral capsids while Rep proteins play roles in viral genome replication, packaging and site-specific viral genome integration at the AAVS1 site in the human chromosome 19 (Kotin et al., 1990, 1992). AAP protein (AAP stands for assembly-activating protein), which was identified in 2010, plays a role in directing VP proteins to nucleolus, the organelle where new AAV virions assemble (Sonntag et al., 2010).

## *The Viral Life Cycle of wtAAV2*

The life cycle of wtAAV2 consists of the lytic and the latent phases. Following infection through its surface receptors including heparan sulfate proteoglycan (Summerford & Samulski, 1998), wtAAV2 viral particles are carried to the nucleus where single-stranded viral genomes are released from virions into nucleoplasm. When an adenovirus co- or super-infects cells that are infected with wtAAV2, adenovirus helper functions are supplied and wtAAV2 enters the lytic phase where productive viral genome replication takes place. Adenoviral E1a, E1b55k, E4orf6, DNA-binding protein (DPB), and virus-associated RNA I (VAI-RNA) have been identified as the helper functions required for the growth of wtAAV2 and rAAV (Berns & Parrish, 2007; Geoffroy & Salvetti, 2005). In the lytic cycle, there are significant interactions between viral components and host DDR mediated by adenoviral E1 and E4 gene products and AAV large Rep proteins (i.e., Rep68/78). The interactions are primarily aimed at blocking the cell cycle and suppressing the NHEJ DNA repair pathway (more details are described in 3.2.).

In the absence of helper virus co-infection, most of the viral genomes are lost during cell division in dividing cells because they do not replicate or segregate together with the cellular genome into daughter cells. However, a certain proportion of AAV genomes establishes a latent phase by integration into the cellular genome, particularly at the AAVS1 site located within the human chromosome 19q13.42 (Kotin et al., 1992; Samulski et al., 1991). The AAVS1 site has a 33-nt DNA sequence within the myosin binding subunit (MBS) 85 gene and this short DNA sequence serves as the target for the site-specific integration (Linden et al., 1996b; Tan et al., 2001). The site-specific integration process requires expression of Rep68/78, which binds to a GCTC repeat element termed Rep binding site (RBS) and creates a nick at a nearby 3'-CCGGT/TG-5', designated terminal resolution site (trs). A set of these two recognition sequences is located within the AAV-ITR and the AAVS1 site (Brister & Muzyczka, 1999; McCarty et al., 1994). Several cellular factors have been shown to modulate Rep68/78-mediated sitespecific integration (Figure 2b), although the experimental observations appear to be in conflict in some aspects. High mobility group protein 1 (HMG1) binds to Rep78, enhances its RBS binding and nicking activities, and promotes site-specific integration at the AAVS1 site (Costello et al., 1997). In addition, the human immunodeficiency virus type 1 (HIV-1) TAR RNA binding protein 185 (TRP-185) binds to the RBS within the AAVS1 site, interact with Rep68, enhances Rep68's helicase activity, and controls selection of AAV genome integration sites within the AAVS1 locus (Figure 2b) (Yamamoto et al., 2007). This mode of latency is unique to wtAAV2 among the animal viruses, and is the case at least in cultured cells. In latently-infected human tissues, a majority of wtAAV2 genomes persist as circular genomes and no site-specific integration has been demonstrated even by sensitive PCR-based assays (Schnepp et al., 2005); therefore, the significance of the site-specific integration of wtAAV2 in natural infection in humans remains elusive.

## *wtAAV2 Viral Components that Evoke DDR*

Among the viral components, large Rep proteins, the cis-acting replication element (CARE) within the p5 promoter (Fragkos et al., 2008; Francois et al., 2005; Nony et al., 2001; Tullis & Shenk, 2000), AAV-ITR, and the unusual single-stranded nature of the viral genome are particularly important in AAV-evoked DDR and interaction with DNA repair machinery. These elements could potentially activate DDR without AAV viral genome replication.

## Recombinant AAV (rAAV) Vectors

Recombinant AAV (rAAV) vectors are genetically-engineered viral agents that carry heterologous DNA to be delivered to target cells and are devoid of all the viral genome sequence except for the 145-nt (ITR) at each genome terminus. Until early 2000s, rAAV vectors were primarily derived from AAV2 due to the limited availability of alternative serotypes at that time. rAAV2 vectors have a broad host range and outstanding ability to deliver genes of interest to both dividing and non-dividing cells of various types in vitro. However, rAAV2 exhibits limited transduction efficiency in tissues and organs in vivo when administered in experimental animals. This drawback of rAAV2 has recently been overcome by the discovery of new serotypes exemplified by AAV serotypes 8 and 9 (AAV8 and AAV9) (Gao et al., 2002, 2004). rAAV vectors derived from serotypes alternative to AAV2 have become widely available at present and been shown to exhibit unprecedented robust transduction in various tissues and organs by intravascular injection of the vector (Figure 1b) (Foust et al., 2009; Ghosh et al., 2007; Inagaki et al., 2006; Nakai et al., 2005a; Sarkar et al., 2006; Vandendriessche et al., 2007; Wang et al., 2005). It should be noted that alternative serotype rAAV vectors are in general those containing a rAAV2 viral genome encapsidated with an alternative serotype viral coat (i.e., pseudoserotyped rAAV2 vectors) (Rabinowitz et al., 2002). Therefore, they are often referred to AAV2/8 and AAV2/9 (genotype/serotype) when rAAV2 genome is contained in AAV8 and AAV9 viral coats, respectively. In addition to the exploitation of AAV capsids derived from various serotypes and variants present in nature, recent advances in genetic engineering of viral capsids aiming for the creation of specific cell type/tissue-targeting vectors have significantly broadened the utility of this vector system (Asokan et al.,

2010; Excoffon et al., 2009; Koerber et al., 2009; Yang et al., 2009). Double-stranded (ds) rAAV vectors and gene targeting rAAV vectors are also worthy of note. Ds rAAV contains a ds viral genome in place of a single-stranded (ss) DNA (McCarty, 2008). Because ds rAAV vectors overcome the rate-limiting step in transduction, i.e., conversion from ss to ds DNA in infected cells, they exhibit a 1-2 log higher transduction efficiency than that achievable with the conventional ss vectors (McCarty et al., 2003; Wang et al., 2003). Gene-targeting rAAV vectors have the ability to introduce genomic alterations precisely and site-specifically at high frequencies of up to 1% (Russell & Hirata, 1998; Vasileva & Jessberger, 2005), which is several logs higher than that achievable by the conventional homologous recombination (HR) approaches (Thomas & Capecchi, 1987) (please refer to 3.5 for more details). Nonetheless, even if rAAV vectors are devoid of several key components that trigger DDR (i.e., large Rep expression and the CARE), establishment of rAAV transduction heavily relies on the interactions between rAAV viral genomes and DNA repair machinery irrespective of serotypes or nature of viral genomes (i.e., ss rAAV or ds rAAV).

# AAV AND DNA REPAIR PATHWAYS

An overview of AAV and DNA repair pathways is summarized in Figure 2.

## AAV-evoked DDR

### *Earlier Evidence for the Role of DDR in the AAV Genome Processing*

Although the interplay between virus and DDR is a relatively new area of research, earlier studies indicated potential roles of DDR in the AAV life cycle or viral genome processing. The first indicative evidence came from the observation that cells treated with a wide variety of genotoxic agents including UV irradiation and carcinogens such as hydroxyurea could support wtAAV2 genome replication in the absence of helper virus co-infection (Yakinoglu et al., 1988; Yakobson et al., 1987, 1989). Subsequently, such treatment was found to augment rAAV2 transduction efficiency in both dividing and non-dividing cells with the latter showing a more dramatic enhancing effect (Alexander et al., 1994; Russell et al., 1995). These earlier observations suggested that activated DNA repair pathways following DNA damage induced by genotoxic treatment somehow facilitated the conversion of rAAV genomes from ss to ds DNA by second-strand synthesis (Figure 2f, g and h) (Ferrari et al., 1996; Fisher et al., 1996). As mentioned earlier, the formation of ds AAV genomes is a critical step for wtAAV to initiate productive infection and for rAAV to undergo abortive infection and express transgene products. A better response to the treatment in non-dividing cells conforms to the idea that DNA repair pathways are constitutively activated to a greater extent in dividing cells. Such activation is required to repair DNA replication errors that occur naturally and unavoidably. Although the underlying mechanism of this effect still remains elusive, one can speculate that up-regulation of DNA repair pathways increases the pool of cellular factors required for AAV genome processing. Alternatively, factors inhibitory for wtAAV genome replication or rAAV transduction may become sequestered from AAV genomes to multiple DNA repair foci formed on the damaged cellular genome (Figure 2f and g). A recent observation that the MRN complex and ATM, the major DDR proteins, have an inhibitory effect on rAAV transduction supports the latter model (Cataldi & McCarty 2010; Cervelli et al., 2008; Choi et al., 2006; Sanlioglu et al., 2000; Schwartz et al., 2007).

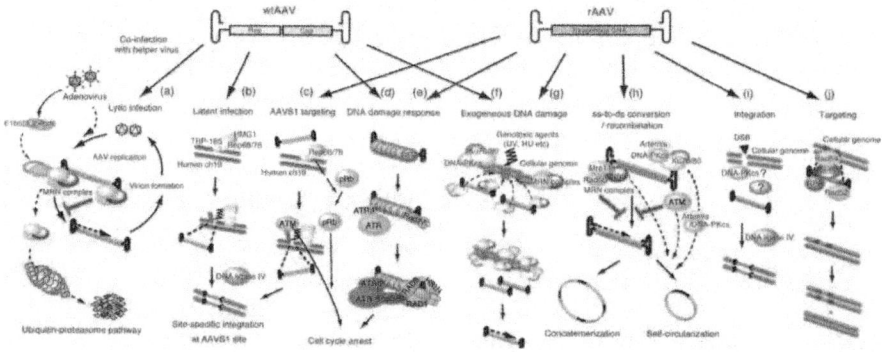

**Figure 2**. A hypothetical model of the interactions between DNA damage responses (DDRs), DNA repair proteins and AAV genomes. (a) In the presence of adenovirus co-infection, adenoviral E1B55k/E4ort6 degrades and E4orf3 mislocates the MRN complex bound to AAV-ITR, allowing efficient second-strand synthesis and subsequent genome replication. (b) In the absence of helper virus co-infection, AAV Rep68/78 introduces a nick at the AAVS1 site in the human chromosome 19q13.42, where wtAAV2 genome integrates. It has been shown that TRP-185 and HMG1 proteins modulate this process. (c) Expression of Rep68/78 supplied in trans, mediates site-specific integration of rAAV (AAVS1 targeting) and/or activates the ATM pathway. Over-expression of Rep68/78 activates both pRb and ATM pathways, resulting in a complete block of the cell cycle in the S phase. (d, e) Activation of the ATR pathway by wtAAV2 or rAAV genome results in the G2/M arrest. (f, g) Genotoxic treatment damages the cellular genome at multiple locations, to which the inhibitory MRN complex and DNA-PK (Ku proteins and DNA-PKcs) are presumed to be sequestered from AAV genomes, allowing processing of viral genomes toward wtAAV2 replication and rAAV transduction. (h) The MRN complex and Ku bind to rAAV genome and suppress rAAV transduction. ATM may also have an inhibitory effect. The MRN complex, ATM

and Artemis/DNA-PKcs promote generation of circular monomers via intramolecular recombination of rAAV genomes. (i) rAAV genomes integrate at pre-existing breaks in the cellular genome via NHEJ pathway(s). DNA-PKcs is dispensable but it could influence the efficiency of viral genome integration. (j) The Rad51/Rad54-mediated HR pathway mediates efficient rAAV gene targeting. In the pathways a, f, g, and h, dashed arrows extending from AAV-ITR indicate leading strands generated by a second-strand synthesis mechanism. In the pathway h, either or both of the second-strand synthesis mechanism and annealing of complementary ss DNA genomes convert ss genomes to ds DNA.

## DNA Repair Proteins Associated with AAV Genomes in Cells

AAV is composed of only two elements, VP proteins that form viral capsids and a singlestranded viral genomic DNA. VP proteins primarily determine biological properties of various AAV serotypes; i.e., how AAV particles reach cells, enter cells, traffic in cytoplasm and nucleoplasm, and uncoat virion shells to release viral genomes. At present there is no evidence indicating that the above-mentioned AAV infection pathways driven by the capsid trigger DDR. AAV-evoked DDR is all about the cellular responses against AAV viral genomes except for wtAAV2, which expresses Rep68/78 proteins that also trigger DDR. It is plausible that the structure of single-stranded DNA with T-shaped hairpin termini, which is unusual and is not present in the cellular genome, is recognized as damaged DNA and triggers DDR. Direct evidence for the association of AAV genomes with DNA repair machinery has obtained in chromatin immunoprecipitation (ChIP) studies where AAV genomes and their associated cellular factors were crosslinked by formalin and precipitated together using antibodies specific to DNA repair proteins. To date, the MRN complex, Ku86, Rad52, RPA and DNA polymerase delta have been identified as factors bound to ss AAV genomes (Cervelli et al., 2008; Jurvansuu et al., 2005; Zentilin et al., 2001). In addition, immunofluorescence microscopy has revealed that the MRN complex, ATR, TopBP1, BLM, Brca1, Rad17, RPA, and Rad51 are recruited to the discrete nuclear foci where AAV genomes accumulate (Cervelli et al., 2008; Jurvansuu et al., 2005). Table 2 summarizes the roles of DNA repair proteins in AAV infection/transduction, AAV genome selfcircularization, and AAV genome integration into the host genome.

## AAV Genome Activates ATR-Mediated DDR

Although not exclusive, the DNA repair proteins found to be associated with AAV genomes described in 3.1.2 are those involved in the ATR-mediated DDR that is triggered by stalled replication forks (Figure 2d and e) (reviewed in Branzei & Foiani, 2010 and Shiotani & Zou, 2009). At stalled replication forks, ss DNA regions become coated with RPA. RPA then recruits ATR-ATRIP and the Rad17 complex to the damaged site. The Rad17 complex subsequently recruits the ring-shaped trimeric Rad9/Rad1/Hus1 (9-1-1) complex, and finally ATR-ATRIP kinase becomes activated by TopBP1 recruited to the site and sends a DNA damage checkpoint signal (Figure 2d and e). In addition, Mre11 has also been reported to relocalize to stalled replication forks to a limited extent (Mirzoeva & Petrini, 2003). AAVITR exhibits a close structural similarity to stalled replication forks in that it contains both ss DNA regions and ss DNA-ds DNA junctions. This strongly supports a model in which AAV-ITR is recognized as a stalled replication fork and triggers the checkpoint response via ATR kinase.

**Table 2**. DNA repair and AAV

| Protein | Effects of deficiency[2] | | | | | | Note | References[3] |
|---|---|---|---|---|---|---|---|---|
| | In vitro | | | In vivo | | | | |
| | T | C | I | T | C | I | | |
| Artemis | | | | ↓ | ↓ | | | 1 |
| ATM | ↑ | ↓/↑ → | ↑ | ↓ | | | | 2, 3, 4, 5 |
| ATR | ↑/→ | → | → | | | | wtAAV2-evoked signal ↑ | 5 |
| BLM | | ↓ | | | | | | 4 |
| Chk1 | | | | | | | wtAAV2-evoked signal ↑ | 14, 15 |
| DNA-PKcs | ↓ | ↓ | ↓/↑ | ↓ | ↓ | ↑/→ | rAAV2 genome replication ↓ when deficient | 1, 4, 5, 6, 7, 8, |
| | | | | | | | wtAAV2 genome replication ↑ when deficient | 15 |
| Ku70/80(86) | ↑ | → | | | | | wtAAV2 genome replication ↓ when deficient | 3, 4 |
| | | | | | | | Targeting efficiency ↑ when deficient | |
| ligase IV | | → | ↓ | | | | | 4, 8 |
| MDC1 | ↑ | | | | | | | 9 |
| MRN | ↑ | ↓ | | | → | | | 9, 4, 10 |
| Rad52 | ↓ | | | → | → | | | 3, 16 |
| Rad54B | | | | | | | Targeting efficiency ↓ when deficient | 11 |
| Rad54L | | | | | | | Targeting efficiency ↓ when deficient | 11 |
| WRN | | ↓ | | | | | | 4 |
| XRCC3 | | → | | | | | Targeting efficiency ↓ when deficient | 4, 11 |
| pRb | | | | | | | Induction of cell death when deficient | 12, 13 |
| p21 | | | | | | | Induction of cell death when deficient | 12, 13 |
| p53 | | | | | | | Induction of cell death when deficient | 12, 13 |

The actual activation of the ATR pathway by AAV genomes has been confirmed by the demonstration that the ATR-downstream effector proteins; i.e., Chk1 and RPA, become phosphorylated in cells infected with wtAAV2 or UV-irradiated wtAAV2, both of which are devoid of the ability to replicate or express viral genes in the system used for the experiment (Fragkos et al., 2008; Ingemarsdotter et al., 2010; Jurvansuu et al., 2005, 2007). Interestingly, rAAV2 genome devoid of the 55-nt CARE within the p5 promoter does not evoke the ATR-mediated checkpoint signal, and it has been shown that co-existence of both ITR and CARE in an AAV genome is essential for the activation (Fragkos et al., 2008). The consequence of the AAV genome-evoked ATR-mediated DDR is G2/M cell cycle arrest in wild type cells, while it leads to cell death in p53-deficient cells (Ingemarsdotter et al., 2010; Jurvansuu et al., 2007) (please see 3.1.4. for more details). Cell cycle arrest in the late S and/or G2 phases following infection of wtAAV2 or UV-irradiated wtAAV2 was observed in an earlier study although how and what DDR is involved was not known at that time (Winocour et al., 1988).

## AAV Genome-Evoked DDR Leading to Cell Death

A unique aspect of AAV-evoked DDR is the ability to induce cell death without productive viral genome amplification, viral gene expression or cellular DNA damage, when cells are devoid of p53 expression. In 2001, Raj et al. reported an unexpected experimental observation that wtAAV2 infection of an osteosarcoma cell line that lacks expression of functional p53 leads to cell death through apoptosis or mitotic catastrophe, whereas the wild type control cells merely undergo a transient cell cycle arrest in the G2 phase (Raj et al., 2001). Mitotic catastrophe is an ill-defined term describing an apoptosis-like cell death during mitosis that takes place even in the presence of unrepaired DNA damage (Castedo et al., 2004; Vakifahmetoglu et al., 2008). This p53 deficiency-dependent cell killing effect was also observed when cells were infected with UV-irradiated wtAAV2 or microinjected with a 145-nt AAV2-ITR oligonucleotide, demonstrating that the unusual structure of the AAV2-ITR sequence itself is the culprit (Raj et al., 2001). Initially it was presumed that infection of wtAAV2 activates the ATM-p53-mediated DDR, which in turn increases and decreases the levels of p21 and CDC25C, respectively, resulting in the G2 arrest (Raj et al., 2001). Although the mechanism of p53 deficiency-dependent cell killing by AAV genomes still remains elusive, a series of subsequent studies on this phenomenon has revealed at least three potentially independent AAV-evoked pathways leading to cell death: the pathways involving (1) p53-p21-pRb, (2) p84N5 via caspase 6, and (3) ATR-Chk1. In the first mechanism, AAV-evoked DDR signal is transduced to a potent antiapoptotic proteins, pRb, via p53 and p21. Therefore, cells defective in this pathway fail to transduce the DDR signal to pRb, leading to apoptosis (Garner et al., 2007). In the second mechanism, functional defect of the p53-p21-pRb pathway allows activation of the nuclear death domain protein p84N5, which otherwise is inhibited by association with pRb (Doostzadeh-Cizeron et al., 1999). The activated p84N5 then induces apoptosis via caspase-6 (Garner et al., 2007). In the third mechanism, AAV genomes activate ATR, which in turn phosphorylates Chk1, causing a transient cell cycle arrest in the G2 phase. In the absence of p53, cells fail to sustain the G2 arrest following degradation of the unstable Chk1, progress suicidally into mitosis, and die via mitotic catastrophe associated with centriole overduplication and the subsequent formation of multipolar mitotic spindles (Ingemarsdotter et al., 2010; Jurvansuu et al., 2007). Whether all of the pathways or only some of them are triggered by AAV genomes remains unknown at present.

## AAV2 Rep68/78-Evoked DDR

In addition to AAV genome as a trigger of DDR, AAV2 large Rep proteins (i.e., Rep68/78) themselves also evoke DDR independent of AAV genome. In the lytic phase of the AAV life cycle where cells are co-infected with a helper virus, Rep proteins are strongly expressed and exert many functions in the network of cellular proteins and viral factors derived from adenovirus or other helpers. Rep proteins can also be expressed without helper virus infection but only to a limited extent due in part to the large Reps' ability to negatively regulate their own promoter (p5) and the promoter for the small Rep proteins (p19) (Beaton et al., 1989; Kyostio et al., 1994). The significance of Rep68/78 expression in the absence of helper viruses reside in a series of the AAVS1-targeting approaches that exploit wtAAV2's ability to introduce exogenously derived DNA into the AAVS1 site in a site-specific manner (Figure 2c) (Henckaerts & Linden, 2010; Linden et al., 1996a). In these approaches, a donor vector in any context (e.g., plasmid DNA, adenoviral vectors or rAAV) containing a gene of interest and RBS is delivered to human cells where AAV2 large Rep expression is supplied by the same vector or a separate one. The AAV2-ITR sequence is commonly used as an RBScontaining cis element; however, the p5 promoter also serves as an alternative (Philpott et al., 2002).

It has been known that Rep68/78 shows significant cellular toxicity due to the strong antiproliferative action of the protein (Yang et al., 1994). Rep78 completely blocks the cell cycle in the S phase (Saudan et al., 2000). Studies have shown that Rep78 exerts two independent but complementing DDR-associated cellular signal transduction pathways to arrest the cell cycle. The two pathways are the pRb pathway and the ATM-Chk2 pathway (Berthet et al., 2005). In the first pRb pathway, Rep78 expression leads to an increased level of the cyclin-dependent kinase inhibitor p21 and accumulation of hypophosphorylated pRb, the active form of pRb protein (Berthet et al., 2005; Saudan et al., 2000) (Figure 2c). Consequently, cellular proteins that control cell cycle progression such as cyclin A, cyclin B1, and Cdc2, are down-regulated, resulting in slowing down the cell cycle (Saudan et al., 2000). Supporting this model, this effect is substantially attenuated in pRb-deficient mouse embryonic fibroblasts (Saudan et al., 2000). An increased amount of p21 might explain the inability to phosphorylate pRb upon large Rep expression, but the transcriptional activation of p21 has been shown to occur via a p53-independent pathway (Hermanns et al., 1997). In the second ATM-Chk2 pathway, the DNA-nicking activity of Rep68/78 creates multiple damaged sites in the cellular genome, which activates the ATM-Chk2 pathway and arrests the cell cycle (Figure 2c) (Berthet et al., 2005). Large Rep proteins create a break in only one strand of two-stranded DNA, which is not a type of damage that usually activates ATM. It is currently unknown how Rep68/78-induced DNA damage triggers this pathway. Worthy to note, activation of either one of the above-mentioned two pathways by itself is not sufficient for the complete block of the cell cycle, which is attainable by Rep68/78 expression (Berthet et al., 2005). It appears that there would be many other Rep68/78-associated DNA repair pathways that have yet to be identified. This is because a recent study using a tandem affinity purification (TAP) approach has demonstrated physical interaction of Rep78 with many DNA repair-associated proteins including DNA-dependent protein kinase catalytic subunit (DNA-PKcs), minichromosome maintenance (MCM) proteins, Ku70/80, proliferating cell nuclear antigen (PCNA), RPA, and structural maintenance of chromosome 2 (SMC2) (Nash et al., 2009).

## wtAAV2 Genome Replication and DNA Repair Pathways

In the lytic phase of the wtAAV2 life cycle, viral genome replication requires co-infection of a helper virus. Since human adenoviruses have been most extensively studied in the context of AAV virology, this section specifically focuses on the interplay between wtAAV2, human adenoviruses, and DNA repair machinery. The adenoviral components required in the lytic infection are E1a, E1b55k, E4orf6, DBP, and VAI. A series of adenovirus/AAV co-infection studies has provided significant insights into how DDR and DNA repair machinery play roles in AAV genome replication in the presence of adenovirus helper functions (Collaco et al., 2009; Schwartz et al., 2009). It has been shown that adenoviral E1b55k/E4orf6 degrades the MRN complex via the ubiquitin-proteasome pathway and E4orf3 mislocalizes the MRN complex to aggresome, abrogating the MRN function in triggering the ATM and ATR pathways (Figure 2a) (Collaco et al., 2009). In addition, E4orf6 dissociates ligase IV from ligase IV/XRCC complex and degrades it (Jayaram et al., 2008). The main consequence of this adenoviral manipulation of DDR is inhibition on NHEJ, which prevents concatemeric adenoviral genome formation and promotes adenoviral genome replication and packaging. Although it remains elusive how beneficial the inhibition of NHEJ is in the AAV lytic life cycle, the significance of E1b/E4-mediated degradation of MRN complex in the wtAAV2 lytic cycle has been revealed by the observation that the MRN complex binds to AAV-ITR and inhibits wtAAV2 genome replication and rAAV vector transduction (Cervelli et al., 2008; Schwartz et al., 2007). Along the same line, the observation that cells deficient in ATM exhibit a higher rAAV transduction efficiency (Sanlioglu et al., 2000) might be explainable on the assumption that lack of ATM would be an equivalent to inactivation of the MRN complex because the MRN complex serves as a damage sensor that activates the ATM pathway (Carson et al., 2003). It is tempting to propose that dislocation of the MRN complex and other inhibitory factors from AAV genomes to the sites in the cellular genome where genome integrity is more severely threatened, is the mechanism for the augmentation of wtAAV2 genome replication and rAAV vector transduction by genotoxic treatment (Figure 2g). Suppression of the NHEJ pathway that involves DNA-PKcs and Ku proteins, however, may or may not be beneficial, because one study has shown that deficiency of these proteins both resulted in impaired rAAV2 genome replication (Choi et al., 2010) whereas another study has reported that siRNA-mediated knockdown of DNA-PKcs enhanced wtAAV2 genome replication (Collaco et al., 2009).

In addition to the adenovirus-evoked DDR, productive wtAAV2 viral genome replication triggers DDR distinct from that observed in adenovirus only infection (Collaco et al., 2009; Schwartz et al., 2009). Adenovirus-wtAAV2 co-infection results in much more pronounced activation of ATM and the checkpoint kinases, Chk1 and Chk2. This activation occurs independently of the MRN complex; therefore, the activation sustains even if MRN complex starts being degraded by adenoviral E1b/E4 proteins (Collaco et al., 2009). Other DDR substrate proteins RPA, NBS1 and H2AX become phosphorylated as the lytic phase progresses (Collaco et al., 2009; Schwartz et al., 2009). It has been shown that AAV genome replication is essential and sufficient to induce the DDR signal transduction cascade observed in the adenovirus co-infection, and Rep proteins does not play a role in the activation of DDR (Collaco et al., 2009; Schwartz et al., 2009). Among the three phosphatidylinositol 3-kinase-like kinases (PIKKs) that initiate signal transduction (i.e., ATM, ATR and DNA-PKcs), ATM and DNA-PKcs are the primary kinases that phosphorylate downstream DDR substrates, and ATR appears to play only a minor role in the lytic phase of the AAV life cycle (Collaco et al., 2009; Schwartz et al., 2009). Although the significance of the DDR in the AAV lytic cycle remains unclear, the activation of the ATM pathway appears to be beneficial for AAV genome replication (Collaco et al., 2009).

## rAAV Genome Recombination and DNA Repair Pathways

### rAAV Genome Processing is Mediated Solely by DNA Repair Machinery

After entering nuclei, rAAV virion shells break down, releasing single-stranded (ss) vector genomes into nucleoplasm, which subsequently convert to various forms of doublestranded (ds) genomes (Deyle & Russell, 2009; Schultz & Chamberlain, 2008). It should be noted that rAAV does not express any viral gene products that can process viral genomes such as recombinases and integrases; therefore the processing of viral genomes must heavily depend on DNA repair machinery. In addition, unlike the battle between adenovirus and the host DNA repair systems as described in 3.2, rAAV has no means to manipulate DNA repair pathways once viral genomes evoke DDR. Unless rAAV genomes have been processed to completion into stable ds DNA with no free ends, DDR would remain activated due to the continued presence of viral DNA in an unusual structure presenting a single strand with free ends. In mammalian cells, extrachromosomal free DNA ends at ds rAAV genome termini as well as those in ds linear plasmid DNA, when exogenously delivered, appear to be removed primarily by ligating two free ends and making a single continuous ds DNA strand via NHEJ and/or occasionally HR rather than by DNA degradation (Nakai et al., 2003b; Nakai, unpublished observation). In this sense, the rAAV genomes processed into various forms in their latency could be viewed as byproducts that have been created and disposed of by a cellular defense mechanism against potentially toxic exogenous agents.

### Single-to-Double-Stranded rAAV Genome Conversion and DNA Repair Machinery

How ss rAAV genomes become ds DNA is not completely understood but the process involves the following two mechanisms; second-strand synthesis (Ferrari et al., 1996; Fisher et al., 1996; Zhong et al., 2008; Zhou et al., 2008) and annealing of plus and minus strands (Hauck et al., 2004; Nakai et al., 2000). It has been shown that, upon rAAV infection, the MRN complex becomes activated, physically associates with AAV2-ITR and inhibits wtAAV2 replication and rAAV transduction (Figure 2a and h) (Cervelli et al., 2008; Schwartz et al., 2007); therefore, MRN appears to have some role in the conversion of ss to ds DNA. ATM has also been suggested to be a cellular factor that inhibits the single-todouble-stranded genome conversion because transduction efficiency with ss rAAV is significantly enhanced in ATM-deficient cells in vitro (Figure 2h) (Sanlioglu et al., 2000). However, a recent study has proposed an ATM-mediated gene silencing model rather than the mechanism involving the second-strand synthesis to explain the ATM's inhibitory effect. This model stems from the observation that, in the absence of ATM, ds rAAV transduction was enhanced as well, indicating that an alternative mechanism other than second-strand synthesis is involved (Cataldi & McCarty, 2010). Another factor that is known to inhibit this process is tyrosine-phosphorylated FKBP52, which binds to AAV-ITR and inhibits secondstrand synthesis (Qing et al., 2001). Its dephosphorylation by T-cell protein tyrosine phosphatase (TC-PTP) dissociates FKBP52 from AAV-ITR, allowing the formation of ds genomes (Qing et al., 2003). In vitro AAV replication studies have identified the DNA polymerase that catalyzes second-strand synthesis as DNA polymerase δ (Nash et al., 2007), which is a polymerase that fills a single-stranded DNA gap created during the nuclear excision repair (Torres-Ramos et al., 1997). Physical association of DNA polymerase δ and AAV genome has also been demonstrated (Jurvansuu et al., 2005). At present it remains elusive whether and how the above-mentioned signal kinases (i.e., MRN and ATM) and effectors (FKBP52 and DNA polymerase δ) are linked in the rAAV-evoked DDR.

## *Extrachromosomal rAAV Genome Recombination and DNA Repair Machinery*

In addition to the above-mentioned single-to-double-stranded genome conversion, rAAV genomes are further processed into the following stable ds forms by intra- or intermolecular DNA recombination mediated solely by DNA repair machinery, and establish the latent infection. The viral genome forms in the latent phase include ds circular monomers, large concatemers (circular and/or linear), and rAAV proviral genomes that are stably integrated into the host cellular genome at low frequencies (Deyle & Russell, 2009; Schultz & Chamberlain, 2008). It has not been determined when the rAAV genome recombination takes place, which may be either before, at, or after completion of the single-to-doublestranded genome conversion. In dividing cells, extrachromosomal genomes are lost because they do not replicate episomally, whereas they can be stabilized and maintained as chromatin in quiescent cells in animal tissues (Penaud-Budloo et al., 2008). Earlier studies indicated that the formation of large concatemeric rAAV genomes is important for transgene expression; however, accumulated observations might favor a model in which extrachromosomal circular monomer genomes, not large concatemers or integrated forms, are primarily responsible for persistent and stable transgene expression in rAAV-transduced animal tissues (Nakai et al., 2001; Nakai et al., 2002; Nathwani et al., 2011).

In extrachromosomal rAAV genome recombination, AAV-ITR plays a pivotal role in mediating recombination. Although it has yet to be elucidated how DDR is evoked by rAAV genomes in the context of rAAV genome recombination, it is not unreasonable to speculate that the T-shaped hairpin structure within the AAV-ITR and/or ss DNA-ds DNA junctions in the stem of the hairpin DNA trigger DDR. A set of DNA repair proteins, which includes DNA-PKcs, Artemis, ATM, MRN, BLM, and WRN (Figure 2h), has been found to be involved in rAAV genome recombination (Cataldi & McCarty, 2010; Choi et al., 2006; Duan et al., 2003; Inagaki et al., 2007b; Nakai et al., 2003b; Sanlioglu et al., 2000; Song et al., 2001). Deficiency of these proteins impairs intramolecular recombination of ss rAAV and/or ds rAAV genomes via the AAV-ITR sequence. DNA-PKcs and Artemis are the two major components in the classical NHEJ pathway of DSB repair. Artemis, when activated by DNAPKcs, possesses an endonuclease activity and resolves DNA hairpin loops and flaps formed at broken DNA ends to facilitate ds DNA end joining (Ma et al., 2005). BLM and WRN are members of the RecQ family of DNA helicases. They unwind ds DNA to ensure the formation of proper recombination intermediates, and mediate a various types of DNA transactions, mainly HR (Bernstein et al., 2010). MRN is a multifaceted protein complex that functions as a primary sensor of DSB, binds DNA lesion, recruits ATM, and processes DNA ends by utilizing the Mre11 endo- and exo-nuclease activity that creates recombinogenic 3' single-stranded tails (Williams et al., 2010). The initial study of the structure of ITR-ITR junction sequences revealed that the majority of the recombination junctions in ds circular monomer genomes exhibited a 165-nt double-D ITR structure, the hallmark of HR (Duan et al., 1999; Xiao et al., 1997). This indicates that HR is the major pathway for intra- and intermolecular genome recombination events. Supporting this view, Rad52, which is a key player in HR, was identified as a protein that binds to rAAV genomes in cultured cells (Zentilin et al., 2001). Interestingly, deficiency of Rad52 does not affect rAAV transduction efficiency or genome processing in murine liver (Nakai, unpublished observation). It remains possible that HR plays a major role in rAAV genome recombination at least under certain cellular environment; however, accumulated observations by us and others rather support a model in which NHEJ is the major pathway for extrachromosomal rAAV genome recombination. In the absence of DNA-PKcs or Artemis, intramolecular recombination is significantly impaired in cultured cells and animal tissues (Cataldi & McCarty, 2010; Duan et al., 2003; Inagaki et al., 2007b; Nakai et al., 2003b; Song et al., 2001), and the footprints on junction DNA are quite consistent with NHEJ-mediated recombination, showing nucleotide deletions of various degrees with occasional microhomology at junctions (Inagaki et al., 2007b). Interestingly, intra- and inter-molecular recombination events that form ds circular monomers and ds concatemers, respectively, are differentially regulated by different DNA repair pathways (Figure 2h). Intramolecular recombination heavily depends on the Artemis/ DNA-PKcs-dependent NHEJ pathway, while the NHEJ pathways that mediate intermolecular recombination are redundant because intermolecular recombination occurs efficiently in the absence of DNA-PKcs or Artemis (Inagaki et al., 2007b). The DNA-PKcs or Artemis-independent NHEJ might be those involving ATM and/or MRN (Cataldi & McCarty, 2010; Choi et al., 2006; Duan et al., 2003; Inagaki et al., 2007b; Nakai et al., 2003b; Sanlioglu et al., 2000; Song et al., 2001). Alternatively, HR might be the major pathway for intermolecular rAAV recombination. This model stems from the observation that recombination between two homologous AAV-ITRs derived from the same serotype is preferred to that between two non-homologous AAV-ITRs derived from different serotypes (Yan et al., 2007). The ATR pathway does not appear to be involved in extrachromosomal rAAV genome recombination (Cataldi & McCarty, 2010).

How DNA-PKcs and Artemis process rAAV genome termini and mediate recombination has been extensively studied in the context of murine tissues. In DNA-PKcs or Artemisdeficient SCID mice, ds linear rAAV genomes with covalently closed hairpin caps at genome termini accumulate in rAAV-transduced tissues (Figure 3b). In SCID mouse thymi, V(D)J recombination is impaired resulting in accumulation of covalently-sealed hairpin intermediates at V(D)J coding ends in the T cell receptor gene (Rooney et al., 2002; Roth et al., 1992) (Figure 3a). These two phenomena are essentially the same in that if hairpin structures at DNA ends are not cleaved by the Artemis/DNA-PKcs endonuclease activity, covalently closed DNA ends accumulate without undergoing further recombination. Therefore, intramolecular recombination most likely uses the same Artemis/DNA-PKcsdependent NHEJ pathway used for V(D)J recombination. It is not easy to determine GC-rich AAV-ITR hairpin DNA structures at sequencing levels; however this shortcoming has been overcome by exploiting the bisulfite PCR technique. Utilizing this method, the primary cleavage site by the Artemis/DNA-PKcs endonuclease activity has been mapped to the 5' end of the 3-base AAA loop at the AAV-ITR hairpin tips (Figure 3b) (Inagaki et al., 2007b). In DNA-PKcs-deficient SCID mouse tissues, the relative proportion of rAAV genome recombination junctions exhibiting the hallmark of HR increases, indicating compensatory activation of HR in the absence of DNA-PKcs in quiescent cells in animal tissues (Nakai, unpublished observation). In this regard, worthy of note are the following observations made by us and others that DNA repair pathways might somehow be linked to epigenetic modifications of rAAV genomes. We have found that the cytomegalovirus (CMV) immediately early gene promoter in rAAV genome can be significantly silenced in Artemisor DNA-PKcs-deficient mouse muscle (Nakai, unpublished observation). Recently, Cataldi et al. reported that the CMV promoter is somewhat silenced in ATM-proficient murine fibroblasts compared to that in ATM-deficient cells (Cataldi & McCarty, 2010). These observations imply that rAAV genome recombination via NHEJ generates more functionally active genomes than HR presumably due to a difference in epigenetic modifications of rAAV genomes (Cataldi & McCarty, 2010).

## rAAV Genome Integration and DNA Repair Pathways

rAAV is devoid of Rep68/78 expression; therefore, it lacks the ability to integrate into the cellular genome site specifically. In addition, rAAV does not harness machinery designed specifically for integration into the cellular genome. rAAV vectors are generally considered as episomal vectors, but they do integrate into the cellular genome of both dividing and non-dividing cells at low frequencies (Deyle & Russell, 2009; McCarty et al., 2004). This process is entirely dependent on the host cellular DNA repair machinery. Although it is not easy to determine the frequency of rAAV genome integration in each case and it may vary depending on the amount of rAAV genomes delivered to cells, integration has been reported to occur at approximately ~0.1% of total input rAAV genomes (Russell et al., 1994) or up to ~4% of cell population in rAAV-infected cultured cells (Cataldi & McCarty, 2010), or at approximately 0.1% of rAAV-transduced hepatocytes when rAAV is injected into newborn mice (Inagaki et al., 2008). rAAV genome integration occurs at nonrandom sites in both cultured cells and somatic cells in animals. The preferred genomic sites for integration include the 45s pre-ribosomal RNA gene, transcriptionally active genes, DNA palindromes, CpG islands, and the neighborhood of transcription start sites (Inagaki et al., 2007a; Miller et al., 2005; Nakai et al., 2003a). Although the mechanism of integration remains largely unknown, it has been presumed that input rAAV genomes are fortuitously captured at pre existing breaks in the cellular genome when the DNA breaks are repaired by DNA repair machinery, which establishes rAAV integration. This model has been supported by the observations that rAAV genome integrations are frequently found at I-SceI-induced DSBs in the cellular genome (Miller et al., 2004) and genotoxic treatments can increase integration rates (Russell et al., 1995). Clinically, rAAV vectors are generally considered to be safe; however, one study has shown that vector genome integration could cause insertional mutagenesis leading to hepatocarcinogenesis in a mouse model (Donsante et al., 2007).

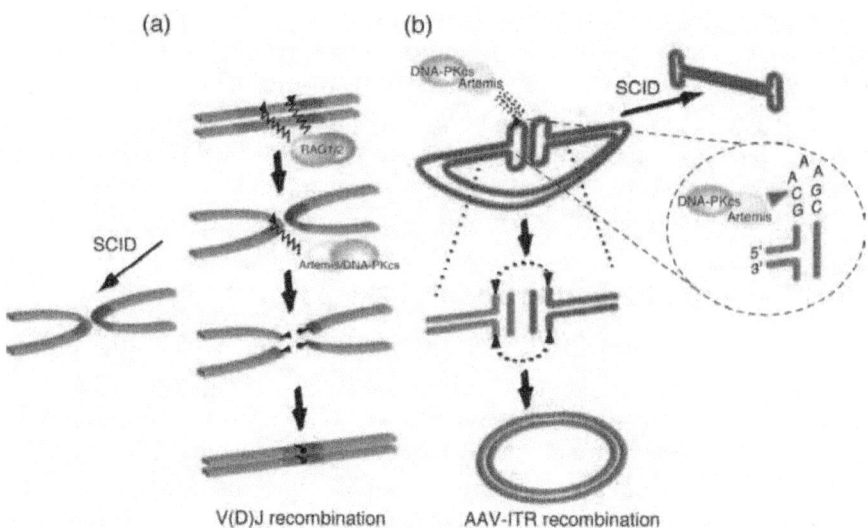

(a)    (b)

V(D)J recombination    AAV-ITR recombination

**Figure 3.** A similarity of Artemis / DNA-PKcs-mediated hairpin cleavage in V(D) J recombination and AAV-ITR recombination. (a) During V(D)J recombination, the recombination activating gene products, Rag1 and Rag2 endonucleases, cleave the immunoglobulin and T cell receptor genes, forming covalently closed hairpin loops at cleaved DNA ends. Artemis/DNA-PKcs complex resolves the hairpin loops, which triggers the subsequent recombination between the two coding ends via the classical NHEJ DNA repair pathway. In cells deficient in either Artemis or DNA-PKcs (SCID phenotype), hairpin coding ends remain unrecombined and accumulate. (b) The same Artemis/DNA-PKcsdependent NHEJ pathway mediates intramolecular AAV-ITR re-combination, forming circular monomer genomes. Intermolecular AAV-ITR recom-bination occurs independently of Artemis/DNA-PKcs. A red arrowhead indicates the primary target for Artemis/DNAPKcs-mediated cleavage. In SCID mouse tissues, ds linear rAAV genomes with covalently closed AAV-ITR hairpin caps accumulate.

The detailed analyses of rAAV vector genome-cellular genome junction sequences in cultured cells and murine tissues have provided significant insights into which and how DNA repair pathways play roles in rAAV integration (Inagaki et al., 2007a; Miller et al., 2005; Nakai et al., 2005b). rAAV integration does not take place in a neat cut-and-paste fashion and always accompanies various degrees of deletions in rAAV genome terminal sequences and the cellular genomes around integration sites. Complex genomic rearrangements are not rare and integration often causes a chromosomal translocation. All of these observations fit very well with a model in which NHEJ mediates rAAV integration. A series of studies has shown that DNA-PKcs has negative or positive effects on integration depending on the experimental systems used (Figure 2i). In a cell-free in vitro rAAV integration system, ss rAAV integration frequency increases and decreases by the addition of DNA-PKcs antibody and purified DNA-PKcs, respectively, leading to a conclusion that DNA-PKcs inhibits rAAV integration (Song et al., 2004). Whereas, in a cell culture system using DNA-PKcs-proficient M059K and deficient M059J cells, DNA-PKcs has been shown to enhance integration of both ss rAAV and ds rAAV (Cataldi & McCarty, 2010; Daya et al., 2009). In the context of animal experiment, Song et al. have exploited a two-thirds partial hepatectomy approach and shown that rAAV genomes integrate in DNA-PKcs-deficient SCID mouse livers at a significantly greater frequency than that of wild type control animals (i.e., >50% in SCID versus

## rAAV-Mediated Gene Targeting and DNA REPAIR PATHWAYS

HR mediated by the conventional vector systems occurs with efficiencies of a range of $10^{-6}$ to $10^{-7}$. In this regard, rAAV has become increasingly popular as the most efficient tool to precisely introduce defined DNA modifications at the target site in the cellular genome with remarkably high efficiencies of up to 1% in the cell population (Hendrie & Russell, 2005; Khan et al., 2011; Russell & Hirata, 1998; Vasileva & Jessberger, 2005). Targeting efficiencies could be increased further by 60-100 fold or more by introducing a DSB at the target site with a site-specific endonuclease (Miller et al., 2003; Porteus et al., 2003). This system, named the gene targeting rAAV vector system, has been applied in various disciplines, not only for gene therapy (Chamberlain et al., 2004) but also for generating knockout animals (Sun et al., 2008) and other types of basic research (Khan et al., 2011). Gene targeting rAAV serves as a donor vector that carries a DNA segment homologous to the chromosomal target sequence with a desired modification being introduced. The length of the homology arms can be 1.7 kb or potentially shorter, which is an advantage over the conventional targeting vectors that require a longer homologous DNA sequence (Hirata & Russell, 2000). Despite significant advance in the applications of the system, the underlying mechanism for rAAV-mediated gene targeting is poorly understood. As described above, rAAV does not harness any machinery designed specifically for mediating highly efficient gene targeting. The unusual structure of viral genome DNA is the only element that makes the system much more efficient than the conventional approaches.

The mechanism of rAAV-mediated gene targeting has just begun to be partly elucidated. Studies have indicated that the single-stranded nature of gene-targeting rAAV is key to efficient gene targeting reactions. Experimental evidence has come from the observation that, when mixtures of gene-targeting ss rAAV and ds rAAV vectors were used, gene correction rates correlated with the amounts of ss rAAV but not ds rAAV within the mixtures (Hirata &

Russell, 2000). Another study took advantage of recombinant minute virus of mice (rMVM), a rAAV-like parvovirus-based vector that predominantly packages viral genomes of minus polarity and does rarely undergo second-strand synthesis to form ds viral genomes. When reporter cells were infected with gene-targeting rMVM vectors containing either the coding or noncoding strand of a transgene cassette, a significant difference in targeting efficiencies was revealed between the two, indicating that ss viral genomes are the substrate (Hendrie et al., 2003). However, a recent study points out limitations in the previously used assay systems and argues against the above model because ds rAAV has also been found to mediate gene targeting at a higher level compared with the ss rAAV control (Hirsch et al., 2010). Although the nature of gene targeting substrates may be a subject of debate, it is clear that rAAV genome integration and rAAVmediated gene targeting use different DNA repair pathways. Genotoxic treatment, which significantly augments rAAV genome integrations, does not affect gene targeting efficiency (Hirata & Russell, 2000). In addition, rAAV gene targeting occurs preferentially in S-phase cells and does not take place at an appreciable level in terminally differentiated murine skeletal muscle fibers (Liu et al., 2004; Trobridge et al., 2005). Moreover, the cell cycle dependence has not clearly been demonstrated in rAAV integration and a study has demonstrated a readily appreciable level of rAAV integration in terminally differentiated cardiomyocytes and skeletal myofibers (Inagaki et al., 2007a). Collectively, NHEJ appears to be the major DNA repair pathway involved in rAAV integration while rAAV-mediated gene targeting uses HR. It has been demonstrated that RAD51/RAD54 pathway of HR is required for efficient rAAV-mediated gene targeting (Figure 2j), and deficiency of either of the NHEJ proteins, DNA-PKcs and Ku70, enhances the targeting rates (Fattah et al., 2008; Vasileva et al., 2006). Although the DNA-PKcs effect appears to be a cell-type dependent phenomenon (Fattah et al., 2008), the observations underscore the significant contribution of the HR pathways in rAAV-mediated gene targeting. Manipulation of HR and NHEJ pathways with small molecules will offer a novel and effective means to further improve rAAV-mediated gene targeting approaches to genetically engineer cellular genomes.

## AAV AS A TOOL FOR STUDYING DAMAGED DNA SITES, DDR, AND DNA REPAIR PATHWAYS

AAV has provided the most powerful means to deliver genetic materials to a broad range of cell and tissue/organ types without toxicity and to introduce sequence modifications at defined locations. What has made AAV more attractive is its utility as an unprecedented research tool to study molecular and cellular biology, where gene delivery is not a primary goal. As described in 3.1.3 and 3.1.4, AAV has been successfully exploited as a refined agent that can trigger DDR toward cell cycle arrest and apoptosis. AAV can deliver an element that triggers DDR (e.g., stalled replication forks) extrachromosomally with minimal transcriptional responses (McCaffrey et al., 2008) and without damaging the cellular genome. Although the phenomena observed in the AAV-based system may not necessarily recapitulate what takes place when the cellular DNA is damaged, it is assumed that molecularly defined extrachromosomal DDR triggers would provide a simple and less complicated means to study cellular responses to DNA damage. In addition, AAV has been exploited to study potential differences in DNA repair pathways among various tissues in the context of living animals. This type of study has demonstrated that, in hepatocytes, there is significant redundancy of Artemis/DNA-PKcs-independent NHEJ pathways that process hairpin DNA ends, while such redundancy is not observed in skeletal myofibers or cardiomyocytes in mice (Inagaki et al., 2007b). Moreover, AAV has recently emerged as a powerful tool to identify DNA sites damaged either endogenously or exogenously by genotoxic treatment or agents. Using rAAV as a tool to label pre-existing damaged DNA sites, a study has shown that DNA palindromes with an arm length of > 20 base pairs in the cellular genome represent the sites susceptible to breakage in mouse tissues (Inagaki et al., 2007a). Another study has taken a similar AAV-based labeling approach and demonstrated frequent off-target cleavage of the cellular genome by a rare cutting endonuclease, I-SceI, following expression of I-SceI in cells (Petek et al., 2010). Perhaps applications of AAV in biological and medical research will not be limited to the disciplines described above and will continue to expand with the advent of novel rAAV vector technologies.

## CONCLUSIONS

The virus-host interaction from a viewpoint of viral components and DNA repair machinery is an emerging research area that would offer unprecedented means to study both virology and molecular and cellular biology. The interaction in this aspect is most studied with adenoviruses, herpesviruses, and retroviruses including human immunodeficiency virus. These viruses have evolved sophisticated machinery to benefit them by manipulating or controlling DDR, DNA repair machinery, and the cell cycle. In this regard, AAV (i.e., wtAAV and rAAV) represents a unique viral agent in that Rep proteins are the sole viral components that interact with DNA repair machinery and rAAV expresses no such component. Despite the seemingly simple nature of AAV, there are significant virus-host interactions that involve DDR and DNA repair machinery in AAV infection, and we have just begun to appreciate them as summarized in this chapter. There has been an increasing interest in AAV primarily as a promising gene delivery vector and more recently as a new tool to study DNA damage, DDR, and DNA repair machinery. Studying AAV from various scientific aspects including virology, immunology, physiology, gene therapy, DNA damage, DDR, DNA repair, genomic instability, carcinogenesis, and so on, would significantly advance our knowledge about AAV and could solve unanswered fundamental biological questions that are difficult to address by the conventional approaches.

## ACKNOWLEDGMENT

Preparation of this chapter is in part supported by the National Institutes of Health (R01 DK078388). The authors are most grateful to Christopher Naitza and Baskaran Rajasekaran for their invaluable assistance in preparation of the manuscript. The authors apologize to investigators whose papers relevant to the topic of this chapter were not cited due to space constraints or inadvertent oversight.

## REFERENCES

1.    Alexander, I. E., Russell, D. W. & Miller, A. D. (1994). DNA-damaging agents greatly increase the transduction of nondividing cells by adeno-associated virus vectors. Journal of Virology, Vol.68, No.12, pp. 8282-8287, ISSN 0022-538X

2.    Asokan, A., Conway, J. C., Phillips, J. L., Li, C., Hegge, J., Sinnott, R., Yadav, S., DiPrimio, N., Nam, H. J., Agbandje-McKenna, M., McPhee, S., Wolff, J. & Samulski, R. J. (2010). Reengineering a receptor footprint of adeno-associated virus enables selective and systemic gene transfer to muscle. Nature Biotechnology, Vol.28, No.1, pp. 79-82, ISSN 1546-1696

3.    Atchison, R. W., Casto, B. C. & Hammon, W. M. (1965). Adenovirus-associated defective virus particles. Science, Vol.149, pp. 754-756, ISSN 0036-8075

4.    Beaton, A., Palumbo, P. & Berns, K. I. (1989). Expression from the adeno-associated virus p5 and p19 promoters is negatively regulated in trans by the rep protein. Journal of Virology, Vol.63, No.10, pp. 4450-4454, ISSN 0022-538X

5.    Berns, K. I. & Parrish C. R.. (2007). Parvoviridae, In Fields VIROLOGY, vol.2, 5th ed., D. M. Knipe, P. M. Howley, D. E. Griffin, R. A. Lamb, M. A. Martin, B. Roizman, and S. E. Straus (eds.), pp. 2437-2477, LIPPINCOTT WILLIAMS & WILKINS, ISBN 0-7817- 1832-5, Philadelphia, PA.

6.      Bernstein, K. A., Gangloff, S. & Rothstein, R. (2010). The recq DNA helicases in DNA repair. Annual Review of Genetics, Vol.44, pp. 393-417, ISSN 1545-2948

7.      Berthet, C., Raj, K., Saudan, P. & Beard, P. (2005). How adeno-associated virus rep78 protein arrests cells completely in s phase. Proceedings of the National Academy of Sciences of the United States of America, Vol.102, No.38, pp. 13634-13639, ISSN 0027-8424

8.      Boutin, S., Monteilhet, V., Veron, P., Leborgne, C., Benveniste, O., Montus, M. F. & Masurier, C. (2010). Prevalence of serum igg and neutralizing factors against adeno-associated virus (aav) types 1, 2, 5, 6, 8, and 9 in the healthy population: Implications for gene therapy using aav vectors. Human Gene Therapy, Vol.21, No.6, pp. 704-712, ISSN 1557- 7422

9.      Branzei, D. & Foiani, M. (2010). Maintaining genome stability at the replication fork. Nature Reviews Molecular Cell Biology, Vol.11, No.3, pp. 208-219, ISSN 1471-0080

10.     Brister, J. R. & Muzyczka, N. (1999). Rep-mediated nicking of the adeno-associated virus origin requires two biochemical activities, DNA helicase activity and transesterification. Journal of Virology, Vol.73, No.11, pp. 9325-9336, ISSN 0022-538X

11.     Burguete, T., Rabreau, M., Fontanges-Darriet, M., Roset, E., Hager, H. D., Koppel, A., Bischof, P. & Schlehofer, J. R. (1999). Evidence for infection of the human embryo with adenoassociated virus in pregnancy. Human Reproduction, Vol.14, No.9, pp. 2396-2401, ISSN 0268-1161

12.     Calcedo, R., Vandenberghe, L. H., Gao, G., Lin, J. & Wilson, J. M. (2009). Worldwide epidemiology of neutralizing antibodies to adeno-associated viruses. Journal of Infectious Diseases, Vol.199, No.3, pp. 381-390, ISSN 0022-1899

13.     Carson, C. T., Schwartz, R. A., Stracker, T. H., Lilley, C. E., Lee, D. V. & Weitzman, M. D. (2003). The mre11 complex is required for atm activation and the g2/m checkpoint. EMBO Journal, Vol.22, No.24, pp. 6610-6620, ISSN 0261-4189

14.     Carter, B. J., Burstein, H. & Peluso, R. W. (2009). Adeno-associated virus and AAV vectors for gene delivery, In: Gene and cell therapy, N. S. Templeton, (ed.), pp. 115-157, CRC press, ISBN 978-0-8493-8768-5, Boca Raton, FL.

15.     Castedo, M., Perfettini, J. L., Roumier, T., Andreau, K., Medema, R. & Kroemer, G. (2004). Cell death by mitotic catastrophe: A molecular definition. Oncogene, Vol.23, No.16, pp. 2825-2837, ISSN 0950-9232

16. Costello, E., Saudan, P., Winocour, E., Pizer, L. & Beard, P. (1997). High mobility group chromosomal protein 1 binds to the adeno-associated virus replication protein (rep) and promotes rep-mediated site-specific cleavage of DNA, atpase activity and transcriptional repression. EMBO Journal, Vol. 16, No. 19, pp. 5943-5954, ISSN 0261- 4189

17. Cataldi, M. P. & McCarty, D. M. (2010). Differential effects of DNA double-strand break repair pathways on single-strand and self-complementary adeno-associated virus vector genomes. Journal of Virology, Vol.84, No.17, pp. 8673-8682, ISSN 1098-5514

18. Cervelli, T., Palacios, J. A., Zentilin, L., Mano, M., Schwartz, R. A., Weitzman, M. D. & Giacca, M. (2008). Processing of recombinant aav genomes occurs in specific nuclear structures that overlap with foci of DNA-damage-response proteins. Journal of Cell Science, Vol.121, No.Pt 3, pp. 349-357, ISSN 0021-9533

19. Chamberlain, J. R., Schwarze, U., Wang, P. R., Hirata, R. K., Hankenson, K. D., Pace, J. M., Underwood, R. A., Song, K. M., Sussman, M., Byers, P. H. & Russell, D. W. (2004). Gene targeting in stem cells from individuals with osteogenesis imperfecta. Science, Vol.303, No.5661, pp. 1198-1201, ISSN 1095-9203

20. Choi, V. W., McCarty, D. M. & Samulski, R. J. (2006). Host cell DNA repair pathways in adenoassociated viral genome processing. Journal of Virology, Vol.80, No.21, pp. 10346- 10356, ISSN 0022-538X

21. Choi, Y. K., Nash, K., Byrne, B. J., Muzyczka, N. & Song, S. (2010). The effect of DNAdependent protein kinase on adeno-associated virus replication. PLoS ONE, Vol.5, No.12, pp. e15073, ISSN 1932-6203

22. Collaco, R. F., Bevington, J. M., Bhrigu, V., Kalman-Maltese, V. & Trempe, J. P. (2009). Adenoassociated virus and adenovirus coinfection induces a cellular DNA damage and repair response via redundant phosphatidylinositol 3-like kinase pathways. Virology, Vol.392, No.1, pp. 24-33, ISSN 1096-0341

23. Daniel, R., Katz, R. A. & Skalka, A. M. (1999). A role for DNA-pk in retroviral DNA integration. Science, Vol.284, No.5414, pp. 644-647, ISSN 0036-8075

24. Daya, S., Cortez, N. & Berns, K. I. (2009). Adeno-associated virus site-specific integration is mediated by proteins of the nonhomologous end-joining pathway. Journal of Virology, Vol.83, No.22, pp. 11655-11664, ISSN 1098-5514

25.  Deyle, D. R. & Russell, D. W. (2009). Adeno-associated virus vector integration. Current Opinion Molecular Therapy, Vol.11, No.4, pp. 442-447, ISSN 2040-3445

26.  Donsante, A., Miller, D. G., Li, Y., Vogler, C., Brunt, E. M., Russell, D. W. & Sands, M. S. (2007). Aav vector integration sites in mouse hepatocellular carcinoma. Science, Vol.317, pp. 477, ISSN 1095-9203

27.  Doostzadeh-Cizeron, J., Evans, R., Yin, S. & Goodrich, D. W. (1999). Apoptosis induced by the nuclear death domain protein p84n5 is inhibited by association with rb protein. Molecular Biology of the Cell, Vol.10, No.10, pp. 3251-3261, ISSN 1059-1524

28.  Duan, D., Yan, Z., Yue, Y. & Engelhardt, J. F. (1999). Structural analysis of adeno-associated virus transduction circular intermediates. Virology, Vol.261, No.1, pp. 8-14, ISSN 0042- 6822

29.  Duan, D., Yue, Y. & Engelhardt, J. F. (2003). Consequences of DNA-dependent protein kinase catalytic subunit deficiency on recombinant adeno-associated virus genome circularization and heterodimerization in muscle tissue. Journal of Virology, Vol.77, No.8, pp. 4751-4759, ISSN 0022-538X

30.  Erles, K., Sebokova, P. & Schlehofer, J. R. (1999). Update on the prevalence of serum antibodies (igg and igm) to adeno-associated virus (aav). Journal of Medical Virology, Vol.59, No.3, pp. 406-411, ISSN 0146-6615

31.  Erles, K., Rohde, V., Thaele, M., Roth, S., Edler, L. & Schlehofer, J. R. (2001). DNA of adenoassociated virus (aav) in testicular tissue and in abnormal semen samples. Human Reproduction, Vol.16, No.11, pp. 2333-2337, ISSN 0268-1161

32.  Evans, J. D. & Hearing, P. (2005). Relocalization of the mre11-rad50-nbs1 complex by the adenovirus e4 orf3 protein is required for viral replication. Journal of Virology, Vol.79, No.10, pp. 6207-6215, ISSN 0022-538X

33.  Excoffon, K. J., Koerber, J. T., Dickey, D. D., Murtha, M., Keshavjee, S., Kaspar, B. K., Zabner, J. & Schaffer, D. V. (2009). Directed evolution of adeno-associated virus to an infectious respiratory virus. Proceedings of the National Academy of Sciences of the United States of America, Vol.106, No.10, pp. 3865-3870, ISSN 1091-6490

34.  Fattah, F. J., Lichter, N. F., Fattah, K. R., Oh, S. & Hendrickson, E. A. (2008). Ku70, an essential gene, modulates the frequency of raav-mediated gene targeting in human somatic cells. Proceedings of the

National Academy of Sciences of the United States of America, Vol.105, No.25, pp. 8703-8708, ISSN 1091-6490

35.   Ferrari, F. K., Samulski, T., Shenk, T. & Samulski, R. J. (1996). Second-strand synthesis is a ratelimiting step for efficient transduction by recombinant adeno-associated virus vectors. Journal of Virology, Vol.70, No.5, pp. 3227-3234, ISSN 0022-538X

36.   Fisher, K. J., Gao, G. P., Weitzman, M. D., DeMatteo, R., Burda, J. F. & Wilson, J. M. (1996). Transduction with recombinant adeno-associated virus for gene therapy is limited by leading-strand synthesis. Journal of Virology, Vol.70, No.1, pp. 520-532, ISSN 0022- 538X

37.   Foust, K. D., Nurre, E., Montgomery, C. L., Hernandez, A., Chan, C. M. & Kaspar, B. K. (2009). Intravascular aav9 preferentially targets neonatal neurons and adult astrocytes. Nature Biotechnology, Vol.27, No.1, pp. 59-65, ISSN 1546-1696

38.   Fragkos, M., Breuleux, M., Clement, N. & Beard, P. (2008). Recombinant adeno-associated viral vectors are deficient in provoking a DNA damage response. Journal of Virology, Vol.82, No.15, pp. 7379-7387, ISSN 1098-5514

39.   Francois, A., Guilbaud, M., Awedikian, R., Chadeuf, G., Moullier, P. & Salvetti, A. (2005). The cellular tata binding protein is required for rep-dependent replication of a minimal adeno-associated virus type 2 p5 element. Journal of Virology, Vol.79, No.17, pp. 11082-11094, ISSN 0022-538X

40.   Gao, G., Vandenberghe, L. H., Alvira, M. R., Lu, Y., Calcedo, R., Zhou, X. & Wilson, J. M. (2004). Clades of adeno-associated viruses are widely disseminated in human tissues. Journal of Virology, Vol.78, No.12, pp. 6381-6388, ISSN 0022-538X

41.   Gao, G. P., Alvira, M. R., Wang, L., Calcedo, R., Johnston, J. & Wilson, J. M. (2002). Novel adeno-associated viruses from rhesus monkeys as vectors for human gene therapy. Proc. Natl. Acad. Sci. U.S.A., Vol.99, No.18, pp. 11854-11859, ISSN 0027-8424

42.   Garner, E., Martinon, F., Tschopp, J., Beard, P. & Raj, K. (2007). Cells with defective p53-p21- prb pathway are susceptible to apoptosis induced by p84n5 via caspase-6. Cancer Research, Vol.67, No.16, pp. 7631-7637, ISSN 0008-5472

43.   Geoffroy, M. C. & Salvetti, A. (2005). Helper functions required for wild type and recombinant adeno-associated virus growth. Current Gene Therapy, Vol.5, No.3, pp. 265-271, ISSN 1566-5232

44.  Ghosh, A., Yue, Y., Long, C., Bostick, B. & Duan, D. (2007). Efficient whole-body transduction with trans-splicing adeno-associated viral vectors. Molecular Therapy, Vol.15, No.4, pp. 750-755, ISSN 1525-0016

45.  Harper, J. W. & Elledge, S. J. (2007). The DNA damage response: Ten years after. Molecular Cell, Vol.28, No.5, pp. 739-745, ISSN 1097-2765

46.  Hauck, B., Zhao, W., High, K. & Xiao, W. (2004). Intracellular viral processing, not singlestranded DNA accumulation, is crucial for recombinant adeno-associated virus transduction. Journal of Virology, Vol.78, No.24, pp. 13678-13686, ISSN 0022-538X

47.  Henckaerts, E. & Linden, R. M. (2010). Adeno-associated virus: A key to the human genome? Future Virology, Vol.5, No.5, pp. 555-574, ISSN 1746-0808

48.  Hendrie, P. C., Hirata, R. K. & Russell, D. W. (2003). Chromosomal integration and homologous gene targeting by replication-incompetent vectors based on the autonomous parvovirus minute virus of mice. Journal of Virology, Vol.77, No.24, pp. 13136-13145, ISSN 0022-538X

49.  Hendrie, P. C. & Russell, D. W. (2005). Gene targeting with viral vectors. Molecular Therapy, Vol.12, No.1, pp. 9-17, ISSN 1525-0016

50.  Hermanns, J., Schulze, A., Jansen-Dblurr, P., Kleinschmidt, J. A., Schmidt, R. & zur Hausen, H. (1997). Infection of primary cells by adeno-associated virus type 2 results in a modulation of cell cycle-regulating proteins. Journal of Virology, Vol.71, No.8, pp. 6020-6027, ISSN 0022-538X

51.  Hirata, R. K. & Russell, D. W. (2000). Design and packaging of adeno-associated virus gene targeting vectors. Journal of Virology, Vol.74, No.10, pp. 4612-4620, ISSN 0022-538X

52.  Hirsch, M. L., Green, L., Porteus, M. H. & Samulski, R. J. (2010). Self-complementary aav mediates gene targeting and enhances endonuclease delivery for double-strand break repair. Gene Therapy, Vol.17, No.9, pp. 1175-1180, ISSN 1476-5462

53.  Inagaki, K., Fuess, S., Storm, T. A., Gibson, G. A., McTiernan, C. F., Kay, M. A. & Nakai, H. (2006). Robust systemic transduction with aav9 vectors in mice: Efficient global cardiac gene transfer superior to that of aav8. Molecular Therapy, Vol.14, No.1, pp. 45- 53, ISSN 1525-0016

54.  Inagaki, K., Lewis, S. M., Wu, X., Ma, C., Munroe, D. J., Fuess, S., Storm, T. A., Kay, M. A. & Nakai, H. (2007a). DNA palindromes with a modest arm length of greater, similar 20 base pairs are a significant target for recombinant adeno-associated virus vector integration in the

liver, muscles, and heart in mice. Journal of Virology, Vol.81, No.20, pp. 11290-11303, ISSN 0022-538X

55.  Inagaki, K., Ma, C., Storm, T. A., Kay, M. A. & Nakai, H. (2007b). The role of DNA-pkcs and artemis in opening viral DNA hairpin termini in various tissues in mice. Journal of Virology, Vol.81, No.20, pp. 11304-11321, ISSN 0022-538X

56.  Inagaki, K., Piao, C., Kotchey, N. M., Wu, X. & Nakai, H. (2008). Frequency and spectrum of genomic integration of recombinant adeno-associated virus serotype 8 vector in neonatal mouse liver. Journal of Virology, Vol.82, No.19, pp. 9513-9524, ISSN 1098-5514

57.  Ingemarsdotter, C., Keller, D. & Beard, P. (2010). The DNA damage response to nonreplicating adeno-associated virus: Centriole overduplication and mitotic catastrophe independent of the spindle checkpoint. Virology, Vol.400, No.2, pp. 271-286, ISSN 1096-0341

58.  Jayaram, S., Gilson, T., Ehrlich, E. S., Yu, X. F., Ketner, G. & Hanakahi, L. (2008). E1b 55kindependent dissociation of the DNA ligase iv/xrcc4 complex by e4 34k during adenovirus infection. Virology, Vol.382, No.2, pp. 163-170, ISSN 1096-0341

59.  Jurvansuu, J., Raj, K., Stasiak, A. & Beard, P. (2005). Viral transport of DNA damage that mimics a stalled replication fork. Journal of Virology, Vol.79, No.1, pp. 569-580, ISSN 0022-538X

60.  Jurvansuu, J., Fragkos, M., Ingemarsdotter, C. & Beard, P. (2007). Chk1 instability is coupled to mitotic cell death of p53-deficient cells in response to virus-induced DNA damage signaling. Journal of Molecular Biology, Vol.372, No.2, pp. 397-406, ISSN 0022-2836

61.  Khan, I. F., Hirata, R. K. & Russell, D. W. (2011). Aav-mediated gene targeting methods for human cells. Nature Protocol, Vol.6, No.4, pp. 482-501, ISSN 1750-2799

62.  Koerber, J. T., Klimczak, R., Jang, J. H., Dalkara, D., Flannery, J. G. & Schaffer, D. V. (2009). Molecular evolution of adeno-associated virus for enhanced glial gene delivery. Molecular Therapy, Vol.17, No.12, pp. 2088-2095, ISSN 1525-0024

63.  Kotin, R. M., Siniscalco, M., Samulski, R. J., Zhu, X. D., Hunter, L., Laughlin, C. A., McLaughlin, S., Muzyczka, N., Rocchi, M. & Berns, K. I. (1990). Site-specific integration by adeno-associated virus. Proceedings of the National Academy of Sciences of the United States of America, Vol.87, No.6, pp. 2211-2215, ISSN 0027-8424

64.  Kotin, R. M., Linden, R. M. & Berns, K. I. (1992). Characterization of a preferred site on human chromosome 19q for integration of adeno-

associated virus DNA by non-homologous recombination. EMBO Journal, Vol. 11, No. 13, pp. 5071-5078, ISSN 0261-4189

65.  Kyostio, S. R., Owens, R. A., Weitzman, M. D., Antoni, B. A., Chejanovsky, N. & Carter, B. J. (1994). Analysis of adeno-associated virus (aav) wild-type and mutant rep proteins for their abilities to negatively regulate aav p5 and p19 mrna levels. Journal of Virology, Vol.68, No.5, pp. 2947-2957, ISSN 0022-538X

66.  Li, L., Olvera, J. M., Yoder, K. E., Mitchell, R. S., Butler, S. L., Lieber, M., Martin, S. L. & Bushman, F. D. (2001). Role of the non-homologous DNA end joining pathway in the early steps of retroviral infection. EMBO Journal, Vol.20, No.12, pp. 3272-3281, ISSN 0261-4189

67.  Lilley, C. E., Schwartz, R. A. & Weitzman, M. D. (2007). Using or abusing: Viruses and the cellular DNA damage response. Trends in Microbiology, Vol.15, No.3, pp. 119-126, ISSN 0966-842X

68.  Linden, R. M., Ward, P., Giraud, C., Winocour, E. & Berns, K. I. (1996a). Site-specific integration by adeno-associated virus. Proceedings of the National Academy of Sciences of the United States of America, Vol.93, No.21, pp. 11288-11294, ISSN 0027-8424

69.  Linden, R. M., Winocour, E. & Berns, K. I. (1996b). The recombination signals for adenoassociated virus site-specific integration. Proceedings of the National Academy of Sciences of the United States of America, Vol.93, No.15, pp. 7966-7972, ISSN 0027-8424

70.  Liu, X., Yan, Z., Luo, M., Zak, R., Li, Z., Driskell, R. R., Huang, Y., Tran, N. & Engelhardt, J. F. (2004). Targeted correction of single-base-pair mutations with adeno-associated virus vectors under nonselective conditions. Journal of Virology, Vol.78, No.8, pp. 4165-4175, ISSN 0022-538X

71.  Ma, Y., Schwarz, K. & Lieber, M. R. (2005). The artemis:DNA-pkcs endonuclease cleaves DNA loops, flaps, and gaps. DNA Repair, Vol.4, No.7, pp. 845-851, ISSN 1568-7864

72.  McCaffrey, A. P., Fawcett, P., Nakai, H., McCaffrey, R. L., Ehrhardt, A., Pham, T. T., Pandey, K., Xu, H., Feuss, S., Storm, T. A. & Kay, M. A. (2008). The host response to adenovirus, helper-dependent adenovirus, and adeno-associated virus in mouse liver. Molecular Therapy, Vol.16, No.5, pp. 931-941, ISSN 1525-0024

73.  McCarty, D. M., Pereira, D. J., Zolotukhin, I., Zhou, X., Ryan, J. H. & Muzyczka, N. (1994). Identification of linear DNA sequences that specifically bind the adeno- associated virus rep protein. Journal of Virology, Vol.68, No.8, pp. 4988-4997, ISSN 0022-538X

74. McCarty, D. M., Fu, H., Monahan, P. E., Toulson, C. E., Naik, P. & Samulski, R. J. (2003). Adeno-associated virus terminal repeat (tr) mutant generates self-complementary vectors to overcome the rate-limiting step to transduction in vivo. Gene Therapy, Vol.10, No.26, pp. 2112-2118, ISSN 0969-7128

75. McCarty, D. M., Young, S. M., Jr. & Samulski, R. J. (2004). Integration of adeno-associated virus (aav) and recombinant aav vectors. Annual Review of Genetics, Vol.38, pp. 819-845, ISSN 0066-4197

76. McCarty, D. M. (2008). Self-complementary aav vectors; advances and applications. Molecular Therapy, Vol.16, No.10, pp. 1648-1656, ISSN 1525-0024

77. Miller, D. G., Petek, L. M. & Russell, D. W. (2003). Human gene targeting by adeno-associated virus vectors is enhanced by DNA double-strand breaks. Molecular and Cellular Biology, Vol.23, No.10, pp. 3550-3557, ISSN 0270-7306

78. Miller, D. G., Petek, L. M. & Russell, D. W. (2004). Adeno-associated virus vectors integrate at chromosome breakage sites. Nature Genetics, Vol.36, No.7, pp. 767-773, ISSN 1061- 4036

79. Miller, D. G., Trobridge, G. D., Petek, L. M., Jacobs, M. A., Kaul, R. & Russell, D. W. (2005). Large-scale analysis of adeno-associated virus vector integration sites in normal human cells. Journal of Virology, Vol.79, No.17, pp. 11434-11442, ISSN 0022-538X

80. Mirzoeva, O. K. & Petrini, J. H. (2003). DNA replication-dependent nuclear dynamics of the mre11 complex. Molecular Cancer Research, Vol.1, No.3, pp. 207-218, ISSN 1541-7786

81. Nakai, H., Storm, T. A. & Kay, M. A. (2000). Recruitment of single-stranded recombinant adeno-associated virus vector genomes and intermolecular recombination are responsible for stable transduction of liver in vivo. Journal of Virology, Vol.74, No.20, pp. 9451-9463, ISSN 0022-538X

82. Nakai, H., Yant, S. R., Storm, T. A., Fuess, S., Meuse, L. & Kay, M. A. (2001). Extrachromosomal recombinant adeno-associated virus vector genomes are primarily responsible for stable liver transduction in vivo. Journal of Virology, Vol.75, No.15, pp. 6969-6976, ISSN 0022-538X

83. Nakai, H., Thomas, C. E., Storm, T. A., Fuess, S., Powell, S., Wright, J. F. & Kay, M. A. (2002). A limited number of transducible hepatocytes restricts a wide-range linear vector dose response in recombinant adeno-associated virus-mediated liver transduction. Journal of Virology, Vol.76, No.22, pp. 11343-11349, ISSN 0022-538X

84.  Nakai, H., Montini, E., Fuess, S., Storm, T. A., Grompe, M. & Kay, M. A. (2003a). Aav serotype 2 vectors preferentially integrate into active genes in mice. Nature Genetics, Vol.34, No.3, pp. 297-302, ISSN 1061-4036

85.  Nakai, H., Storm, T. A., Fuess, S. & Kay, M. A. (2003b). Pathways of removal of free DNA vector ends in normal and DNA-pkcs-deficient scid mouse hepatocytes transduced with raav vectors. Human Gene Therapy, Vol.14, No.9, pp. 871-881, ISSN 1043-0342

86.  Nakai, H., Fuess, S., Storm, T. A., Muramatsu, S., Nara, Y. & Kay, M. A. (2005a). Unrestricted hepatocyte transduction with adeno-associated virus serotype 8 vectors in mice. Journal of Virology, Vol.79, No.1, pp. 214-224, ISSN 0022-538X

87.  Nakai, H., Wu, X., Fuess, S., Storm, T. A., Munroe, D., Montini, E., Burgess, S. M., Grompe, M. & Kay, M. A. (2005b). Large-scale molecular characterization of adeno-associated virus vector integration in mouse liver. Journal of Virology, Vol.79, No.6, pp. 3606- 3614, ISSN 0022-538X

88.  Nash, K., Chen, W., McDonald, W. F., Zhou, X. & Muzyczka, N. (2007). Purification of host cell enzymes involved in adeno-associated virus DNA replication. Journal of Virology, Vol.81, No.11, pp. 5777-5787, ISSN 0022-538X

89.  Nash, K., Chen, W., Salganik, M. & Muzyczka, N. (2009). Identification of cellular proteins that interact with the adeno-associated virus rep protein. Journal of Virology, Vol.83, No.1, pp. 454-469, ISSN 1098-5514

90.  Nathwani, A. C., Rosales, C., McIntosh, J., Rastegarlari, G., Nathwani, D., Raj, D., Nawathe, S., Waddington, S. N., Bronson, R., Jackson, S., Donahue, R. E., High, K. A., Mingozzi, F., Ng, C. Y., Zhou, J., Spence, Y., McCarville, M. B., Valentine, M., Allay, J., Coleman, J., Sleep, S., Gray, J. T., Nienhuis, A. W. & Davidoff, A. M. (2011). Long-term safety and efficacy following systemic administration of a self-complementary aav vector encoding human fix pseudotyped with serotype 5 and 8 capsid proteins. Molecular Therapy, pp., ISSN 1525-0024

91.  Nony, P., Tessier, J., Chadeuf, G., Ward, P., Giraud, A., Dugast, M., Linden, R. M., Moullier, P. & Salvetti, A. (2001). Novel cis-acting replication element in the adeno-associated virus type 2 genome is involved in amplification of integrated rep-cap sequences. Journal of Virology, Vol.75, No.20, pp. 9991-9994, ISSN 0022-538X

92.  Penaud-Budloo, M., Le Guiner, C., Nowrouzi, A., Toromanoff, A., Cherel, Y., Chenuaud, P., Schmidt, M., von Kalle, C., Rolling, F., Moullier, P. & Snyder, R. O. (2008). Adenoassociated virus vector genomes persist

as episomal chromatin in primate muscle. Journal of Virology, Vol.82, No.16, pp. 7875-7885, ISSN 1098-5514

93. Pereira, C. C., de Freitas, L. B., de Vargas, P. R., de Azevedo, M. L., do Nascimento, J. P. & Spano, L. C. (2010). Molecular detection of adeno-associated virus in cases of spontaneous and intentional human abortion. Journal of Medical Virology, Vol.82, No.10, pp. 1689-1693, ISSN 1096-9071

94. Petek, L. M., Russell, D. W. & Miller, D. G. (2010). Frequent endonuclease cleavage at off-target locations in vivo. Molecular Therapy, Vol.18, No.5, pp. 983-986, ISSN 1525-0024

95. Philpott, N. J., Gomos, J., Berns, K. I. & Falck-Pedersen, E. (2002). A p5 integration efficiency element mediates rep-dependent integration into aavs1 at chromosome 19. Proceedings of the National Academy of Sciences of the United States of America, Vol.99, No.19, pp. 12381-12385, ISSN 0027-8424

96. Porteus, M. H., Cathomen, T., Weitzman, M. D. & Baltimore, D. (2003). Efficient gene targeting mediated by adeno-associated virus and DNA double-strand breaks. Molecular and Cellular Biology, Vol.23, No.10, pp. 3558-3565, ISSN 0270-7306

97. Qing, K., Hansen, J., Weigel-Kelley, K. A., Tan, M., Zhou, S. & Srivastava, A. (2001). Adenoassociated virus type 2-mediated gene transfer: Role of cellular fkbp52 protein in transgene expression. Journal of Virology, Vol.75, No.19, pp. 8968-8976, ISSN 0022- 538X

98. Qing, K., Li, W., Zhong, L., Tan, M., Hansen, J., Weigel-Kelley, K. A., Chen, L., Yoder, M. C. & Srivastava, A. (2003). Adeno-associated virus type 2-mediated gene transfer: Role of cellular t-cell protein tyrosine phosphatase in transgene expression in established cell lines in vitro and transgenic mice in vivo. Journal of Virology, Vol.77, No.4, pp. 2741-2746, ISSN 0022-538X

99. Rabinowitz, J. E., Rolling, F., Li, C., Conrath, H., Xiao, W., Xiao, X. & Samulski, R. J. (2002). Cross-packaging of a single adeno-associated virus (aav) type 2 vector genome into multiple aav serotypes enables transduction with broad specificity. Journal of Virology, Vol.76, No.2, pp. 791-801, ISSN 0022-538X

100. Raj, K., Ogston, P. & Beard, P. (2001). Virus-mediated killing of cells that lack p53 activity. Nature, Vol.412, No.6850, pp. 914-917, ISSN 0028-0836

101. Rooney, S., Sekiguchi, J., Zhu, C., Cheng, H. L., Manis, J., Whitlow, S., DeVido, J., Foy, D., Chaudhuri, J., Lombard, D. & Alt, F. W. (2002).

Leaky scid phenotype associated with defective v(d)j coding end processing in artemis-deficient mice. Molecular Cell, Vol.10, No.6, pp. 1379-1390, ISSN 1097-2765

102. Roth, D. B., Menetski, J. P., Nakajima, P. B., Bosma, M. J. & Gellert, M. (1992). V(d)j recombination: Broken DNA molecules with covalently sealed (hairpin) coding ends in scid mouse thymocytes. Cell, Vol.70, No.6, pp. 983-991, ISSN 0092-8674

103. Russell, D. W., Miller, A. D. & Alexander, I. E. (1994). Adeno-associated virus vectors preferentially transduce cells in s phase. Proceedings of the National Academy of Sciences of the United States of America, Vol.91, No.19, pp. 8915-8919, ISSN 0027-8424

104. Russell, D. W., Alexander, I. E. & Miller, A. D. (1995). DNA synthesis and topoisomerase inhibitors increase transduction by adeno-associated virus vectors. Proceedings of the National Academy of Sciences of the United States of America, Vol.92, No.12, pp. 5719- 5723, ISSN 0027-8424

105. Russell, D. W. & Hirata, R. K. (1998). Human gene targeting by viral vectors [see comments]. Nature Genetics, Vol.18, No.4, pp. 325-330, ISSN 1061-4036

106. Samulski, R. J., Berns, K. I., Tan, M. & Muzyczka, N. (1982). Cloning of adeno-associated virus into pbr322: Rescue of intact virus from the recombinant plasmid in human cells. Proceedings of the National Academy of Sciences of the United States of America, Vol. 79, No. 6, pp. 2077-2081,ISSN 0027-8424

107. Samulski, R. J., Zhu, X., Xiao, X., Brook, J. D., Housman, D. E., Epstein, N. & Hunter, L. A. (1991). Targeted integration of adeno-associated virus (aav) into human chromosome 19 [published erratum appears in embo j 1992 mar;11(3):1228]. EMBO Journal, Vol.10, No.12, pp. 3941-3950, ISSN 0261-4189

108. Sanlioglu, S., Benson, P. & Engelhardt, J. F. (2000). Loss of atm function enhances recombinant adeno-associated virus transduction and integration through pathways similar to uv irradiation. Virology, Vol.268, No.1, pp. 68-78, ISSN 0042-6822

109. Sarkar, R., Mucci, M., Addya, S., Tetreault, R., Bellinger, D. A., Nichols, T. C. & Kazazian, H. H., Jr. (2006). Long-term efficacy of adeno-associated virus serotypes 8 and 9 in hemophilia a dogs and mice. Human Gene Therapy, Vol.17, No.4, pp. 427-439, ISSN 1043-0342

110. Saudan, P., Vlach, J. & Beard, P. (2000). Inhibition of s-phase progression by adeno-associated virus rep78 protein is mediated by

hypophosphorylated prb. EMBO Journal, Vol.19, No.16, pp. 4351-4361, ISSN 0261-4189

111. Schnepp, B. C., Jensen, R. L., Chen, C. L., Johnson, P. R. & Clark, K. R. (2005). Characterization of adeno-associated virus genomes isolated from human tissues. Journal of Virology, Vol.79, No.23, pp. 14793-14803, ISSN 0022-538X

112. Schultz, B. R. & Chamberlain, J. S. (2008). Recombinant adeno-associated virus transduction and integration. Molecular Therapy, Vol.16, No.7, pp. 1189-1199, ISSN 1525-0024

113. Schwartz, R. A., Palacios, J. A., Cassell, G. D., Adam, S., Giacca, M. & Weitzman, M. D. (2007). The mre11/rad50/nbs1 complex limits adeno-associated virus transduction and replication. Journal of Virology, Vol.81, No.23, pp. 12936-12945, ISSN 1098-5514

114. Schwartz, R. A., Carson, C. T., Schuberth, C. & Weitzman, M. D. (2009). Adeno-associated virus replication induces a DNA damage response coordinated by DNA-dependent protein kinase. Journal of Virology, Vol.83, No.12, pp. 6269-6278, ISSN 1098-5514

115. Shiotani, B. & Zou, L. (2009). Atr signaling at a glance. Journal of Cell Science, Vol.122, No.Pt 3, pp. 301-304, ISSN 0021-9533

116. Smith, R. H. & Kotin, R. M. (2002). Adeno-associated virus, In: Mobile DNA II, N. L. Craig, R. Craigie, M. Gellert, A. M. Lambowitz (eds.), pp. 905-923, ASM press, ISBN 1-55581- 209-0, Herndon, VA.

117. Song, S., Laipis, P. J., Berns, K. I. & Flotte, T. R. (2001). Effect of DNA-dependent protein kinase on the molecular fate of the raav2 genome in skeletal muscle. Proceedings of the National Academy of Sciences of the United States of America, Vol.98, No.7, pp. 4084-4088, ISSN 0027-8424

118. Song, S., Lu, Y., Choi, Y. K., Han, Y., Tang, Q., Zhao, G., Berns, K. I. & Flotte, T. R. (2004). DNA-dependent pk inhibits adeno-associated virus DNA integration. Proceedings of the National Academy of Sciences of the United States of America, Vol.101, No.7, pp. 2112- 2116, ISSN 0027-8424

119. Sonntag, F., Schmidt, K. & Kleinschmidt, J. A. (2010). A viral assembly factor promotes aav2 capsid formation in the nucleolus. Proceedings of the National Academy of Sciences of the United States of America, Vol.107, No.22, pp. 10220-10225, ISSN 1091-6490

120. Stracker, T. H., Carson, C. T. & Weitzman, M. D. (2002). Adenovirus oncoproteins inactivate the mre11-rad50-nbs1 DNA repair complex. Nature, Vol.418, No.6895, pp. 348-352, ISSN 0028-0836

121. Strickler, H. D., Viscidi, R., Escoffery, C., Rattray, C., Kotloff, K. L., Goldberg, J., Manns, A., Rabkin, C., Daniel, R., Hanchard, B., Brown, C., Hutchinson, M., Zanizer, D., Palefsky, J., Burk, R. D., Cranston, B., Clayman, B. & Shah, K. V. (1999). Adenoassociated virus and development of cervical neoplasia. Journal of Medical Virology, Vol.59, No.1, pp. 60-65, ISSN 0146-6615

122. Su, P. F. & Wu, F. Y. (1996). Differential suppression of the tumorigenicity of hela and siha cells by adeno-associated virus. British Journal of Cancer, Vol.73, No.12, pp. 1533-1537, ISSN 0007-0920

123. Summerford, C. & Samulski, R. J. (1998). Membrane-associated heparan sulfate proteoglycan is a receptor for adeno-associated virus type 2 virions. Journal of Virology, Vol.72, No.2, pp. 1438-1445, ISSN 0022-538X

124. Sun, X., Yan, Z., Yi, Y., Li, Z., Lei, D., Rogers, C. S., Chen, J., Zhang, Y., Welsh, M. J., Leno, G. H. & Engelhardt, J. F. (2008). Adeno-associated virus-targeted disruption of the cftr gene in cloned ferrets. Journal of Clinical Investigation, Vol.118, No.4, pp. 1578-1583, ISSN 0021-9738

125. Tan, I., Ng, C. H., Lim, L. & Leung, T. (2001). Phosphorylation of a novel myosin binding subunit of protein phosphatase 1 reveals a conserved mechanism in the regulation of actin cytoskeleton. Journal of Biological Chemistry, Vol.276, No.24, pp. 21209-21216, ISSN 0021-9258

126. Thomas, K. R. & Capecchi, M. R. (1987). Site-directed mutagenesis by gene targeting in mouse embryo-derived stem cells. Cell, Vol.51, No.3, pp. 503-512, ISSN 0092-8674

127. Torres-Ramos, C. A., Prakash, S. & Prakash, L. (1997). Requirement of yeast DNA polymerase delta in post-replicational repair of uv-damaged DNA. Journal of Biological Chemistry, Vol.272, No.41, pp. 25445-25448, ISSN 0021-9258

128. Trobridge, G., Hirata, R. K. & Russell, D. W. (2005). Gene targeting by adeno-associated virus vectors is cell-cycle dependent. Human Gene Therapy, Vol.16, No.4, pp. 522-526, ISSN 1043-0342

129. Tullis, G. E. & Shenk, T. (2000). Efficient replication of adeno-associated virus type 2 vectors: A cis-acting element outside of the terminal repeats and a minimal size. Journal of Virology, Vol.74, No.24, pp. 11511-11521, ISSN 0022-538X

130. Vakifahmetoglu, H., Olsson, M. & Zhivotovsky, B. (2008). Death through a tragedy: Mitotic catastrophe. Cell Death and Differentiation, Vol.15, No.7, pp. 1153-1162, ISSN 1350-9047

131. Vandendriessche, T., Thorrez, L., Acosta-Sanchez, A., Petrus, I., Wang, L., Ma, L., L, D. E. W., Iwasaki, Y., Gillijns, V., Wilson, J. M., Collen, D. & Chuah, M. K. (2007). Efficacy and safety of adeno-associated viral vectors based on serotype 8 and 9 vs. Lentiviral vectors for hemophilia b gene therapy. Journal of Thrombosis Haemostasis, Vol.5, No.1, pp. 16-24, ISSN 1538-7933

132. Vasileva, A. & Jessberger, R. (2005). Precise hit: Adeno-associated virus in gene targeting. Nature Reviews: Microbiology, Vol.3, No.11, pp. 837-847, ISSN 1740-1526

133. Vasileva, A., Linden, R. M. & Jessberger, R. (2006). Homologous recombination is required for aav-mediated gene targeting. Nucleic Acids Res, Vol.34, No.11, pp. 3345-3360, ISSN 1362-4962

134. Walz, C. & Schlehofer, J. R. (1992). Modification of some biological properties of hela cells containing adeno-associated virus DNA integrated into chromosome 17. Journal of Virology, Vol.66, No.5, pp. 2990-3002, ISSN 0022-538X

135. Wang, Z., Ma, H. I., Li, J., Sun, L., Zhang, J. & Xiao, X. (2003). Rapid and highly efficient transduction by double-stranded adeno-associated virus vectors in vitro and in vivo. Gene Therapy, Vol.10, No.26, pp. 2105-2111, ISSN 0969-7128

136. Wang, Z., Zhu, T., Qiao, C., Zhou, L., Wang, B., Zhang, J., Chen, C., Li, J. & Xiao, X. (2005). Adeno-associated virus serotype 8 efficiently delivers genes to muscle and heart. Nature Biotechnology, Vol.23, No.3, pp. 321-328, ISSN 1087-0156

137. Weitzman, M. D., Carson, C. T., Schwartz, R. A. & Lilley, C. E. (2004). Interactions of viruses with the cellular DNA repair machinery. DNA Repair, Vol.3, No.8-9, pp. 1165-1173, ISSN 1568-7864

138. Weitzman, M. D., Lilley, C. E. & Chaurushiya, M. S. (2010). Genomes in conflict: Maintaining genome integrity during virus infection. Annual Review of Microbiology, Vol.64, pp. 61- 81, ISSN 1545-3251

139. Williams, G. J., Lees-Miller, S. P. & Tainer, J. A. (2010). Mre11-rad50-nbs1 conformations and the control of sensing, signaling, and effector responses at DNA double-strand breaks. DNA Repair, Vol.9, No.12, pp. 1299-1306, ISSN 1568-7856

140. Winocour, E., Callaham, M. F. & Huberman, E. (1988). Perturbation of the cell cycle by adenoassociated virus. Virology, Vol.167, No.2, pp. 393-399, ISSN 0042-6822

141. Xiao, X., Xiao, W., Li, J. & Samulski, R. J. (1997). A novel 165-base-pair terminal repeat sequence is the sole cis requirement for the adeno-

associated virus life cycle. Journal of Virology, Vol.71, No.2, pp. 941-948, ISSN 0022-538X

142. Yakinoglu, A. O., Heilbronn, R., Burkle, A., Schlehofer, J. R. & zur Hausen, H. (1988). DNA amplification of adeno-associated virus as a response to cellular genotoxic stress. Cancer Research, Vol.48, No.11, pp. 3123-3129, ISSN 0008-5472

143. Yakobson, B., Koch, T. & Winocour, E. (1987). Replication of adeno-associated virus in synchronized cells without the addition of a helper virus. Journal of Virology, Vol.61, No.4, pp. 972-981, ISSN 0022-538X

144. Yakobson, B., Hrynko, T. A., Peak, M. J. & Winocour, E. (1989). Replication of adeno-associated virus in cells irradiated with uv light at 254 nm. Journal of Virology, Vol.63, No.3, pp. 1023-1030, ISSN 0022-538X

145. Yamamoto, N., Suzuki, M., Kawano, M. A., Inoue, T., Takahashi, R. U., Tsukamoto, H., Enomoto, T., Yamaguchi, Y., Wada, T. & Handa, H. (2007). Adeno-associated virus site-specific integration is regulated by trp-185. Journal of Virology, Vol. 81, No. 4, pp. 1990-2001, ISSN 0022-538X

146. Yan, Z., Lei-Butters, D. C., Zhang, Y., Zak, R. & Engelhardt, J. F. (2007). Hybrid adenoassociated virus bearing nonhomologous inverted terminal repeats enhances dualvector reconstruction of minigenes in vivo. Human Gene Therapy, Vol.18, No.1, pp. 81- 87, ISSN 1043-0342

147. Yang, L., Jiang, J., Drouin, L. M., Agbandje-McKenna, M., Chen, C., Qiao, C., Pu, D., Hu, X., Wang, D. Z., Li, J. & Xiao, X. (2009). A myocardium tropic adeno-associated virus (aav) evolved by DNA shuffling and in vivo selection. Proceedings of the National Academy of Sciences of the United States of America, Vol.106, No.10, pp. 3946-3951, ISSN 1091-6490

148. Yang, Q., Chen, F. & Trempe, J. P. (1994). Characterization of cell lines that inducibly express the adeno- associated virus rep proteins. Journal of Virology, Vol.68, No.8, pp. 4847- 4856, ISSN 0022-538X

149. Zentilin, L., Marcello, A. & Giacca, M. (2001). Involvement of cellular double-stranded DNA break binding proteins in processing of the recombinant adeno-associated virus genome. Journal of Virology, Vol.75, No.24, pp. 12279-12287, ISSN 0022-538X

150. Zhong, L., Zhou, X., Li, Y., Qing, K., Xiao, X., Samulski, R. J. & Srivastava, A. (2008). Singlepolarity recombinant adeno-associated virus

2 vector-mediated transgene expression in vitro and in vivo: Mechanism of transduction. Molecular Therapy, Vol.16, No.2, pp. 290-295, ISSN 1525-0024

151. Zhou, X., Zeng, X., Fan, Z., Li, C., McCown, T., Samulski, R. J. & Xiao, X. (2008). Adenoassociated virus of a single-polarity DNA genome is capable of transduction in vivo. Molecular Therapy, Vol.16, No.3, pp. 494-499, ISSN 1525-0024

# Chapter 4

# RECOMBINANT VIRAL VECTORS FOR INVESTIGATING DNA DAMAGE RESPONSES AND GENE THERAPY OF XERODERMA PIGMENTOSUM

Carolina Quayle[1], Carlos Frederico Martins Menck[1], and Keronninn Moreno Lima-Bessa[2]

[1]Dept. of Microbiology, Institute of Biomedical Sciences, University of Sao Paulo, Brazil

[2]Dept. of Cellular Biology and Genetics, Institute of Biosciences Federal University of Rio Grande do Norte, Brazil

## INTRODUCTION

### The Dark Side of the Sun

The genome of all living organisms is constantly threatened by a number of endogenous and exogenous DNA damaging agents. Such damage may disturb essential cellular processes, such as DNA replication and transcription, thereby resulting in double-strand breaks (referred to as 'replication fork collapse'), which can lead to chromosomal aberrations and/or cell death, ultimately contributing to mutagenesis, early aging and tumorigenesis (Ciccia & Elledge, 2010). One of the most important exogenous sources of DNA damage is the ultraviolet radiation (UV) component of sunlight, since it is responsible for a wide range of biological effects, including alteration in the structure of biologically essential molecules, such as proteins and nucleic acids. Indeed, UV is one of the most effective and carcinogenic exogenous agents that act on DNA, threatening the genome integrity and affecting normal life processes in different aquatic and terrestrial organisms, ranging from prokaryotes to mammals (Rastogi et al., 2010). In addition, UV is the major etiologic agent in the development of human skin cancers (Narayanan et al., 2010).

Sunlight is the primary UV source, whose spectrum is usually classified according to its wavelength in UVA (320-400 nm; lowest energy), UVB (280-320 nm) and UVC (200-280 nm; highest energy). Although these three UV bands are present in sunlight, the stratospheric ozone layer entirely blocks the

UVC and most of UVB, thus the solar UV spectrum that reaches the Earth's ground is composed by UVA and some UVB, even though ozone layer depletion can cause changes in this spectral distribution (Kuluncsics et al., 1999).

The chemical nature and efficiency in the formation of DNA lesions greatly depend on the wavelength of the incident photons. Despite its lowest energy, UVA light can deeply penetrate into the cells, mostly damaging DNA by indirect effects caused by the generation of reactive oxygen species which may react with nitrogen bases, resulting in base alterations and breaks in the DNA molecule. On the other hand, UVB can be directly absorbed by DNA bases, producing two main types of DNA damage, the cyclobutane pyrimidine dimers (CPDs) and pyrimidine-pyrimidone-(6-4)-photoproducts (6-4PPs), both resulting from covalent linkages between adjacent pyrimidines located on the same DNA strand, which leads to severe structural distortions in the DNA double helix. Interestingly, it has been recently demonstrated that UVA can also be directly absorbed by the DNA molecule, efficiently generating both CPDs and 6-4PPs (Schuch et al, 2009).

CPDs correspond to the formation of a four-member ring structure involving carbons C5 and C6 of both neighboring bases, whereas 6-4PPs are formed by a non-cyclic bond between C6 (of the 5'-end) and C4 (of the 3'-end) of the involved pyrimidines. Since those lesions induce strong distortions in the DNA molecule, they may lead to severe consequences to the cell if not properly removed, such as transcription arrest and replication blockage, thus disturbing cell metabolism, interfering with the cell cycle and, eventually, inducing cell death. DNA mutations can also result from misleading DNA processing. Long term consequences may include even more deleterious events, such as photoaging and cancer (Sinha & Häder, 2002; Narayanan et al., 2010; Rastogi et al., 2010).

## DNA Repair of UV Lesions and Related Human Syndromes

To ensure the maintenance of the genome integrity, several mechanisms that counteract DNA damage have emerged very early in evolution, including an intricate machinery of DNA repair, damage tolerance, and checkpoint pathways (Figure 1).

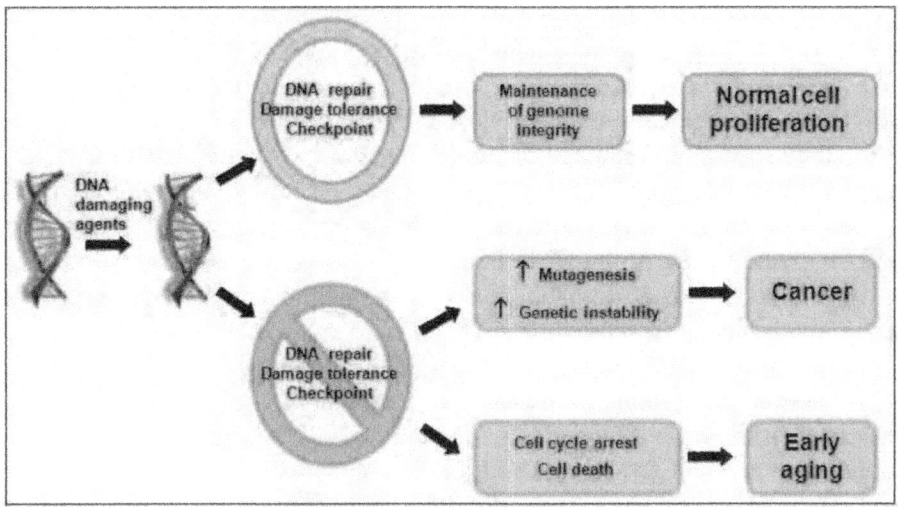

**Figure 1**. Main consequences of DNA damage. DNA damage can be induced by a variety of endogenous and exogenous agents. Several mechanisms, including an intricate machinery of DNA repair, damage tolerance, and checkpoint pathways, counteract DNA damage, aiming for the maintenance of genome stability, and guaranteeing normal cell proliferation. When these mechanisms fail, errors in DNA replication and/ or aberrant chromosomal segregations take place, increasing mutagenesis and genetic instability and contributing to a higher risk of cancer development. Alternatively, these damages may disturb the transcription and/or cause replication blockage, leading to cell death , thus contributing to early aging.

The nucleotide excision repair (NER) is one of the most versatile and flexible DNA repair systems, removing a wide range of structurally unrelated DNA double-helix distorting lesions, including UV photoproducts, bulky chemical adducts, DNA-intrastrand crosslinks, and some forms of oxidatively generated damage by orchestrating the concerted action of over 30 proteins, including the seven that are functionally impaired in xeroderma pigmentosum patients (XPA to XPG) (Costa et al., 2003; de Boer & Hoeijmakers, 2000). The NER pathway has been extensively studied at the molecular level in both prokaryotic and eukaryotic organisms. Depending on whether the damage is located in a transcriptionally active or inactive domain in the genome, its repair will be processed by one of two NER subpathways: global genome repair (GG-NER) or transcription-coupled repair (TC-NER). Indeed, while GG-NER is a random process, removing distorting lesions over the entire genome, TC-NER focus on those lesions which block RNA polymerases elongation, thus being highly specific and efficient (Fousteri & Mullenders, 2008; Hanawalt, 2002).

Briefly, the NER pathway involves a sequential cascade of events that starts with damage recognition, which defines the major difference between GG-NER and TC-NER. The latter is triggered upon blockage of RNA polymerase translocation at the DNA damage site, whereas GG-NER is evoked by specialized damage recognition factors, including the XPChHR23B heterodimer, and also XPE for certain lesions. The subsequent steps are carried out by a common set of NER factors that are shared by both subpathways and involve opening of the DNA helix around the lesion site by the concerted action of two helicases; dual incision of the damaged strand at both sides of the lesion by two endonucleases; removal of the damaged oligonucleotide (24-32 mer); gap filling of the excised patch using the undamaged strand as a template by the action of the replication machinery; and ligation of the new fragment to the chromatin by DNA ligase (Cleaver et al., 2009; Costa et al., 2003). Even though the core NER proteins that carry out damage recognition, excision, and repair reactions have been identified and extensively characterized, the regulatory pathways which govern the threshold levels of NER have not been fully elucidated (Liu et al., 2010). A schematic representation of this repair mechanism in humans is illustrated in Figure 2.

Several human autosomal recessive diseases are caused by dysfunction of the NER pathway, xeroderma pigmentosum (XP) being the prototype. Although this chapter will mainly focus on the XP syndrome, deficiencies in NER can also lead to other genetic diseases, such as trichothiodystrophy (TTD), Cockayne syndrome (CS), cerebro–oculo– facial–skeletal syndrome (COFS) and UV-sensitive syndrome (UVsS), all of which have photosensitivity as a common feature.

Xeroderma pigmentosum (XP) is a rare human disorder transmitted in an autosomal recessive fashion characterized by severe UV light photosensitivity, pigmentary changes, premature skin aging and a greater than 1,000-fold increase incidence of skin and mucous membrane cancer, including squamous and basal cell carcinomas and melanomas, with a 30-year reduction in life span (Cleaver et al., 2009; Karalis et al., 2011; Narayanan et al., 2010). In addition to cutaneous features, patients often develop ocular abnormalities, including neoplasms which may cause blindness. For most patients, often referred to as classical XP, this syndrome is caused by an impaired GG-NER activity, with or without deficiencies in TC-NER, determined by mutations in one of seven NER genes (XPA to XPG). When TC-NER is also affected (mutations in XPA, XPB, XPD and XPG genes), accelerated neurodegeneration may also occur in a substantial number of patients, suggesting increased neuronal cell death due to accumulated endogenous damage (Gerstenblith et al., 2010; Hoeijmakers, 2009). The eighth complementation group corresponds to the

XP-variant (XPV) patients, whose XP phenotype is related to mutations in the POLH gene, which encodes the translesion synthesis DNA polymerase eta responsible for the replication process on UV-irradiated DNA templates (Johnson et al., 1999; Masutani et al., 1999).

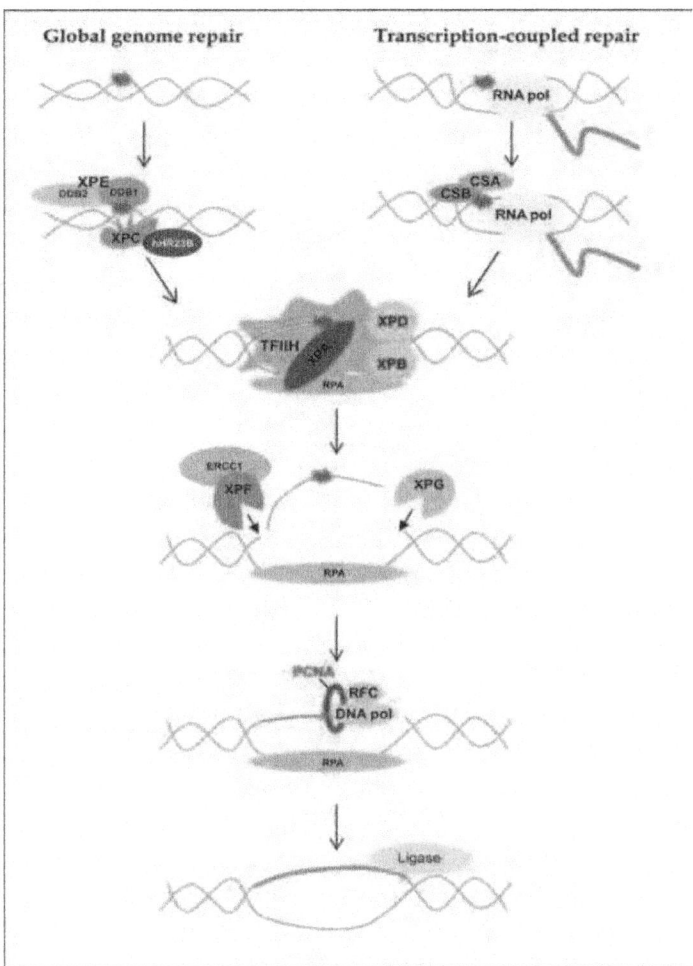

**Figure 2**. Schematic representation of repair of DNA lesions by nucleotide excision repair (NER). Depending on where the DNA damage is located in the genome, it will be processed by one of the two NER subpathways: the global genome repair (GG-NER) or the transcription-coupled repair (TC-NER), that basically differ in the lesion recognition step. Lesions occurring randomly in the genome are recognized by the XPC-HR23B complex, with the participation of XPE (DDB1-DDB2) for certain lesions, both complexes are GGNER-specific. On the other hand, lesions present in the transcribed strand of active genes that lead to the RNA polymerase arrest trigger the

TC-NER subpathway, which involves the CSA and CSB proteins. The following steps are common to both subpathways. The DNA double helix around the lesion is opened by XPB and XPD (helicases belonging to the TFIIH complex) and the single strand region is stabilized by RPA, allowing damage verification by the XPA protein. The DNA around the damaged site is then cleaved by the XPF–ERCC1 and XPG endonucleases, excising an oligonucleotide of 24-32 mer, and this patch is resynthesized by the replication machinery using the undamaged strand as a template. Finally, the new fragment is sealed to the chromatin by the DNA ligase.

A list of NER genes, which are related to XP syndrome, with their specific functions is given in Table 1.

**Table 1**. List of NER genes related to xeroderma pigmentosum and their roles in human DNA repair.

| Gene | Protein | Protein size (A.A.) | Function | Pathway |
|------|---------|---------------------|----------|---------|
| XPA | XPA | 273 | Interacts with RPA and other NER proteins, stabilizing ssDNA regions and also facilitating the repair complex assembly. | GG-NER TC-NER |
| XPB | XPB | 782 | Belongs to TFIIH complex, working as a 3` → 5` helicase. | GG-NER TC-NER |
| XPC | XPC | 940 | Responsible for lesion recognition in GG-NER. | GG-NER |
| XPD | XPD | 760 | Belongs to the TFIIH complex, working as a 5` → 3` helicase. | GG-NER TC-NER |
| DDB2 | XPE/p48 subunit | 428 | Forms a complex with XPE/p127 subunit, which is believed to facilitate the identification of lesions that are poorly recognized by XPC-hHR23B. | GG-NER |
| XPF | XPF | 905 | Found as a complex with ERCC1, which functions as an endonuclease 5' of the lesion. | GG-NER TC-NER |
| XPG | XPG | 1186 | Functions as an endonuclease 3' of the lesion. | GG-NER TC-NER |

*GG-NER- global genome repair; TC-NER- transcription-coupled repair.

The Cockayne syndrome (CS) is predominantly a developmental and neurological disorder, caused by mutations leading to a defective TC-NER, which prevents recovery from blocked transcription after DNA damage. CS patients are characterized by early growth and development cessation, severe and progressive neurodysfunction associated with demyelination, sensorineural hearing loss, cataracts, cachexia, and frailty (Weidenheim et al., 2009). Curiously, although severe photosensitivity is a common feature reported for most CS patients, it is not linked to an increased frequency of skin cancers, like it is in XP patients. Interestingly, specific mutations in one of three XP genes (XPB, XPD and XPG) may result in a clinical phenotype which reflects a combination of the traits associated with XP and CS (XP/CS patients). This observation indicates that simultaneous defects in GGNER and TC-NER can cause mutagenesis and cancer in some tissues and accelerated cell death and premature aging in others (Hoeijmakers, 2009).

The hallmark of trichothiodystrophy (TTD) is sulfur-deficient brittle hair, caused by a greatly reduced content of cysteine-rich matrix proteins in the hair shafts. In severe cases, mental abilities are also affected. Abnormal characteristics at birth and pregnancy complications are also common features of TTD, which may imply a role for DNA repair genes in normal fetal development (Stefanini et al., 2010). As CS patients, TTD patients do not present a high incidence of skin cancers. Genetically, three genes were identified for this disease (XPB, XPD and TTDA), but most TTD patients exhibit mutations on the two alleles of the XPD gene (Itin et al., 2001).

Cerebro–oculo–facial–skeletal syndrome (COFS) is a disorder determined by mutations in CSB, XPD, XPG and ERCC1 genes, leading to a defective TC-NER (Suzumura & Arisaka, 2010). It is characterized by congenital microcephaly, congenital cataracts and/or microphthalmia, arthrogryposis, severe developmental delay, an accentuated postnatal growth failure and facial dysmorphism.

Photosensitivity and freckling are the main features of patients with UV-sensitive syndrome (UVsS), but these patients have mild symptoms and no neurological or developmental abnormalities or skin tumors. Although other genes may be involved, mutations in the CSB gene were found in some of these patients, leading to defective TC-NER of UV damage (Horibata et al., 2004; Spivak, 2005).

Therefore, the general relationship between defects in NER genes and clinical disease phenotypes is complex, since mutations in several genes can cause the same phenotype, and different mutations in the same gene can cause different phenotypes (Kraemer et al., 2007).

Even though DNA repair malfunctions are autosomal recessive diseases and their incidence is therefore relatively low (~1/100,000), many of the individuals with DNA repair deficiencies die in early childhood since there is no effective treatment, only palliative care. Therefore, the search for a long-term treatment has been intense. Several strategies using recombinant viral vectors are being used in order to improve the resistance of cells from these patients to DNA damaging agents (Lima-Bessa et al., 2009; Menck et al., 2007). Also, the studies of DNA repair mechanisms have yielded a better understanding of specific cell processes which lead to human diseases such as cancer, neurodegeneration and aging (Hoeijmakers, 2009). This review will focus on the use of recombinant viral vectors for the purposes of investigating both the cellular responses to DNA damage and the perspectives of providing therapy for XP patients.

# RECOMBINANT VIRAL VECTORS AS GENE DELIVERY TOOLS

An ideal gene delivery tool should have the ability to transduce proliferating and fully differentiated cells with high efficiency; mediate high-level, prolonged and controlled transgene expression; have little toxicity (both at cellular and organism levels); elicit small immune responses in vivo; and be able to accommodate large DNA fragments for transgene transduction (Howarth et al., 2010). Unfortunately, there is no single tool that fulfills all these criteria.

Viruses have had million of years to improve their capacity to infect cells with the aid of evolutionary pressures. Researchers have been trying to take advantage of this ability creating recombinant viral vectors. In general, for that purpose, the viral genome is manipulated and sequences needed to form the infective virion are deleted, opening space to insert the transgene of interest.

Several viral vectors have been created and the most widely used are: adenovirus, retrovirus (including lentivirus) and adeno-associated virus. The main characteristics of these vectors are presented in Table 2.

**Table 2**. Main features of viruses currently used as recombinant vectors for gene delivery.

| Virus | Nucleic acid | Genome size (Kb) | Envelope | Virion size (nm) | Integration | Transgene size (Kb) | Immune response | Transgene expression |
|---|---|---|---|---|---|---|---|---|
| Adenovirus | dsDNA linear | 36 | | 90 | episomal | 8 - 25 | *** | days - months |
| Adeno-associated virus | ssDNA linear | 4.7 | | 25 | site-specific | 4.7 - 9 | * | months-years |
| Retrovirus | ssRNA (homodimer) | 7 - 12 | * | 100 | random | <10 | ** | years |
| Lentivirus | ssRNA (homodimer) | 9 | * | 100 | random | 10 - 16 | ** | years |

Searching for the perfect gene delivery tool, intense modifications have been added to the vectors' genomes, nucleocapsid and envelopes, always searching for less immunogenic vectors, with higher and more specific transduction properties. Currently, recombinant viruses are the vector of choice for research and clinical trials worldwide, but still only few phase II or III trials are being conducted (Atkinson & Chalmers, 2010). All viral vectors cited here have already been used in in vitro, ex vivo and in vivo experiments and in clinical trials.

## Recombinant Adenoviral Vectors

Adenoviruses (Ad) are non-enveloped double-stranded DNA viruses with tropism for the respiratory and ocular tissues. The first generation recombinant vector can carry up to 8 Kbp of DNA, while the last generation, in which the viral DNA sequence is completely deleted (also named gutless), is able to efficiently transduce over 25 Kbp of DNA (Atkinson & Chalmers, 2010).

Despite the fact that the gutless vector needs the aid of helper viral proteins supplied in trans, adenoviral vectors are easily produced in high titers. Once the transgene has been delivered inside the nucleus it remains episomal, reducing the risk of tumorigenesis induced by insertional mutagenesis. On the other hand, the episomal DNA is not replicated and its segregation in mitosis leads to the eventual loss of the transgene in the daughter cells. Thus, the transgene expression is short-lived. A possible solution is to add a site-specific integration sequence next to the transgene, leading to a prolonged transgene expression (Atkinson & Chalmers, 2010). Another advantage of the adenoviral vectors is their ability to transduce post mitotic cells since the transgene is already delivered in its active form, as a double-stranded DNA. This property is of particular interest when aiming for gene therapy in neurons (Atkinson & Chalmers, 2010).

The biggest challenge for the use of adenoviral vectors in vivo is the immunological response it elicits. This strong response is not only due to the natural immunogenicity of its components, but also to pre-existing immunity caused by previous contact with at least one of the over 50 serotypes of human infecting adenovirus (Seregin & Amalfitano, 2009). Taking into consideration that these vectors are only capable of a transient expression of the transgene and that repeated dosage might be necessary, a strong immune response is very undesirable. Possible alternatives to circumvent this issue are: manipulation of the viral capsid proteins and DNA, making them less immunogenic; the usage of a different serotype on each application; and the use of immunosuppressants (Atkinson & Chalmers, 2010; Seregin & Amalfitano, 2009).

The great importance of the immunological response against a gene therapy vector was brought to attention when, in 1999, a patient suffering from an ornithine transcarbamylase deficiency, died due to an unexpected inflammatory response reaction to the adenoviral vector used in a clinical trial (Edelstein et al., 2007). Still, adenoviral vectors are currently the most widely used viral vectors in clinical trials, accounting for approximately 24% of all vectors used in gene therapy clinical trials (Edelstein et al., 2007; Hall et al., 2010).

## Recombinant Adeno-Associated Viral (AAV) Vectors

Adeno-associated viruses (AAV) are non enveloped, single-stranded DNA, with serotypespecific tropism viruses. To date, 12 serotypes have been identified in primate or human tissues (Schmidt et al., 2008) in a total of over 100 known serotypes (Wang et al., 2011). Their productive lytic infection depends on the presence of a helper virus, adeno or herpesvirus, that provide in trans the necessary genes for the AAV replication and virion production. In the absence of a helper virus, the AAV establishes its latent cycle integrating specifically in the 19q13.4 region of the human genome (Daya & Berns, 2008). The site-specific integration is mainly dependent on the virus internal terminal repeats (ITRs), the integration efficiency element (IEE) and Rep 68 and Rep 78 genes. In the 19q13.4 region, several muscle-related genes are present, including some responsible for actin organization. No significant side effects have been observed due to AAV genome integration in this chromosome region (Daya & Berns, 2008).

The onset of transgene expression delivered by an AAV vector is delayed, usually starting several days after the transduction, probably due to the time invested in the synthesis of the DNA second strand (Michelfelder & Trepel, 2009). Although late, the transgene expression is long lasting and there is a very low humoral response, mainly related to previous exposure to the viral antigens (Daya & Berns, 2008). Despite the small size of the AAV nucleocapsid and genome, it has been shown that transgenes up to 7.2 Kb can be delivered by AAV vectors, but the oversized genomes reduce at least 10 fold the transduction efficiency (Dong et al., 2010). Several strategies have been developed seeking to optimize the vector capacity, such as the trans-splicing vector. With the simultaneous usage of two AAV vectors, this technology takes advantage of the concatamers formed by the ITRs that can recombine to form the desired transgene inside the transduced cell. These trans-splicing vectors allow the final transgene to have up to 9 Kb (Daya & Berns, 2008).

Only recently adeno-associated viral vectors started being used in gene therapy research and account for less than 4% of all vectors used in gene therapy clinical trials (Edelstein et al., 2007; Hall et al., 2010). Although these vectors do not behave as the parental virus, since they do not integrate in the genome (due to the lack of the REP protein), gene expression can be very long and elicit low immunological responses, making AAV vectors promising in gene therapy investigations.

## Recombinant Retroviral Vectors

The Retroviridae family is characterized by a single-stranded RNA genome which can only replicate inside the host cell with the aid of an RNA-dependent DNA polymerase, the reverse transcriptase. This enzyme transcribes the virus' RNA into a DNA sequence that the host cell machinery can transcribe and translate (Froelich et al., 2010).

Retroviral vectors are capable of transducing a wide range of cell types, are able to accommodate extensive changes in their genome, accept long transgenes, have low immunogenicity, can be produced in high titers, and promote a prolonged transgene expression due to their ability to integrate into the host cell genome (Froelich et al., 2010). On the other hand, most retroviral vectors can only transduce replicating cells since the transport of the transcribed viral DNA to the nucleus is mitosis-dependent. Additionally, there is always the risk of insertional mutagenesis due to the semi-random integration of the vector genome in the host cell's genome (Froelich et al., 2010). Nowadays, the most widely used retroviruses as gene therapy tools are the lentiviruses (LVs), such as the human immunodeficiency virus (HIV). These vectors have the same advantages as other retroviral vectors and are capable of transducing post mitotic cells. Moreover, the LVs tend not to integrate by transcription initiation sites, reducing the risk of insertional tumorigenesis (Froelich et al., 2010).

The retroviral vectors were the first vectors used in gene therapy clinical trials in 1989 (Edelstein et al., 2007, Rosenberg et al., 1990) and are extensively used in fundamental biological research, functional genomics and gene therapy (Mátrai et al., 2010). In 2004, 28% of the clinical trials involving viral vectors included retroviral vectors (Edelstein et al., 2007); in 2010 that number dropped to approximately 23% (Voigt et al., 2008). This drawback is due to the unfortunate events of the French severe combined immunodeficiency (SCID) trial in 2002, where two out of ten children died in consequence of a leukemia, which was related to the insertional mutagenesis of the retroviral vector used (Edelstein et al., 2007).

Since then, special attention has been paid to the safety of these vectors as many are known to derive from viruses that cause severe diseases, such as the acquired immunodeficiency syndrome (AIDS). Strategies are constantly developed to prevent the risk of insertional mutagenesis. For that purpose, in addition to the virions being replication-defective, generated by trans-complementation, several further manipulations of the viral genome were made. The development of a self-inactivating (SIN) vector (Iwakuma et al., 1999) prevents horizontal and vertical gene transfer and diminishes the probability of the production of a replicating virion or over-expression of a host cell oncogene (Edelstein et al., 2007).

# INVESTIGATING DNA DAMAGE RESPONSES WITH | ADENOVIRAL VECTORS IN HUMAN CELLS

## *In Vitro* and *In Vivo* Adenoviral Gene Transduction for the Correction of DNA Repair Defects

The knowledge of the molecular defects in XP cells was the starting point for understanding how human cells handle lesions in their genome. So far, different techniques have been used to study DNA repair mechanisms and reverse malfunctions in this essential system. One powerful tool employed in these studies has been the use of recombinant adenoviral vectors to transduce DNA repair genes directly into human skin cells, aiming to improve the knowledge of basic mechanisms that cells use to protect their genome.

Experiments using first generation recombinant adenoviral vectors have been successfully employed in the transduction of both SV40-transformed and primary fibroblasts derived from XP-A, XP-C, XP-D and XP-V patients (Armelini et al., 2007). The expression of the respective functional proteins in all transduced defective cell populations was significantly increased, reaching levels even higher than seen for wild type cells (Armelini et al., 2005; Lima-Bessa et al., 2006; Muotri et al., 2002). Moreover, different phenotypical analyses, including cell cycle, apoptosis and cell survival assays, have been carried out, all indicating that the protein expression mediated by the recombinant adenoviruses was clearly accompanied by the recovery of the DNA repair ability and increased resistance to UV radiation, thereby demonstrating functional correction of the XP phenotype. It is worth mentioning that, even though transgene expression mediated by adenoviruses is typically short-lived, sustainable high expression of XPA and XPC proteins with parallel increased UV-irradiation resistance was obtained even two months after cell transduction (Muotri et al., 2002).

For XP-A, XP-C and XP-D transduced cell lines, phenotypic analyses also involved assays aiming to investigate their ability to perform DNA repair after UV irradiation. This has been measured through determination of unscheduled DNA repair synthesis (UDS), which corresponds to the incorporation of [methyl-$^3$H] thymidine in cells that are not in S-phase, and is visualized by autoradiography as the presence of radioactive grains inside nuclei. Interestingly, UDS activity in all transduced deficient cell lines was restored to levels comparable to NER proficient cell lines, indicating those cells became able to efficiently remove UV lesions by restoring NER activity.

It is well known that UV radiation promotes DNA elongation delay as a result of replication blockage by UV photolesions (Cleaver et al., 1983), which can be easily seen by running pulse-chase experiments in alkaline sucrose gradients. Using this approach, it has been possible to show that XP-V transduced cells were able to elongate nascent DNA on UVdamaged DNA templates as efficiently as wild type cells (Lima-Bessa et al., 2006), once again demonstrating the great potential of recombinant adenoviruses in the transduction and expression of functional proteins.

One interesting conclusion came from the observation that even though XPA, XPC and XPD genes were over-expressed in all transduced cell lines when compared to NER proficient cells, this had no impact in the UV-resistance or NER capability, suggesting that neither of these proteins is limiting for NER in human cells. Another possible explanation is that once the NER pathway requires a coordinated action of several proteins, increasing only one of these proteins does not result in speeding up removal of the DNA lesions. Similarly, the excess of polη (XPV) mediated by adenoviral transduction has not affected cell survival nor elongation of replication products in UV-treated XP-A human cells, suggesting not only that polη is not a limiting factor for the efficient replication of the UV-damaged DNA in XP-A cells, but also demonstrating that the deleterious effects caused by the remaining DNA lesions in the genome cannot be mitigated by an efficient bypass mediated by polη.

However, the potential of such vectors is not restricted to in vitro assays. Indeed, another real perspective is their use to investigate the molecular mechanisms of DNA repair and their consequences in vivo, thus opening new avenues for a better understanding of cellular and physiologic responses to DNA damage. In vivo experiments may also help to establish the relationship between DNA repair, cancer and aging, as mice models for different DNA repair syndromes have been developed by different groups worldwide. Despite the extensive use of these models to broaden the understanding of several DNA repair related disorders, little work has been done in vivo testing gene therapy strategies for these diseases. Indeed, up to the present moment, only one study showed an efficient in vivo gene therapy protocol for complementation of the XP phenotype (Marchetto et al., 2004).

Exciting results by Marchetto and co-workers showed that the administration of subcutaneous injections of an adenoviral vector carrying the XPA human gene directly into the dorsal region of XP-A knockout mice led to an extensive expression of the heterologous protein in different skin cells, including dermal fibroblasts, cells of the hair follicle and basal replicating keratinocytes, which are believed to be the starting point of most skin tumors. As a result, the repair capability of these transduced cells was restored, thus preventing UVBinduced deleterious skin effects, such as persistent scars, skin hyperkeratosis and, ultimately, avoiding the formation of squamous cell carcinomas (Marchetto et al., 2004).

Despite the promising results of this work, no others followed. Researchers are now aware of several possible limitations and complications of gene therapy after some unexpected severe events in clinical trials (Edelstein et al., 2007) and are spending more time improving gene targeting tools and techniques before risking in vivo approaches. In that sense, extreme progress has been made with experiments in vitro, as previously presented. A general panel showing the main uses of the recombinant adenoviral vectors carrying DNA repair genes is presented in Figure 3.

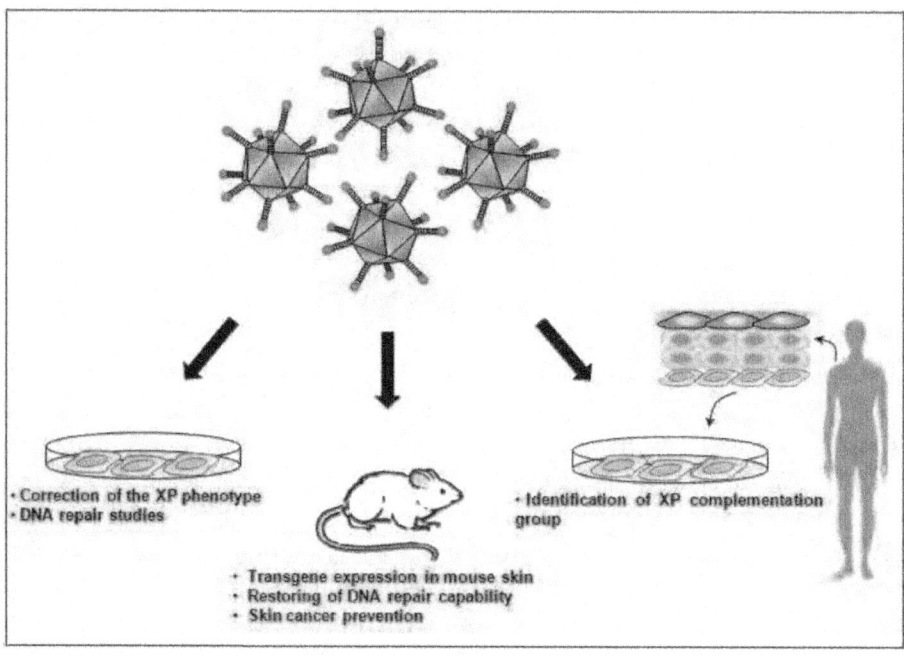

**Figure 3**. DNA repair gene transduction by recombinant adenoviruses. Adenoviral vectors have been successfully employed to transduce human XP genes directly into established human cell lines (left), XP knockout mice skin (center), and fibroblasts from the skin of XP patients (right ). Endpoints are indicated for each particular case.

Based on the successful complementation of the XP phenotype both in vitro and in vivo, adenoviral vectors could be proposed as an efficient tool for diagnosis and identification of XP patients' complementation groups. This hypothesis was recently tested and confirmed: with the use of adenoviruses carrying DNA repair genes, it has been possible to determine the complementation group of three Brazilian XP patients, now characterized as XP-C patients. To that end, adenoviral transduced cells from these patients have been submitted to UV treatment and then analyzed by simple assays, such as cell survival and UDS (Leite et al., 2009). This diagnosis has been performed using the patients' skin fibroblasts but the potential use of adenoviral vectors for this purpose becomes even more exciting, considering that the adenoviral transduction could be held in cells present in the patients' blood, thus becoming a faster and less invasive technique. Besides scientific and epidemiological goals, the identification of the gene defect may help to predict clinical prognosis for the XP patients and guide appropriate genetic counseling for their families. Direct gene sequencing can be performed to identify the mutated genes, but as there are eight potential candidate genes for XP, functional complementation assays are still used for the genetic diagnosis of these patients.

## Investigating UV-Induced Cell Responses Employing Photolyases

Photoreactivation is a very efficient DNA repair mechanism, which specifically removes the two main UV photoproducts. Photoreactivation is carried out by flavoproteins known as photolyases. These enzymes recognize and specifically bind to UV lesions, thus reverting them back to the undamaged monomers, using a blue-light photon as energy source (Brettel & Byrdin, 2010; Sancar, 2008). Interestingly, photolyases demonstrate a great efficiency for discriminating the target lesion, either CPDs or 6-4PPs, and so far no photolyase has been shown to be able to repair both lesions. Thus, enzymes that repair CPDs are referred to as CPD-photolyases, while 6-4PP-photolyases specifically repair 6-4PPs (Müller & Carell, 2009). Both classes of photolyases are evolutionarily related, but functionally distinct (LucasLledó & Lynch, 2009). Curiously, genes encoding genuine photolyases have been lost somehow in the course of the evolution of placental mammals, including humans. Instead, these organisms retain cryptochromes, photolyase-homologous proteins that participate in the maintenance of circadian rhythm, but that do not keep any residual activity related to DNA repair (Partch & Sancar, 2005).

Previous studies have confirmed that the CPD-photolyase is active when delivered to human cells, reducing mutagenesis (You et al., 2001), preventing UV-induced apoptosis (Chiganças et al., 2000) and recovering RNA transcription driven by RNA polymerase II (Chiganças et al., 2002). These successful studies have motivated the adenoviruses-mediated expression of the CPD-photolyase from the rat kangaroo Potorous tridactylus and the plant 6-4PP-photolyase from Arabidopsis thaliana in human cells aiming to discriminate the precise role of UV-induced cellular responses in both NER-deficient and NER-proficient human cells. Employing immunofluorescence, immunoblot and local UV experiments, it has been possible to see that these enzymes are not only very specific for their lesions, but are also really fast to find them, colocalizing with regions of damaged DNA and other DNA repair enzymes in less than two minutes (Chiganças et al., 2004; Lima-Bessa et al., 2008).

Adenoviral-mediated photorepair of CPDs substantially prevented apoptosis in all UVirradiated cell lines (both NER-deficient and NER-proficient cells), confirming the involvement of these lesions in cell death signaling, as previously reported. On the other hand, 6-4PP repair by the 6-4PP-photolyase decreased UV-induced apoptosis only in those cell lines deficient for both NER subpathways, causing minimal effect, if any, in NERproficient cells, including those lacking polη. These results suggest that, when not efficiently repaired, 6-4PPs also have important biological consequences, triggering cell responses leading to the activation of apoptotic cascades. Interestingly, in CS-A cells (TC-NER deficient), a substantial attenuation of apoptotic levels could be again detected when CPDs were removed from the genome by the means of CPD-photolyase, while no detectable effect was observed as a consequence of photorepair of 6-4PPs, indicating that CPD lesions are the major UV-induced DNA damage leading to cell death, also in cells that are only proficient in GG-NER, the main subpathway of NER responsible for the removal of 6-4PPs in humans (Lima-Bessa et al., 2008).

These results suggest that CPDs and 6-4PPs may play different roles in UV-induced apoptosis depending on the repair capacity of human cells. In GG-NER proficient cells, the harmful effects of UV light seem to be predominantly due to the prolonged remaining CPDs in the genome caused by their slow removal by NER, with the minor participation of 6-4PPs (Lima-Bessa et al., 2008). Indeed, it has been reported that about 80–90% of 6-4PPs are removed from the human genome in the first 4 hours following UV exposure, whereas 40– 50% of CPDs still remain to be repaired 24 hours later, probably due to the higher affinity of the XPC/hHR23B complex for 6-4PPs (Kusumoto et al., 2001). Thus, the lack of noticeable effects on UV-induced apoptosis in NER-proficient cells after 6-4PPs photorepair may be simply due to their fast repair by GG-NER. On the other hand, as for CPDs, the remaining of 6-4PPs in the genome seems to cause major disturbances in cell metabolism that lead to cell death. A summary of these results is shown in Figure 4.

To further confirm the idea that the roles of CPDs and 6-4PPs in UV-killing are related to the cellular repair capacity, authors have expressed these photolyases in TTD1V1 cells, a particular TTD cell line with a slower kinetics of 6-4PPs repair, eliminating about 50% and 70% of 6-4PPs at 6 and 24 hours post-UV treatment, respectively. Once again, repair of both lesions by the respective photolyase notably reduced apoptosis in these cells, even though the 6-4PP photorepair was less effective than seen for NER-deficient cell lines (Lima-Bessa et al., 2008). These photolyases were also used to identify a defect in the recruitment of downstream NER factors on certain XPD/TTD mutated cells, slowing down the removal of UV-induced lesions. As this recruitment was recovered by treatment with the histone deacetylase inhibitor trichostatin A, the data indicated that this defect is partially related to the accessibility of DNA damage in closed chromatin regions (Chiganças et al., 2008).

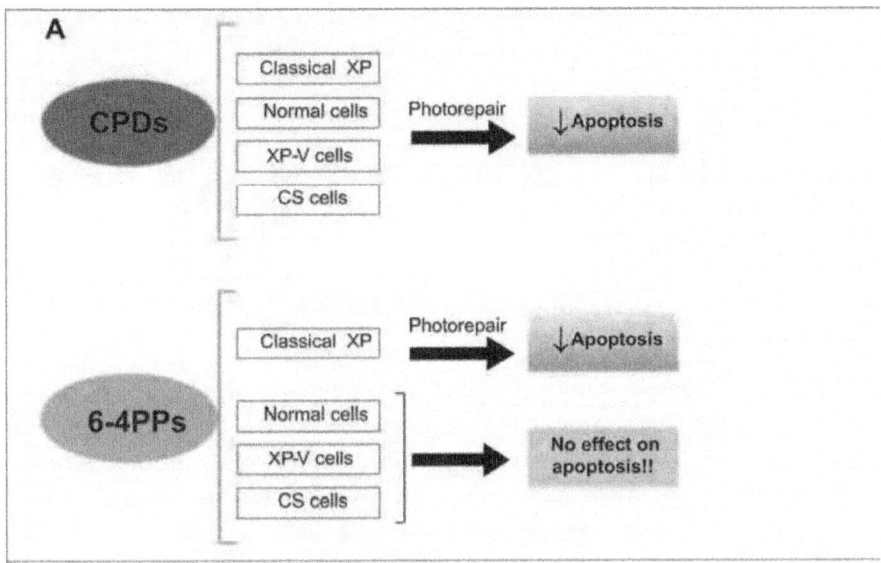

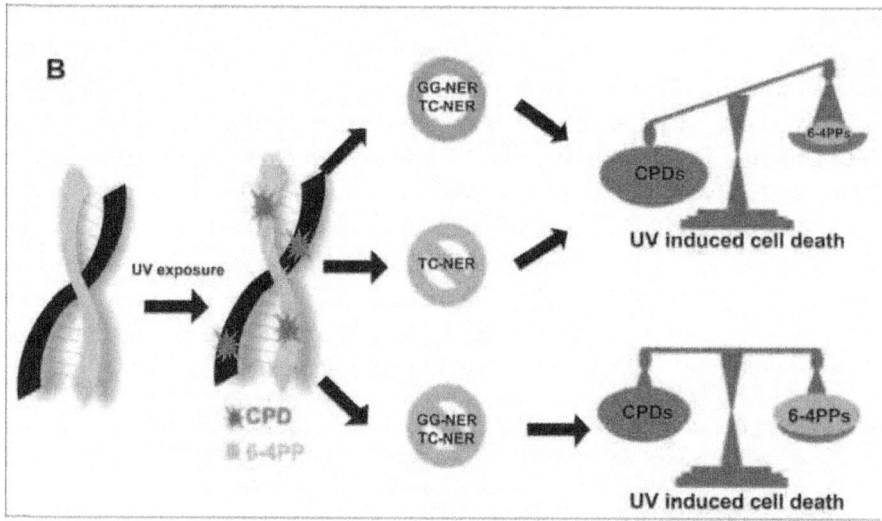

**Figure 4**. Effects of photorepair of CPDs and 6-4PPs on UV-induced apoptosis. (A) Summary of the impact of the specific removal of CPDs and 6-4PPs by photorepair in human cell lines with different DNA repair capabilities. (B) Schematic representation of the main conclusions of the results shown in panel A. Those results clearly implicate that CPDs and 6-4PPs play different roles on UV-induced apoptosis depending on the cellular repair capacity.

Another interesting finding came from assays investigating the time-dependent kinetics of the apoptosis commitment after UV treatment. Transduced XP-A cells were UV-treated and photoreactivated (to allow photorepair of the respective UV lesions) at increasing periods of time. Surprisingly, the data suggests that the initial trigger event to cell death after UV irradiation is relatively delayed, since photorepair of CPDs or 6-4PPs was able to reduce apoptosis even when photoreactivation was performed up to 8 hours after UV irradiation. After that, photoreactivation did not prevent UV-killing in these cells, indicating a commitment by events that irreversibly lead to cell death. These results are also in agreement with the indications that fast removed lesions (such as 6-PPs) do not activate apoptosis in NER-proficient human cells (Lima-Bessa et al., 2008). The main implication of all these findings is the fact that skin carcinogenesis in XP patients may also have 6-4PP lesions as important players, suggesting that tumors from these individuals are not only quantitatively different from those of normal people, but may also have different causative lesions. Transduction of XP knockout mice with adenoviral vectors carrying photolyase genes may help to address this question.

# EMPLOYING RETROVIRAL VECTORS FOR CORRECTING XP PHENOTYPE

The first genetic analysis of XP patients was performed through somatic cell fusion followed by analysis of restoration of normal UDS. If somatic cell fusion complements XP genetic deficiency, it will then be positive for UDS activity. These experiments were able to identify the seven classical XP complementation groups and the variant group (Zeng et al., 1998). This implies that DNA repair deficiencies can, in fact, be corrected by the introduction of a normal copy of the affected gene, giving hope for the development of gene therapy protocols for XP patients. In fact, the introduction of a normal copy of the defective gene in XP cells can complement the DNA repair ability, as demonstrated by the delivery of conventional expression vectors, via calcium precipitation and microneedle injection (Mezzina et al., 1994).

In 1995, viral vectors were first used as gene delivery tools in DNA repair experiments (Carreau et al., 1995a). In this study, a LXPDSN retroviral vector carrying the wild-type XPD gene was capable of complementing primary fibroblasts of XPD patients with a long-term expression. A subsequent study showed that this complementation was gene-specific and that there was a long-term expression of the transgene (Quilliet et al, 1996). The use of retroviral vectors for DNA repair genes delivery was further validated in 1996 and 1997, when XP-A, XP-B, XP-C and TTD-D cells were also complemented with the aid of genespecific retroviral vectors (Marionnet et al., 1996; Zeng et al., 1998).

The compilation of these results shows that the retroviral delivery of several DNA repair genes was able to specifically complement several deficiencies presented by XP, CS and TTD patients such as UDS, reduced catalase activity, UV-sensitivity, recovery of RNA synthesis, increased mutation frequency, stabilization of p53 (Dumaz et al., 1998) and deregulation of ICAM-1 (Ahrens et al., 1997).

Since XP patients already receive autologous graft transplants after massive skin tissue removal surgery (Atabay et al., 1991; Bell et al., 1983), most researches in the field of XP gene therapy focus on the three-dimensional skin reconstruction in vitro, using the patients' cells genetically corrected ex vivo. In this technique, the patients' fibroblasts and keratinocytes are cultured in vitro after a skin biopsy of a non-UV-irradiated area. Then, retroviral vectors are used to stably complement the genetic deficiency of these cells. Finally, the keratinocytes and the fibroblasts are used to three-dimensionally reconstruct the epidermis and dermis, respectively. This construct can then be used as a graft when the part of patient's damaged skin is removed in a necessary surgery. To that end, Arnaudeau-Bégard and co-workers managed to complement XP-C keratinocytes, recovering a wild-type phenotype and UVresistance with the aid of a retroviral vector carrying a normal copy of the XPC gene (Arnaudeau-Bégard et al., 2003). Furthermore, Bergoglio and co-workers have also developed a selection method for genetically corrected keratinocytes that does not involve particles derived from microorganisms which could lead to immunological clearance of the transgene, using CD24 as an ectopic marker (Bergoglio et al., 2007).

In 2005, Bernerd and co-workers were able to reconstruct a three-dimensional skin model in vitro using fibroblasts and keratinocytes from a donor XP-C patient. With this model, they were able to see that the XP skin has peculiar characteristics: hypoplastic horny layers, decreased and delayed keratinocyte differentiation, epidermal invaginations, a generally altered proliferation control and fibroblasts with distinct morphology and orientation. Furthermore, the epidermal invaginations were proven to be related to alterations of both keratinocytes' and fibroblasts' functions and were characterized as epidermoid carcinomalike structures (Bernerd et al., 2005). It is important to keep in mind that an XP skin biopsy might give us further and more precise knowledge of the XP skin physiology, but this is a delicate procedure which requires the patients' agreement

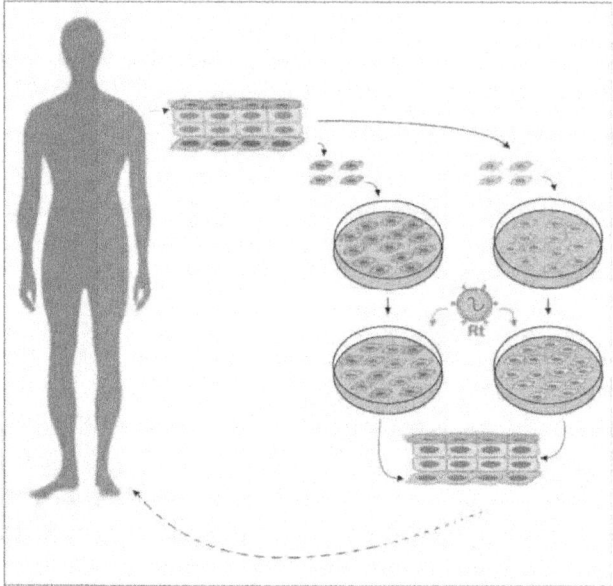

**Figure 5**. Schematic representation of ex vivo gene therapy for XP patients using recombinant retrovirus (Rt). Skin-derived fibroblasts and keratinocytes from an XP patient are cultivated in vitro, and transduced with retroviral vector carrying the wild

type XP cDNA. Transduced cells are then used to reconstruct the human skin in vitro, with a normal phenotype. Dashed line raises the possibility of engraftment of the reconstructed skin directly on XP patients.

Since the use of common retroviral vectors in gene therapy can be dangerous due to semirandom insertional mutagenesis, researchers have developed several self-inactivatinglentiviral vectors carrying DNA repair genes. These vectors were shown to efficiently transduce primary and transformed fibroblasts, complementing in a gene-specific manner XP-A, XP-C and XP-D cells. Furthermore, the recovery of normal levels of UV-resistance in the transduced cells was shown to be persistent for at least 3 months (Marchetto et al., 2006). The reconstruction of a genetically corrected, three-dimensional XP skin followed by the implantation of the graft on a patient (Figure 5) is still an ongoing chore that has to be taken very cautiously, always prioritizing the patient's well-being.

It is also important to keep in mind that these grafts do not include melanocytes, responsible for the very common melanomas in these patients (Khavari, 1998), and that the skin will only be genetically complemented in the areas that receive the grafts, all the other areas of the body will still be extremely photosensitive since no paracrine effect is known for DNA repair proteins and that immunological clearance or gene silencing by cellular methylation can always prohibit a long-term transgene expression (Magnaldo & Sarasin, 2002). Importantly, several XP complementation groups also present other relevant symptoms, such as neurodegeneration, which will not be improved by the skin grafts. For those patients, another kind of gene therapy might be more efficient, such as the development of genetically corrected stem cells (ESs) (Magnaldo & Sarasin, 2002) or induced pluripotent cells (iPSCs, see below (Alison, 2009). Unfortunately, there is still no reference on that kind of research for xeroderma pigmentosum.

## HOST CELL REACTIVATION (HCR) AS A TOOL FOR DNA REPAIR RESEARCH

The host cell reactivation (HCR) technique was first described in human cells by ProticSabljic and co-workers in 1985 (Protic-Sabljic et al., 1985). In this first work, the technique consisted of transducing cells with a plasmid containing a putative cDNA with a selective gene into XP cells to look for a reversion of the UV sensitivity due to gene complementation, allowing identification of the genes responsible for that phenotype.

Other studies have refined the technique which is now widely used as an indirect measure of cellular DNA repair capacity. Mostly, a plasmid containing a reporter gene such as luciferase (LUC) or chloramphenicol acetyltransferse (CAT) is treated with a genotoxic agent such as UV radiation and introduced in the cell where DNA repair capacity is to be evaluated. If the cell is able to remove the lesions from the plasmid, the reporter gene will be expressed. Different DNA repair rates can be addresses by differences on the amount of gene reporter expression at a certain time (Merkle et al., 2004). A schematic representation of HCR is shown in Figure 6.

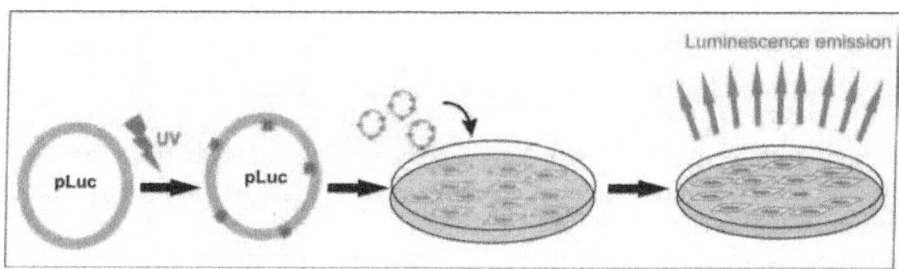

**Figure 6**. Schematic representation of host cell reactivation (HCR) assay. A plasmid carrying a reporter gene (in this case, the luciferase gene) is UV-irradiated in vitro and then transfected into host cells. 48 hours later, the cellular DNA repair capacity is indirectly estimated by measurement of the reporter gene activity in the cellular extract.

In 1995, the HCR assay was further used to visualize the genetic complementation of mammalian expression vectors carrying the DNA repair genes XPA, XPB, XPC, XPD and CSB. In this study, plasmids containing LUC or CAT were UV irradiated and co-transfected with the plasmids containing each of the complementing genes of the DNA-repair deficient cells. Again, only the cells with the correct complementation were capable of removing the DNA damage in the reporter gene, allowing the expression of that protein. This technique facilitates the identification of the complementation group of a given patient, being particularly useful in cases of CS, TTD and some XP patients such as XP-E that present a normal UDS after UV treatment (Carreau et al., 1995b).

Recent data using the HCR assay has shown that the CS proteins are essential for the reversion of oxidated lesions (Pitsikas et al., 2005; Spivak & Hanawalt, 2006; Leach & Rainbow, 2011) and evidence obtained with HCR suggests that, unlike what was previously shown with UDS assays, DNA repair capacity in fibroblasts does not decrease with aging (Merkle et al., 2004). This reduction may however be cell type-specific and DNA repair pathway-specific since blood cells repair capacity decreases approximately 0.6% per year of age (Moriwaki et al., 1996). This technique is still widely used and its great advantage is that the DNA plasmids or the viral vectors are treated in a controlled manner, not being subject to the cell's global response to the same treatment. Further technique improvements will surely allow HCR to be used in different assays such as in vivo, yielding a better knowledge of the DNA repair pathways and their interactions with other pathways and physiological events.

# OTHER TREATMENTS FOR XERODERMA PIGMENTOSUM

## General Care

There is no treatment that has been proven so far to be 100% effective in all XP cases. The only palliative measure that patients can rely on is complete sun avoidance. This includes not only avoiding going out even on cloudy days and covering all exposed body areas such as skin and eyes, but also using special artificial lights that emit no UV wavelengths (Kraemer, 2008). Premalignant lesions, such as actinic keratoses, and malignant lesions must be quickly treated with topical 5-fluoracil or liquid nitrogen, imiquimod cream, electrodesiccation and curettage, surgical excision or chemosurgery, as needed. When extensive areas are damaged and have to be surgically removed, skin grafts from sun unexposed areas of the same patient should be used. When eyes are affected, methylcellulose eye drops or contact lenses can help prevent trauma and corneal transplantations might be needed in extreme cases (Kraemer, 2008).

When caring for XP patients, it is very important to keep in mind that the total sun avoidance also prevents the production of vitamin D in the skin, so dietary supplementation might be needed. Furthermore, the DNA repair deficiencies which prevent the repair of photolesions may also make individuals sensitive to other mutagens such as cigarette smoke, so patients should be protected against these agents (Kraemer, 2008).

Aside from removal of local lesions and total sun avoidance, two other palliative treatments might help improving the XP patient's quality of life: topical use of T4 endonuclease (Yarosh et al., 2001) and oral intake of retinoids (Campbell & DiGiovanna, 2006).

## Topical Use of T4 Endonuclease

In 1975, Tanaka and co-workers demonstrated that the bacteriophage T4 endonuclease V is capable of making an incision 5' within a CPD lesion. The resulting DNA flap is recognized and removed by a 5'◊3' exonuclease, leaving a gap that is filled by a DNA polymerase, using the undamaged single strand as a template. A DNA ligase then joins the repaired fragment to the parental DNA (Tanaka et al., 1975).

In the 80's, Yarosh's laboratory discovered that the T4 endonuclease V could be delivered into cells using 200 nm liposomes as a delivery vehicle. The anionic liposomes not only protect the cationic enzyme inside, but also promote the escape from a clatrin-coated endosome after cellular intake by destabilizing the vesicle's membrane with an acid pH. By cleaving DNA at the site of UV-induced lesions, the enzyme reverses the DNA repair defect of XP cells (Yarosh, 2002). Further work by the same group also showed that these T4N5 liposomes in a 1% hydrogel lotion when applied in cultured human fibroblasts, mouse dorsal back or cultured human breast skin is capable of delivering the enzyme into cells in less than one hour, being almost entirely restricted to the epidermis (Ceccoli et al., 1989; Kibitel et al., 1991).

An inverse correlation was later shown between the T4N5 dose and the level of CPDs that remained in the epidermis. This curve reached a plateau (at 0.5 µg/ml), probably due to saturation of the cell machinery for further repairing the damage after the initial incision by the T4 enzyme (Yarosh et al., 1994). These studies also showed that even in the higher dose of T4N5 liposomes, only ~50% of the CPD lesions were removed but that was capable of reducing the mutagenesis rate by 99% in transformed fibroblasts and 30% in primary fibroblast cell culture. These numbers are probably not only related to the number of remaining lesions, but also to the smaller size of the repair patch filled by BER compared to that needed in NER (Yarosh, 2002; Cafardi & Elmets, 2008).

Finally, after two phase I clinical trials (Yarosh et al., 1996 as cited in Cafardi & Elmets, 2008) and three phase II clinical trials (Wolf et al., 2000 and Yarosh et al., 1996 as cited in Cafardi & Elmets, 2008), in 2001 the T4N5 liposomes were tested in XP patients. The patients were instructed to apply 4-5 ml of the lotion containing 1 mg/ml of endonuclease everyday for a year. Except for lesion removal when necessary, and daily use of sunscreens of 15 SPF or higher, no concomitant treatments were allowed. The treatment was shown to be efficient, reducing the rate of actinic keratoses and basal-cell carcinomas to 68% and 30% respectively in the placebo and treatment groups, reducing tumor promotion and progression. The treatment was also capable of reducing some immunosuppressant molecules, such as interleukin-10 (IL-10) and tumor necrosis factor-α (TNF- α). Unfortunately, the treatment was only effective for patients under 18 years-old. This might be because XP patients older than that already had too much DNA damage in their cells that could not be reversed (Yarosh et al., 2001). Despite the promising results, there are currently no topical DNA repair enzymes approved by the FDA. Clinical trials are still being conducted to analyze the application of T4N5 liposomes in other deficiencies and immunosuppressed patients (Cafardi & Elmets, 2008).

## Oral Use of Retinoids

Despite interventions such as sunlight avoidance and tumor removal, most of the XP patients continue to develop a large number of skin cancers. These high-risk patients may suffer from field cancerization that may happen when a wide field of the epithelium has been exposed to the same genotoxic agent and adjacent but not contiguous areas present genetic and morphological alterations that may lead to a carcinogenesis process. As the whole skin area has been exposed to sunlight, inducing independent tumors with different growth rates, this hypothesis may explain why the patients have a 30% increase in the chances of having a second basal cell carcinoma (BCC) and then a 50% increase of a third BCC (Campbell & DiGiovanna, 2006).

In XP patients, the oral use of retinoids might be beneficial, regardless of the strong side effects. In chemoprevention the goal is to identify early biological events in the epithelium which may lead to a carcinogenesis process and intervene with chemicals which will help stop or reverse the process (Campbell & DiGiovanna, 2006). Retinoids, also known as vitamin A, are the most studied chemopreventive agent for skin cancers, upper aerodigestive tract and breast and cervical cancers. The exact mechanisms through which the retinoids are capable of reducing cancer incidence are still unclear, but it has been shown that they are capable of altering keratinocytes' growth, increasing their differentiation status, affecting their cell surface and immune modulation. Retinoids mediate gene transcription by binding to two families of nuclear receptors, the retinoid acid receptors (RARs) and the retinoid X receptors (RXRs). Retinoids have only a mild effect on existing tumors, but can suppress the development of new lesions (Campbell & DiGiovanna, 2006). In 1988, it was shown in a three year study that isotretinoin in a dose up to 2 mg/day/Kg was able to reduce skin cancers in XP patients by 63%. Unfortunately, in the year following the discontinuation of the treatment there was an increase of 8.5% of cancer incidence in those patients with reference to the two years of treatment (Kraemer et al., 1988).

Furthermore, the constant use of retinoids can have severe side effects ranging from inflammation in existing tumors, dry skin and mucosa and hair loss to pancreatitis, osteoporosis, hyperostosis and myalgia among others. The retinoids' toxicity is dose related and cumulative, but most of the side effects can be prevented with constant check-ups and use of local special moisturizers (Campbell & DiGiovanna, 2006). Indeed, several retinoids can be used as chemopreventives. The two most common are isotretinoin and acitretin, the first having a shorter half-life and being the drug of choice for women due to retinoids' theratogenic potential, especially in fetuses (Campbell & DiGiovanna, 2006).

## Potential Effects of DNA Repair Adjuvants

The use of DNA repair adjuvants and antioxidants may also help reducing skin cancer incidence in XP patients. Some known DNA repair adjuvants are selenium, aquosum extract of Urcaria tomentosa and Interleukin-12 (IL-12) (Emanuel & Scheinfeld, 2007).

Selenium seems to interact with Ref-1, activating p53, inducing the DNA repair branch of the p53 pathway, in a BRCA1-dependent manner, dealing mainly with oxidative stress (Fisher et al., 2006). On the other hand, it has been already reported that high levels of selenium can be mutagenic, carcinogenic and possibly teratogenic (Shamberger, 1985), probably due to non-specific sulfur substitution on proteins and consequent TC-NER activity decrease (Abul-Hassan et al., 2004). Thus, special attention should be taken regarding the dose of dietary selenium supplementation.

The aquosum extract of Urcaria tomentosa (cat's claw) seems to increase the removal of CPDs and reduce oxidative damage, either by an increase in base excision repair (BER) or by an antioxidant property, reducing erythema and blistering after UV. Despite several studies in vitro and in vivo, the precise mechanisms are still unknown (Emanuel & Scheinfeld, 2007).

Another interesting finding is that, besides IL-12 being a strong immunomodulatory molecule, able to prevent UV-induced immunosuppression through IL-10 inhibition (de Gruijl, 2008), it is also capable of increasing DNA repair by inducing NER, as shown by the RNA level increase in some NER molecules (Schwarz et al., 2005).

## Gene Therapy Targeted Approaches: The Use of Meganucleases for Correcting XP-C Cells

There are several techniques to specifically target, substitute, or correct a gene, diminishing the chances of insertional recombination, such as the use of recombinases, transposons, zincfinger nucleases, endonucleases and meganucleases (Silva et al., 2011). Meganucleases can function as RNA maturases, facilitating the maturation of their own intron or as specific endonucleases that can recognize and cleave the exon-exon junction sequence wherein their intron resides, creating a specific double strand break (DSB), giving rise to the moniker "homing endonuclease". The meganuclease function is probably related to the current status of its lifecycle (Silva et al., 2011).

Meganucleases can be used as gene targeting tools in several ways. Ideally, they can provide a true reversion of the mutation, but the efficacy of correction is invertionally correlated to the distance of the initial DNA DSB. Alternatively, it can insert a functional gene upstream of the mutated one or in a safe location where it will not induce insertional mutagenesis. Also, meganucleases can be used for introducing specific mutations for research purposes such as understanding the role of a gene or of a specific point mutation. Furthermore, meganucleases capable of targeting viral sequences are being researched as antiviral agents (Silva et al., 2011).

Recently, the design of a specific I-CreI meganuclease targeting for the XPC gene was able to specifically target two XPC sequences, showing in vitro for the first time that extensive redesign of homing endonuclease can modify a specific chromosome region without lost of specificity or efficiency (Arnould et al., 2007). These results are very promising for the development of future gene therapy strategies for XP patients.

## Induced Pluripotent Cells (iPSCs) as Gene Therapy Agents

In 2006, the induction of pluripotent cells (iPSCs) by the expression of Oct3/4, Sox2, c-Myc and Klf4 in fibroblasts gave hope for a new gene therapy using pluripotent cells that would not elicit an immunological response in the patient, since his own cells would be used to induce the iPSCs and that would not be confronted by ethical issues like the use of embryonic stem cells (Takahashi & Yamanaka, 2006). Since then, this technology has been improved and iPSCs have been induced in a variety of cell types from different species. Also, iPSCs have been differentiated to several different cell types, from fibroblasts to neurons (Sidhu, 2011).

Fanconi Anemina (FA) is a DNA repair related disease, where mutations in one of fourteen genes lead to extreme sensitivity to interstrand crosslinking agents. Patients show progressive bone marrow failure, congenital developmental abnormalities and early onset of cancers, mostly acute myelogenous leukemia and squamous cell carcinomas. Bone marrow transplantation is a palliative treatment for the secondary leukemia but no cure is currently available for FA patients (Kitao &.Takata, 2011). In 2009, Raya and coworkers were able to use lentiviral vectors to genetically correct fibroblast and keratinocytes from patients with various FA complementation group deficiencies and then induce their dedifferentiation into pluripotent stem cells. Interestingly, uncorrected FA cells did not generate iPSCs, indicating a role for DNA repair in nuclear reprogramming. Thus, the generated iPSCs had normal FA genes and have the potential of being used for gene therapy of the donor patients, with no risk of inducing immunological rejection (Raya et al., 2009). Hopefully soon FA patients and others will be able to benefit from this technology as a safe gene therapy approach.

## CONCLUDING REMARKS

Recombinant viral vectors were developed more than thirty years ago, and they have provided extremely useful tools to understand cell metabolism. This chapter focuses on their use to understand cells' responses to DNA damage, especially UV-irradiated DNA repair-deficient cells. These vectors provide means to interfere in these responses, affecting DNA metabolism and revealing important aspects of the DNA repair mechanisms. The discovery of RNA interference mechanisms in human cells offer still more opportunities to modify cells' responses by silencing specific DNA repair genes. Several libraries of viral vectors for the expression of small double-stranded RNA molecules (shRNA) targeting human genes are commercially available, and are already being used for understanding gene function. The use of such vectors to make cells deficient in more than one DNA repair pathway, using cells deficient in XP genes as hosts, for example, may help us to reveal the intricate network of interactions between the different metabolic pathways that contribute to genome maintenance after damage induction (Moraes, et al., 2011; in press). Moreover, the progress that has been made towards gene therapy for xeroderma pigmentosum, using these recombinant viral vectors is also discussed. Although the results indicate a series of limitations, and it is clear that there is still a long way to go, they make researchers go forward, giving a gleam of hope to these patients and their families.

## ACKNOWLEDGEMENTS

KMLB has a post-doctoral fellowship from CAPES (Brasília, Brazil) and CQ has a PhD fellowship from FAPESP (São Paulo, Brazil). This research was supported by FAPESP (São Paulo, Brazil) and CNPq (Brasília, Brazil).

## REFERENCES

1.    Abul-Hassan, K.S., Lehnert, B.E., Guant, L., & Walmsley, R. (2004). Abnormal DNA repair in selenium-treated human cells. Mutation research, V. 565, N. 1, pp. 45-51, ISSN 0027- 5107

2.    Alison, M.R. (2009). Stem cells in pathobiology and regenerative medicine. The Journal of pathology, V. 217, N. 2, pp. 141-143, ISSN 1096-9896

3.    Ahrens, C., Grewe, M., Berneburg, M., Grether-Beck, S., Quilliet, X., Mezzina, M., Sarasin, A., Lehmann, A.R., Arlett, C.F., & Krutmann, J. (1997). Photocarcinogenesis and inhibition of intercellular adhesion molecule 1 expression in cells of DNA-repairdefective individuals. Proceedings of the National Academy of Sciences of the United States of America (PNAS)., V. 94, N. 13, pp. 6837-6841, ISSN 1091-6490

4.    Armelini, M.G., Muotri, A.R., Marchetto, M.C., de Lima-Bessa, K.M., Sarasin, A., & Menck, C.F. (2005). Restoring DNA repair capacity of cells from three distinct diseases by XPD gene-recombinant adenovirus. Cancer gene therapy, V. 12, N. 4, pp. 389-396, ISSN 0929-1903

5.    Armelini, M.G., Lima-Bessa, K.M., Marchetto, M.C., Muotri, A.R., Chiganças, V., Leite, R.A., Carvalho, H., & Menck, C.F. (2007). Exploring DNA damage responses in human cells with recombinant adenoviral vectors. Human , & experimental toxicology, V. 26, N. 11, pp. 899-906, ISSN 0960-3271

6.    Arnaudeau-Bégard, C., Brellier, F., Chevallier-Lagente, O., Hoeijmakers, J., Bernerd, F., Sarasin, A., & Magnaldo, T. (2003). Genetic correction of DNA repairdeficient/cancer-prone xeroderma pigmentosum group C keratinocytes. Human gene therapy, V. 14, N. 10, pp. 983-996, ISSN 1043-0342

7.    Arnould, S., Perez, C., Cabaniols, J.P., Smith, J., Gouble, A., Grizot, S., Epinat, J.C., Duclert, A., Duchateau, P., & Pâques, F. (2007). Engineered I-CreI derivatives cleaving sequences from the human XPC gene can induce highly efficient gene correction in mammalian cells. Journal of molecular biology, V. 371, N. 1, pp 49-65, ISSN 0022-2836

8.    Atabay, K., Celebi, C., Cenetoglu, S., Baran, N.K., & Kiymaz, Z. (1991). Facial resurfacing in xeroderma pigmentosum with monoblock full-thickness skin graft. Plastic and reconstructive surgery, V. 87, N. 6, pp. 1121-1125, ISSN 0032-1052

9.    Atkinson, H., & Chalmers, R. (2010). Delivering the goods: viral and non-viral gene therapy systems and the inherent limits on cargo DNA and internal sequences. Genetica, V. 138, N. 5, pp. 485-498, ISSN 0016-6707

10.   Bell, E., Sher, S., Hull, B., Merrill, C., Rosen, S., Chamson, A., Asselineau, D., Dubertret, L., Coulomb, B., Lapiere, C., Nusgens, B., & Neveux, Y. (1983). The reconstitution of living skin. Journal of investigative dermatology, V. 81, N. 1 (suppl)., pp. 2s-10s, ISSN 0022-202X

11.   Bergoglio, V., Larcher, F., Chevallier-Lagente, O., Bernheim, A., Danos, O., Sarasin, A., Rio, M.D., & Magnaldo, T. (2007). Safe selection of genetically manipulated human primary keratinocytes with very high

growth potential usind CD24. Molecular therapy, V. 15, N. 12, pp. 2186-2193, ISSN 1525-0016

12. Bernerd, F., Asselineau, D., Frechet, M., Sarasin, A., & Magnaldo, T. (2005). Reconstruction of DNA repair-deficient xeroderma pigmentosum skin in vitro: a model to study hypersensitivity to UV light. Photochemistry and photobiology, V. 81, N. 1, pp. 19-24, ISSN 0031-8655

13. Brettel, K., & Byrdin, M. (2010). Reaction mechanisms of DNA photolyase. Current opinion in structural biology, V. 20, N. 6, pp. 693-701, ISSN 0959-440X

14. Cafardi, J.A., & Elmets, C.A. (2008). T4 endonuclease V: review and application to dermatology. Expert opinion on biological therapy, V. 8, N. 6, pp. 829-838, ISSN 1471- 2598

15. Campbell, R.M., & DiGiovanna, J.J. (2006). Skin cancer chemoprevention with systemic retinoids: an adjunct in the management of selected high-risk patients. Dermatologic therapy, V. 19, N. 5, pp. 306-314, ISSN 13960296

16. Carreau, M., Quilliet, X., Eveno, E., Salvetti, A., Danos, O., Heard, J.M., Mezzina, M., & Sarasin, A. (1995a). Functional retroviral vector for gene therapy of xeroderma pigmentosum group D patients. Human gene therapy, V. 6, N. 10, pp. 1307-1315, ISSN 1043-0342

17. Carreau, M. Eveno, E., Quilliet, X., Chevalier-Lagente, O., Benoit, A., Tanganelli, B., Stefanini, M., Vermeulen, W., Hoeijmakers, J.H., Sarasin, A., & Mezzina, M. (1995b). Development of a new easy complementation assay for DNA repair deficiency human syndromes using cloned repair genes. Carcinogenesis, V. 16, N. 5, pp. 1003- 1009, ISSN 0143-3334

18. Ceccoli, J., Rosales, N., Tsimis, J., & Yarosh, D.B. (1989). Encapsulation of the UV-DNA repair enzyme T4 endonuclease V in liposomes and delivery to human cells. Journal of investigative dermatology, V. 93, N. 2, pp. 190-194, ISSN 0022-202X

19. Chiganças, V., Miyaji, E.N., Muotri, A.R., de Fátima-Jacysyn, J., Amarante-Mendes, G.P., Yasui, A., & Menck, C.F. (2000). Photorepair prevents ultraviolet-induced apoptosis in human cells expressing the marsupial photolyase gene. Cancer research, V. 60, N. 9, pp. 2458-2463, ISSN 1538-7445

20. Chiganças, V., Batista, L.F., Brumatti, G., Amarante-Mendes, G.P., Yasui, A., & Menck, C.F. (2002). Photorepair of RNA polymerase arrest and apoptosis after ultraviolet irradiation in normal and XPB deficient rodent cells. Cell death and differentiation, V. 9, N. 10, pp. 1099-1107, ISSN 1350-9047

21. Chiganças, V., Sarasin, A., & Menck, C.F. (2004). CPD-photolyase adenovirus-mediated gene transfer in normal and DNA-repair-deficient human cells. Journal of cell science, V. 117, N. Pt 16, pp. 3579-3592, ISSN 0021-9533

22. Chiganças, V., Lima-Bessa, K.M., Stary, A., Menck, C.F., & Sarasin, A. (2008). Defective transcription/repair factor IIH recruitment to specific UV lesions in trichothiodystrophy syndrome. Cancer research, V. 68, N. 15, pp. 6074-6083, ISSN 1538-7445

23. Ciccia, A., & Elledge, S.J. (2010). The DNA damage response: making it safe to play with knives. Molecular cell, V. 40, N. 2, pp. 179-204, ISSN 1538-7445

24. Cleaver, J.E., Kaufmann, W.K., Kapp, L.N., & Park, S.D. (1983). Replicon size and excision repair as factors in the inhibition and recovery of DNA synthesis from ultraviolet damage. Biochimica et biophysica Acta, V. 739, N. 2, pp. 207-215, ISSN 0304-4165

25. Cleaver, J.E., Lam, E.T., & Revet, I. (2009). Disorders of nucleotide excision repair: the genetic and molecular basis of heterogeneity. Nature reviews.Genetics, V. 10, N. 11, pp. 756-768, ISSN 1471-0056

26. Costa, R.M., Chiganças, V., Galhardo, Rda.S., Carvalho, H., & Menck, C.F. (2003). The eukaryotic nucleotide excision repair pathway. Biochimie, V. 85, N. 11, pp. 1083- 1099, ISSN 0300-9084

27. Daya, S., & Berns, K.I. (2008). Gene therapy using adeno-associated virus vectors. Clinical microbiology reviews, V. 21, N. 4, pp. 583-593, ISSN 0893-8512

28. de Boer, J., & Hoeijmakers, J.H. (2000). Nucleotide excision repair and human syndromes. Carcinogenesis, V. 21, N. 3, pp. 453-460, ISSN 0143-3334

29. de Gruijl, F. (2008). UV-induced immunosuppression in the balance. Photochemistry and photobiology, V. 84, N. 1, pp. 2-9, ISSN 0031-8655

30. Dong, B., Nakai, H., & Xiao, W. (2010). Characterization of genome integrity for oversized recombinant AAV vector. Molecular therapy, V. 18, N. 1, pp. 87-92, ISSN 1525-0016

31. Dumaz, N., Drougard, C., Quilliet, X., Mezzina, M., Sarasin, A., & Daya-Grosjean, L. (1998). Recovery of the normal p53 response after UV treatment in DNA repair-deficient fibrobalsts by retroviral-mediated correction with the XPD gene. Carcinogenesis, V. 19, N. 9, pp. 1701-1704, ISSN 0143-3334

32. Edelstein, M.L., Abedi, M.R., & Wixon, J. (2007). Gene therapy clinical trials worldwide to 2007- an update. The journal of gene medicine, V. 9, N. 10, pp. 833-842, ISSN 1521-2254

33. Emanuel, P., & Scheinfeld, N. (2007). A review of DNA repair and possible DNA-repair adjuvants and selected natural anti-oxidants. Dermatology online journal, V. 13, N. 3, pp. 10, ISSN 1087-2108

34. Fisher, J.L., Lancia, J.K., Mathur, A., & Smith, M.L. (2006). Selenium protection from DNA damage involves Ref1/p53/Brca1 protein complex. Anticancer research, V. 26, N. 2A, pp. 899-904, ISSN 0250-7005

35. Froelich, S., Tai, A., & Wang, P. (2010). Lentiviral vectors for immune cells targeting. Immunopharmacology and immunotoxicology, V. 32, N. 2, pp. 208-218, ISSN 0892-3973

36. Fousteri, M., & Mullenders, L.H. (2008). Transcription-coupled nucleotide excision repair in mammalian cells: molecular mechanisms and biological effects. Cell research, V. 18, N. 1, pp. 73-84, ISSN 1001-0602

37. Gerstenblith, M.R., Goldstein, A.M., & Tucker, M.A. (2010). Hereditary genodermatoses with cancer predisposition. Hematology/oncology clinics of North America, V. 24, N. 5, pp. 885–906, ISSN 0889-8588

38. Hall, K., Blair Zajdel, M.E., & Blair, G.E. (2010). Unity and diversity in the human adenoviruses: exploiting alternative entry pathways for gene therapy. The Biochemical journal, V. 431, N. 3, pp. 321-336, ISSN 0264-6021

39. Hanawalt, P.C. (2002). Subpathways of nucleotide excision repair and their regulation. Oncogene, V. 21, N. 58, pp. 8949-8956, ISSN 0950-9232

40. Hoeijmakers, J.H. (2009). Molecular origins of cancer: DNA damage, aging, and cancer. The New England journal of medicine, V. 361, N. 15, pp. 1475-1485, ISSN 0028-4793

41. Horibata, K., Iwamoto, Y., Kuraoka, I., Jaspers, N.G.J., Kurimasa, A., Oshimura, M., Ichihashi, M., & Tanaka, K. (2004). Complete absence of Cockayne syndrome group B gene product gives rise to UV-sensitive syndrome but not Cockayne syndrome. PNAS, V. 101, N. 43, pp. 15410–15415, ISSN 1091-6490

42. Howarth, J.L., Lee, Y.B., & Uney, J.B. (2010). Using viral vectors as gene transfer tools. Cell biology and toxicology, V. 26, N. 1, pp. 1-20, ISSN 0742-2091

43.  Itin, P.H., Sarasin, A., & Pittelkow, M.R. (2001). Trichothiodystrophy: update on the sulfurdeficient brittle hair syndromes. Journal of the American Academy of Dermatology, V. 44, N. 6, pp. 891-920, ISSN 0190-9622

44.  Iwakuma, T., Cui, Y., & Chang, L.J. (1999). Self-inactivating lentiviral vectors with U3 and U5 modifications. Virology, V. 261, N. 1, pp. 120-132, ISSN 0042-6822

45.  Johnson, R.E., Kondratick, C.M., Prakash, S., & Prakash, L. (1999). hRAD30 mutations in the variant form of xeroderma pigmentosum. Science, V. 285, N. 5425, pp. 263–265, ISSN 0036-8075

46.  Karalis, A., Tischkowitz, M., & Millington, G.W. (2011). Dermatological manifestations of inherited cancer syndromes in children. The British journal of dermatology, V. 164, N. 2, pp. 245–256, ISSN 0007-0963

47.  Khavari, P.A. (1998). Gene therapy for genetic skin disease. Journal of investigative dermatology, V. 110, N. 4, pp. 462-467, ISSN 0022-202X

48.  Kibitel, J.T., Yee, V., & Yarosh, D.B. (1991). Enhancement of ultraviolet-DNA repair in denV gene transfectants and T4 endonuclease V-liposome recipients. Photochemistry and photobiology, V. 54, N. 5, pp. 753-60, ISSN 0031-8655

49.  Kitao, H., & Takata, M. (2011). Fanconi anemia: a disorder defective in the DNA damage response. International journal of hematology, V. 93, N. 4, pp. 417-424, ISSN 0925- 5710

50.  Kraemer, K.H., DiGiovanna, J.J., Moshell, A.N., Tarone, R.E , & Peck, G.L. (1988). Prevention of skin cancer in xeroderma pigmentosum with the use of isotretinoin. The New England journal of medicine, V. 318, N. 25, pp. 1633-1637, ISSN 0028-4793

51.  Kraemer, K.H., Patronas, N.J., Schiffmann, R., Brooks, B.P., Tamura, D., & DiGiovanna, J.J. (2007). Xeroderma pigmentosum, trichothiodystrophy and Cockayne syndrome: a complex genotype-phenotype relationship. Neuroscience, V. 145, N. 4, pp. 1388– 1396, ISSN 0306-4522

52.  Kraemer, K. H. (2008). Xeroderma pigmentosum. Gene reviews-NCBI Bookshelf, NBK1397, PMID: 20301571.

53.  Kuluncsics, Z., Perdiz, D., Brulay, E., Muel B., & Sage, E. (1999). Wavelength dependence of ultraviolet-induced DNA damage distribution: involvement of direct or indirect mechanisms and possible artefacts. Journal of Photochemistry and photobiology.B, Biology, V. 49, N. 1, pp. 71–80, ISSN 1011-1344

54. Kusumoto, R., Masutani, C., Sugasawa, K., Iwai, S., Araki, M., Uchida, A., Mizukoshi, T., & Hanaoka, F. (2001). Diversity of the damage recognition step in the global genomic nucleotide excision repair in vitro. Mutation research, V. 485, N. 3, pp. 219–227, ISSN 0027-5107

55. Leach, D.M., & Rainbow, A.J. (2011). Early host cell reactivation of an oxidatively damaged adenovirus-encoded reporter gene requires the Cockayne syndrome proteins CSA and CSB. Mutagenesis, V. 26, N. 2, pp. 315-321, ISSN 1383-5718

56. Leite, R.A., Marchetto, M.C., Muotri, A.R., Vasconcelos, Dde.M., de Oliveira, Z.N., Machado, M.C., & Menck, C.F. (2009). Identification of XP complementation groups by recombinant adenovirus carrying DNA repair genes. The journal of investigative dermatology, V. 129, N. 2, pp. 502-506, ISSN 0022-202X

57. Lima-Bessa, K.M., Chigancas, V., Stary, A., Kannouche, P., Sarasin, A., Armelini, M.G., de Fatima Jacysyn, J., Amarante-Mendes, G.P., Cordeiro-Stone, M., Cleaver, J.E., & Menck, C.F. (2006). Adenovirus mediated transduction of the human DNA polymerase eta cDNA. DNA repair, V. 5, N. 8, pp. 925-934, ISSN 1568-7864

58. Lima-Bessa, K.M., Armelini, M.G., Chiganças, V., Jacysyn, J.F., Amarante-Mendes, G.P., Sarasin, A., & Menck, C.F. (2008). CPDs and 6-4PPs play different roles in UVinduced cell death in normal and NER-deficient human cells. DNA repair, V. 7, N. 2, pp. 303-312, ISSN 1568-7864

59. Lima-Bessa, K.M., Soltys, D.T., Marchetto, M.C., & Menck, C.F.M. (2009). Xeroderma pigmentosum: living in the dark but with hope in therapy. Drugs of the Future, V. 34, N. 8, pp. 665-672, ISSN 0377-8282

60. Liu, L., Lee, J., & Zhou, P. (2010). Navigating the nucleotide excision repair threshold. Journal of cellular physiology, V. 224, N. 3, pp. 585-589, ISSN 1097-4652

61. Lucas-Lledó, J.I., & Lynch, M. (2009). Evolution of mutation rates: phylogenomic analysis of the photolyase/cryptochrome family. Molecular biology and evolution, V. 26, N. 5, pp. 1143-1153, ISSN 0737-4038

62. Magnaldo, T., & Sarasin, A. (2002). Genetic reversion of skin disorders. Mutation research, V. 509, N. 1-2, pp. 211-220, ISSN 0027-5107

63. Marchetto, M.C., Muotri, A.R., Burns, D.K., Friedberg, E.C., & Menck, C.F. (2004). Gene transduction in skin cells: preventing cancer in xeroderma pigmentosum mice. PNAS, V. 101, N. 51, pp. 17759-17764, ISSN 1091-6490

64. Marchetto, M.C., Correa, R.G., Menck, C.F., & Muotri, A.R. (2006). Functional lentiviral vectors for xeroderma pigmentosum gene therapy. Journal of biotechnology, V. 126, N. 4, pp, 424-430, ISSN 0168-1656

65. Masutani, C., Kusumoto, R., Yamada, A., Dohmae, N., Yokoi, M., Yuasa, M., Araki, M., Iwai, S., Takio, K., & Hanaoka, F. (1999). The XPV (xeroderma pigmentosum variant). gene encodes human DNA polymerase eta. Nature, V. 399, N. 6737, pp. 700-704, ISSN 0028-0836

66. Menck, C.F., Armelini, M.G., & Lima-Bessa, K.M. (2007). On the search for skin gene therapy strategies of xeroderma pigmentosum disease. Current gene therapy, V. 7, N. 3, pp. 163-174, ISSN 1566-5232

67. Marionnet, C., Quilliet, X., Benoit, A., Armier, J., Sarasin, A., & Stary, A. (1996). Recovery of normal DNA repair and mutagenesis in trichothiodistrophy cells after transduction of the XPD human gene. Cancer research, V. 56, N. 23, pp. 5450-5456, ISSN 1538-7445

68. Mátrai, J., Chuah, M.K., & VandenDriessche, T. (2010). Recent advances in lentiviral vector development and applications. Molecular therapy, V. 18, N. 3, pp. 477-490, ISSN 1525-0016

69. Merkle, T.J., O'Brien, K., Brooks, P.J., Tarone, R.E., & Robbins, J.H. (2004). DNA repair in human fibroblasts, as reflected by host-cell reactivation of a transfected UVirradiated luciferase gene, is not related to donor age. Mutation research, V. 554, N. 1-2, pp. 9-17, ISSN 0027-5107

70. Mezzina, M., Eveno, E., Chevallier-Lagente, O., Benoit, A., Carreau, M., Vermeulen, W., Hoeijmakers, J.H., Stefanini,M, Lehmann,A.R., Weber, C.A., & Sarasin, A. (1994). Correction by the ERCC2 gene of UV-sensitivity and repair deficiency phenotype in a subset of trichothiodistrophy cells. Carcinogenesis, V. 15, N. 8, pp. 1493-1498, ISSN 0143-3334

71. Michelfelder, S., & Trepel, M. (2009). Adeno-associated viral vectors and their redirection to cell-type specific receptors. Advances in genetics, V. 67, pp. 29-60, ISSN 0065 -2660

72. Moraes, M.C.S., Cabral-Neto, J.B., & Menck, C.F. (2011). DNA repair mechanisms protect our genome from carcinogenesis. Frontiers in bioscience, in press, ISSN 0143-3334

73. Moriwaki, S., Ray, S., Tarone, R.E., Kraemer, K.H., & Grossman, L. (1996). The effect of donor age on the processing of UV-damaged DNA by cultured human cells: reduced DNA repair capacity and increased DNA mutability. Mutation research, V. 364, N. 2, pp. 117-123, ISSN 0027-5107

74. Müller, M., & Carell, T. (2009). Structural biology of DNA photolyases and cryptochromes. Current opinion in structural biology, V. 19, N. 3, pp. 277-285, ISSN 0959-440X

75. Muotri, A.R., Marchetto, M.C., Zerbini, L.F., Libermann, T.A., Ventura, A.M., Sarasin, A., & Menck, C.F. (2002). Complementation of the DNA repair deficiency in human xeroderma pigmentosum group A and C cells by recombinant adenovirusmediated gene transfer. Human gene therapy, V. 13, N. 15, pp. 1833-1844, ISSN 1043- 0342

76. Narayanan, D.L., Saladi, R.N., & Fox, J.L. (2010). Ultraviolet radiation and skin cancer. International journal of dermatology, V. 49, N. 9, pp. 978–986, ISSN 00119059

77. Partch, C.L., & Sancar, A. (2005). Cryptochromes and circadian photoreception in animals. Methods in enzymology, V. 393, pp. 726-745, ISSN 0076-6879

78. Pitsikas, P., Francis, M.A., & Rainbow, A.J. (2005). Enhanced host cell reactivation of a UVdamaged reporter gene in pre-UV-treated cells is delayed in Cockayne syndrome cells. Journal of photochemestry and photobiology. B, Biology, V. 81, N. 2, pp. 89-97, ISSN 1011-1344

79. Protic-Sabljic, M., Whyte, D., Fagan, J., Howard, B.H., & Gorman, C. M., Padmanabhan, R., & Kraemer, K.H. (1985). Quantification of expression of linked cloned genes in a simian virus 40-transformed xeroderma pigmentosum cell line. Molecular and cellular biology, V. 5, N. 7, pp. 1685-1693, ISSN 0270- 7306

80. Quilliet, X., Chevallier-Lagente, O., Eveno, E., Stojkovic, T., Destée, A., Sarasin, A., & Mezzina, M. (1996). Long-term complementation of DNA repair deficient human primary fibroblasts by retroviral transduction of the XPD gene. Mutation research, V. 364, N. 3, pp. 161-169, ISSN 0027-5107

81. Rastogi, R.P., Richa, Kumar, A., Tyagi, M.B., & Sinha, R.P. (2010). Molecular mechanisms of ultraviolet radiation-induced DNA damage and repair. Journal of nucleic acids, ISSN 2090-0201

82. Raya, A., Rodríguez-Pizà, I., Guenechea, G., Vassena, R., Navarro, S., Barrero, M.J., Consiglio, A., Castellà, M., Río, P., Sleep, E., González, F., Tiscornia, G., Garreta, E., Aasen, T., Veiga, A., Verma, I.M., Surrallés, J., Bueren, J., & Izpisúa Belmonte, J.C. (2009). Disease-corrected haematopoietic progenitors from Fanconi anaemia induced pluripotent stem cells. Nature, V. 460, N. 7251, pp. 53-59, ISSN 0028-0836

83. Rosenberg, S.A., Aebersold, P., Cornetta, K. Kasid, A., Morgan, R.A., Moen, R., Karson, E.M., Lotze, M.T. Yang, J.C., Topalian, S.L., Merino,

M.J., Culver, K., Miller, D., Blaese, M., & Anderson, W.F. (1990). Gene transfer into humans – immunotherapy of patients with advanced melanoma, using tumor-infiltrating lymphocytes modified by retroviral gene transduction. The New England journal of medicine, V. 323, N. 9, pp. 570-578, ISSN 0028-4793

84. Sancar, A. (2008). Structure and function of photolyase and in vivo enzymology: 50th anniversary. The Journal of biological chemistry, V. 283, N. 47, pp. 32153-32157, ISSN 0021-9258

85. Schmidt, M., Voutetakis, A., Afione, S., Zheng, C., Mandikian, D., & Chiorini, J.A. (2008). Adeno-associated virus type 12 (AAV12).: a novel AAV serotype with sialic acidand heparin sulfate proteoglycan-independent transduction activity. Journal of virology, V. 82, N. 3, pp. 1399-1406, ISSN 0022-538X

86. Schuch, A.P., da Silva Galhardo, R., de Lima-Bessa, K.M., Schuch, N.J., & Menck C.F. (2009). Development of a DNA-dosimeter system for monitoring the effects of solarultraviolet radiation. Photochemical , & photobiological sciences, V. 8, N. 1, pp. 111-120, ISSN 1474-905X

87. Schwarz, A., Maeda, A., Kernebeck, K., van Steeg, H., Beissert, S., & Schwarz, T. (2005). Prevention of UV radiation-induced immunosuppression by IL-12 is dependent on DNA repair. The Journal of experimental medicine, V. 201, N. 2, pp. 173-179, ISSN 0022-1007

88. Seregin, S.S., & Amalfitano, A. (2009). Overcoming pre-existing adenovirus immunity by genetic engineering of adenovirus-based vectors. Expert opinion on biological therapy, V. 9, N. 12, pp. 1521-1531, ISSN 1471-2598

89. Shamberger, R.J. (1985). The genotoxicity of selenium. Mutation research, V. 154, N. 1, pp. 29- 48, ISSN 0027-5107

90. Sidhu, K.S. (2011). New approaches for the generation of induced pluripotent stem cells. Expert opinion on biological therapy, V. 11, N. 5, pp. 569-579, ISSN 1471-2598

91. Silva, G., Poirot, L., Galetto, R., Smith, J., Montoya, G., Duchateau, P., & Pâques, F. (2011). Meganucleases and other tools for targeted genome engineering: perspectives and challenges for gene therapy. Current gene therapy, V. 11, N. 1, pp. 11-27, ISSN 1566- 5232

92. Sinha, R.P., & Häder, D.P. (2002). UV-induced DNA damage and repair: a review. Photochemical , & photobiological sciences, V. 1, N. 4, pp. 225–236, ISSN 1474-905X

93. Spivak, G. (2005). UV-sensitive syndrome. Mutation research, V. 577, N. 1-2, pp. 162-169, ISSN 0027-5107

94. Spivak G., & Hanawalt, P.C. (2006). Host cell reactivation of plasmids containing oxidative DNA lesions is defective in Cockayne syndrome but normal in UV-sensitive syndrome fibroblasts. DNA repair, V. 5, N. 1, pp. 13-22, ISSN 1568-7864

95. Stefanini, M., Botta, E., Lanzafame, M., & Orioli, D. (2010). Trichothiodystrophy: from basic mechanisms to clinical implications. DNA repair, V. 9, N. 1, pp. 2-10, ISSN 1568-7864

96. Suzumura, H., & Arisaka, O. (2010). Cerebro-oculo-facio-skeletal syndrome. Advances in experimental medicine and biology, V. 685, pp. 210-214, ISSN 0065-2598

97. Takahashi, K., & Yamanaka, S. (2006). Induction of pluripotent stem cells from mouse embryonic and adult fibroblast cultures by defined factors. Cell, V. 126, N. 4, pp. 663-676, ISSN 0092-8674

98. Tanaka, K., Sekiguchi, M., & Okada, Y. (1975). Restoration of ultraviolet-induced unscheduled DNA synthesis of xeroderma pigmentosum cells by the concomitant treatment with bacteriophage T4 endonuclease V and HJV (Sendai virus). PNAS, V. 72, N. 10, pp. 4071-4075, ISSN 1091-6490

99. Voigt, K., Izsvák, Z., & Ivics, Z. (2008). Targeted gene insertion for molecular medicine. Journal of molecular medicine, V. 86, N. 11, pp. 1205-1219, ISSN 0946-2716

100. Wang, J., Faust, S.M., & Rabinowitz, J.E. (2011). The next step in gene delivery: molecular engineering of adeno-associated virus serotypes. Journal of molecular and cellular cardiology, V. 50, N. 5, pp. 793-802, ISSN 0022-2828

101. Weidenheim, K.M., Dickson, D.W., & Rapin, I. (2009). Neuropathology of Cockayne syndrome: evidence for impaired development, premature aging, and neurodegeneration. Mechanisms of ageing and development, V. 130, N. 9, pp. 619-636, ISSN 0047-6374

102. Wolf, P. Maier, H., Mullegger, R.R., Chadwick, C.A., Hofmann-Wellenhof, R., Soyer, H.P., Hofer, A., Smolle, J., Horn, M., Cerroni, L., Yarosh, D., Klein, J., Bucana, C., Dunner K. Jr., Potten, C S., Hönigsmann, H., Kerl, H., & Kripke, M.L. (2000). Topical treatment with liposomes containing T4 endonuclease V protects human skin in vivo from ultraviolet-induced upregulation of interleukin-10 and tumor necrosis factor-α. Journal of investigative dermatology, V. 114, N. 1, pp. 149-156, ISSN 0022- 202X

103. Yarosh, D, Bucana, C., Cox, P., Alas, L., Kibitel, J., & Kripke, M. (1994). Localization of liposomes containing a DNA repair enzyme in murine

skin. Journal of investigate dermatology, V. 103, N. 4, pp. 461-468, ISSN 0022-202X

104. Yarosh, D., Klein, J., Kibitel, J., Alas, L., O'Connor, A., Cummings, B., Grob, D., Gerstein, D., Gilchrest, B.A., Ichihashi, M., Ogoshi, M., Ueda, M., Fernandez, V., Chadwick, C., Potten, C.S., Proby, C.M., Young, A.R., & Hawk, J.L. (1996). Enzyme therapy of xeroderma pigmentosum: safety and efficacy testing of T4N5 liposome lotion containing a prokaryotic DNA repair enzyme. Photodermatology, photoimmunology , & photomedicine, V. 12, N. 3, pp. 122-130, ISSN 0905-4383

105. Yarosh, D., Klein, J., O'Connor, A.O., Hawk, J., Rafal, E., & Wolf, P. (2001). Effect of topically applied T4 endonuclease V in liposomes on skin cancer in xeroderma pigmentosum: a randomized study. Lancet, V. 357, N. 9260, pp. 926-929, ISSN 0140- 6736

106. Yarosh, D.B. (2002). Enhanced DNA repair of cyclobutane pyrimidine dimers changes the biological response to UV-B radiation. Mutation research, V. 509, N. 1-2, pp. 221-226, ISSN 0027-5107

107. You, Y.H., Lee, D.H., Yoon, J.H., Nakajima, S., Yasui, A., & Pfeifer GP. (2001). Cyclobutane pyrimidine dimers are responsible for the vast majority of mutations induced by UVB irradiation in mammalian cells. The Journal of biological chemistry, V. 276, N. 48, pp. 44688-44694, ISSN 0021-9258

108. Zeng, L., Sarasin, A., & Mezzina, M. (1998). Retrovirus-mediated DNA repair gene transfer into xeroderma pigmentosum cells: perspectives for a gene therapy. Cell biology and toxicology, V. 14, N. 2, pp. 105-110, ISSN 0742-2091

# Chapter 5

# RECOMBINANT ANTIBODIES AND NON-ANTIBODY SCAFFOLDS FOR IMMUNOASSAYS

Bhupal Ban and Diane A. Blake

Department of Biochemistry and Molecular Biology, Tulane University School of Medicine, New Orleans, Louisiana, USA

## INTRODUCTION

The measurement of trace amounts of physiologically active small molecules (for example, lipids, drugs, other synthetic chemicals and metals) is critical for both clinical and environmental analyses. Most small molecules can be analyzed using highly sophisticated analytical techniques, including high pressure liquid chromatography (HPLC), gas chromatography (GC), and inductively coupled plasma atomic emission spectroscopy (ICPAES). However, these methods require extensive purification, experienced technicians, and expensive instruments and reagents. Immunoassays offer an alternative to these instrument-intensive methods. Immunoassays rely on an antibody (Ab), or mixture of antibodies, for recognition of the molecule being analyzed (the analyte). Immunoassays are frequently applied to the analysis of both low molecular ligands and macromolecular drugs, and are also applied in such important areas as the quantitation of biomarkers that indicate disease progression and immunogenicity of therapeutic drug candidates. The performance of immunoassays is critically dependent on the binding properties of the antibody used in the analysis, and identification of suitable antibodies is often a major hurdle in assay development. Recombinant antibodies will play a major role in future immunoassay development.

## NATURAL AND RECOMBINANT ANTIBODY FRAGMENTS

The antibody is the key reagent of an immunoassay and it can be produced by animal immunization, hybridoma technology, and/or recombinant techniques. Most, but not all, production methods require immunization of an animal with an antigen. An antigen is a molecule that can be recognized by the immune system (immunogenicity) and that can be bound specifically to an antibody (reactogenicity). Molecules with both immunogenicity and reactogenicity are called "complete antigens" and molecules that possess only reactogenicity are called "incomplete antigens". Incomplete antigens, also called haptens, encompass a wide variety of molecules, including drugs, explosives, pesticides, herbicides, polycyclic aromatic hydrocarbons, and metal ions. These haptens can induce the immune system to produce antibodies only when they are covalently conjugated to a larger carrier molecule such as a protein.

Although polyclonal antibodies hold their place as the reagents of choice for generalpurpose applications in the biological sciences, the volume of serum that can be obtained from immunized animals and batch-to-batch differences in affinity and cross-reactivity make them less attractive for quantitative immunoassays. The first milestone for the generalized the use of immunoassays was the development of hybridoma technology, which overcame problems of heterogeneity and supply (Kohler & Milstein, 1975). While traditional monoclonal antibodies are used throughout biological research, many potential applications remain unfulfilled. The production of monoclonal antibodies requires considerable time, expense and expertise, as well as specialized cell culture facilities. The use of animal immunization means that the selection for relevant binding specificities occurs in the uncontrolled serum environment. This technology is adequate for stable antigens but not for molecules that are highly toxic, not immunogenic in mammals or not stable enough to withstand the immune processing steps required for the in vivo immune response. Most importantly, when working with monoclonal antibodies, it is not possible to alter or improve an antibody's binding properties without cumbersome procedures that convert the molecules to recombinant forms that can be engineered. All these reasons urged the development of strategies aimed at the production of recombinant antibodies (rAbs) and alternative scaffolds (Gebauer & Skerra, 2009) of smaller dimensions that can be easily selected, manipulated and produced using standard molecular biology techniques.

There are several distinct classes of natural antibodies (IgG, IgM, IgA, and IgE) that provide animals with key defenses against pathogenic organisms and toxins. Most immunoassay systems rely upon IgG as the immunoglobulin of choice. IgG is bivalent, and its ability to bind to two antigenic sites greatly increases its functional affinity and confers high retention time on cell surface receptors. The basic structure of an IgG molecule is shown in figure 1. Most IgG molecules are composed of two heavy chains (HC) and two light chains (LC), which are stabilized and linked by inter- and intra-chain disulfide bonds. The HC and LC can be further subdivided into variable regions and constant regions. The antigen binding site is formed by the combination of the variable region of the HC and LC. Most IgG molecules have two identical antigen binding sites, which are usually flat and concave for protein antigens, but which may form a pocket when the antibody has been selected against a hapten. Within the HC and LC variable regions are 3 hypervariable regions, also called complementary determining regions (CDRs), and 4 frameworks regions (FRs). The greatest sequence variation among individual antibodies occurs within the CDRs, while the FRs are more conserved. In general, it is assumed that the CDR regions from the LC and HC associate to form the antigen binding site. The lower part of the IgG molecule contains the heavy chain domains (crystallizable fragment, Fc) that are responsible for important biological effector functions. In additional to these conventional antibodies, camelids and sharks produce unusual antibodies composed only of heavy chains, also shown in figure 1. These peculiar heavy chain antibodies lack light chains (and, in the case of camelid antibodies also CH1 domain). Therefore, the antigen binding site of heavy chain antibodies is formed only by a single domain that is linked directly via a hinge region to the Fc domain. Intact IgG molecules, the bivalent (Fab')2, or the monovalent (Fab), all of which contain the antigen binding site(s), can be used in immunoassays.

Recombinant antibody forms have also been developed to facilitate antibody engineering. The single chain fragment variable (scFv) molecule is a small antibody fragment of 26-27 kDa. It contains the complete variable domain of the HC and LC, typically linked by a 15 aa long hydrophilic and flexible polypeptide linker. The scFv fragments can also include a His tag for purification, an immunodetection epitope and a protease-specific cleavage site.

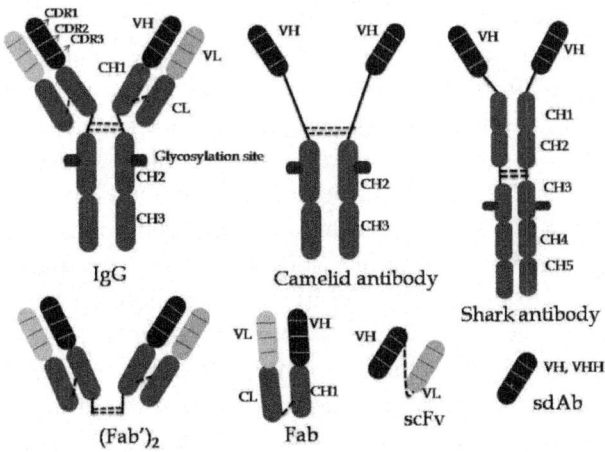

**Figure 1.** Structure of conventional, camelid and shark antibodies and of antibody fragments.

The orientation of the HC and LC domains is critical for binding activity, expression and proteolytic stability. Although a vast number of recombinant antibody (rAb) structures have been proposed (Holliger & Hudson, 2005), scFv fragments derived from mammalian IgGs and the single domain antibodies (sdAbs), which include the VHH from camelid and llama and the VH from shark, are the antibody fragments most widely used for both research and industrial applications (Kontermann, 2010; Wesolowski et al., 2009).

# PRINCIPLES AND SELECTION PLATFORMS OF RABS

Powerful combinatorial technologies have enabled the development of in vitro immune repertoires and selection methodologies that can be used to derive antibodies with or without the direct immunization of a living host (Hoogenboom, 2005; Marks & Bradbury, 2004). Recombinant antibody technology has provided an alternative method to engineer antibody fragments with the desired specificity and affinity within inexpensive and relatively simple host systems. Effective in vitro libraries have been constructed using either the entire antigen-binding fragment (Fab) or the single chain variable fragment (scFv), which represents the smallest domain capable of mediating antigen recognition. The simplest and most widely used antibody libraries utilize the scFv format, although single domain heavy chain libraries (VH and VHH) have also been constructed. The construction of in vitro libraries using different sources will be reviewed herein.

## Antibodies from Immune Antibody Libraries

The first rAbs were derived from pre-existing hybridomas; now, however, rAbs are mostly isolated from immune antibody libraries, i.e., antibody libraries generated from genetic material derived from immunized animals or naturally infected animals or humans. These libraries are biased for binding to the antigen. Thus, affinity maturation takes place in vivo and the chances of isolating the high–affinity antibodies are increased. Immune libraries are constructed using HC and LC variable domain gene pools amplified directly from immune sources; lymphoid sources include peripheral blood, bone marrow, spleen and tonsil (Huse et al., 1989; Schoonbroodt et al., 2005). In contrast to hybridoma technology, which can sample no more ~10% of the immune repertoire of an animal, a recombinant immune library, when prepared with the appropriate primers, can sample >80% of the immune repertoire and the diversity of antibodies that can be derived from a single immunized donor is much higher than what is possible using hybridomas. Selection is performed in vitro, which enhances the ability to select for rare antibody specificities. In addition, the immune repertoires of almost any species can be trapped, even those where hybridoma technology has not been described (chicken and llama), is not freely available (rabbit), or is not very robust (sheep). Immune libraries can provide higher-affinity binders than nonimmune libraries. Immunizations are generally required for each targeted antigen, although multi-antigen immunizations have been performed successfully (Li et al., 2000). Advantages and disadvantages of immune libraries include: (1) the ease of preparation compared to naïve libraries; (2) the time requirement for animal immunization; (3) the unpredictability of the immune response of the animal to an antigen of interest; (4) lack of immune response to some antigens; and (5) the necessity of construction of new libraries for each new antigen.

## Antibodies from Nonimmune, Synthetic, and Semi-Synthetic Libraries

Non-immune (naïve) libraries are derived from normal, unimmunized, rearranged V gene from the IgM/IgG mRNA of B cells, peripheral blood lymphocytes, bone marrow, spleen or tonsil. These libraries are not explicitly biased to contain clones binding to antigens; as such they are useful for selecting antibodies against a wide variety of antigens. Using specific sets of primers and PCR, IgM and IgG variable regions are amplified and cloned into specific vectors designed for selection and screening (Bradbury & Marks, 2004; Marks et al., 1991, 2004). An ideal naïve library is expected to contain a representative sample of the primary repertoires of the immune system, although it will not contain a large proportion of antibodies with somatic hypermutations produced by natural immunization. The major advantages and disadvantages of using very large naïve libraries are: (1) the large antibody repertoire, which can be selected for binders for all antigens including non-immunogenic and toxic agents; (2) a shorter time period to binding proteins, because selection is performed on an already existing library; (3) low affinity antibodies are obtained from these libraries; and (4) it is technically demanding to construct these large non-immune repertoires. Many of these disadvantages may be bypassed by using synthetic antibody libraries.

Synthetic antibody libraries are created by introducing degenerate, synthetic DNA into the regions encoding CDRs of the defined variable-domain frameworks. Synthetic diversity bypasses the natural biases and redundancies of antibody repertoires created in vivo and allows control over the genetic makeup of V genes and the introduction of diversity (Hoogenboom & Winter, 1992). A synthetic library has been described that was constructed on the basis of existing information on the structure of the antigenic site of proteins and small molecules (Persson et al., 2006; Sidhu & Fellouse 2006).

Semi-synthetic libraries have been constructed by incorporating CDR loops with both natural and synthetic diversity into one or more of the antibody framework regions. High diversity semi-synthetic repertoires have been generated by introducing partially or completely randomized sequences mainly into the CDR3 region of the heavy chain. This process generates highly complex libraries and facilitates the selection of antibodies against self-antigens, which are normally removed by the negative selection of the immune system (Barbas et al.,1992). An efficient cloning system (in vivo Cre/loxP site specific recombination) combined with dual antibody cloning strategies allows construction of very large repertoires with about 109-11 individual clones (Sblattero & Bradbury, 2000). Semi-synthetic libraries, however, have the disadvantage of always containing a certain number of non-functional clones, stemming from PCR errors, stop codons in the random sequence, or improperly folded protein products.

# IN VITRO SELECTION PROCEDURES FOR RABS FROM COMBINATORIAL LIBRARIES

Recombinant antibody technologies provide the investigator with a great deal of control over selection and screening conditions and thus permit the generation of antibodies against highly specialized antigen conformations or epitopes. The most powerful methods, phage, yeast, and ribosomal display technologies, are complementary in their properties and can be used with naïve, immunized or synthetic antibody repertoires.

## Phage Display Libraries for the Isolation of Antibodies

Phage display-based selections are now a relatively standard procedure in many molecular biology laboratories. The generation of antibody fragments with high specificity and affinity for virtually any antigen has been made possible using phage display. Phage display libraries are produced by cloning the pool of genes coding for antibody fragments into vectors that can be packed into the viral genome. The rAb is then expressed as an antibody fragment on the surface of mature phage particles. Selection of specific antibody fragments involves exposure to antigen, which allows the antigen-specific phage antibodies to bind their target during the bio-panning. The binding is followed by extensive wash steps and subsequent recovery of antigen-specific phage. The phage particles can then be used to infect E. coli bacteria. Different display systems can lead to monovalent (single copy) or to multivalent (multiple copy) display of the antibody fragment, depending on the type of anchor protein and display vector used (Sidhu et al., 2000). The most popular system uses a monovalent display vector system, which is convenient for selecting antibodies with higher affinity. Monovalent display is achieved by using a direct fusion to a minor viral coat protein (pIII). The vector into which most antibody libraries are cloned is a phagemid vector that requires a helper phage for the production of phage particles. Use of a phagemid vector makes propagation in bacteria much easier to accomplish than would be possible with a phage vector (Hust & Dubel, 2005). A general scheme for the isolation of antibody fragments by phage display is shown in figure 2. Libraries with 106-11 individual clones can be made using recombinant-based protocols. Due to limitations of the E. coli folding machinery, complete IgG molecules are very difficult to express in E. coil and display on the surface of phage. Therefore, smaller antibody fragments such as Fab, scFv and sdAb are primarily used for antibody phage display.

## Yeast Surface Display

Yeast surface display is a powerful method for isolating and engineering antibody fragments (Fab, scFv) from immune and non-immune libraries, and has been used to isolate recombinant antibodies with binding specificity to variety of proteins, peptides, and small molecules (Boder & Wittrup, 2000; Chao et al., 2006). In this system, antibodies are displayed on the surface of yeast Saccharomyces cerevisiae via fusion to an ⅃-agglutinin yeast adhesion receptor, which is located in the yeast cell wall.

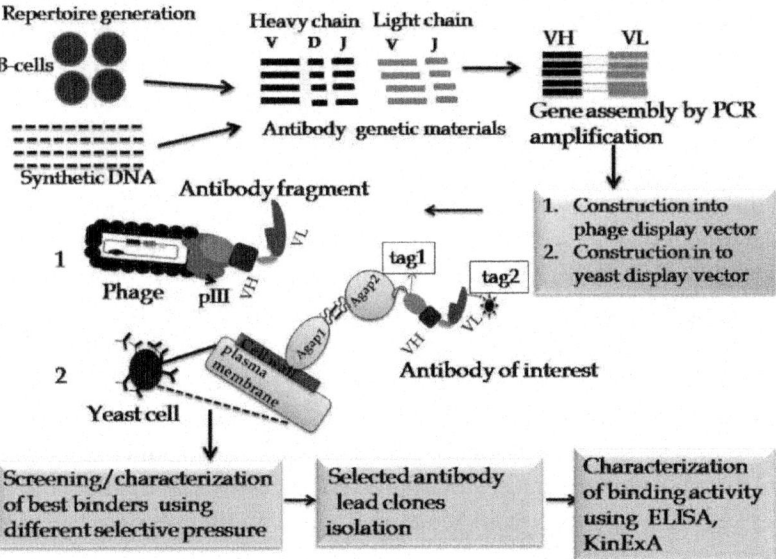

**Figure 2**. Schematic diagram for construction of antibody libraries and in vitro display system; phage and yeast display.

Like phage display, yeast display provides a direct connection between genotype and phenotype; a plasmid containing the gene of interest is contained within yeast cells, while the encoded antibody is expressed on the surface. The display level of each yeast cell is variable, with each cell displaying $1x10^4$ to $1x10^5$ copies of the scFv. Variation of surface expression and avidity can be quantified using fluorescence activated cell sorting (FACS), which measures both antigen binding and antibody expression on the yeast cell surface (Feldhaus et al., 2003). The main advantage of yeast surface display over other display technologies is the eukaryotic expression bias of yeast, which contains post-translational modification and processing machinery similar to that of mammalian cells. Thus, yeast may be better suited for the expression of antibodies as compared to prokaryotes such as E. coli. Yeast display libraries have been used during the affinity maturation of scFvs from mutagenic libraries (Boder et al., 2000; Lou et al., 2010; Orcutt et al., 2011). Limiting factors of yeast display include a more limited transforming efficacy of yeast as compared to bacteria, which can lead to a smaller functional library size (about 107-109 ) than is possible with other display technologies.

## Ribosomal Display

Ribosomal display is an in vitro selection and evolution technology for proteins and peptides from large libraries (Dreier & Pluckthun, 2011; Hanes & Pluckthun, 1997). The general scheme of ribosomal display is shown in figure 3. This display system was developed from a peptide-display approach that was extended to screen scFv and scaffold proteins having very high affinity for antigen ($K_d$s as low as $10^{-11}$ M) from very large libraries (Binz et al., 2004; Zahnd et al., 2007). The DNA library coding for proteins such as antibodies and scaffolds are transcribed in vitro. The mRNA has been engineered without a stop codon; therefore, the translated protein remains attached to the peptidyl tRNA and occupies the ribosomal tunnel. This allows the protein of interest to protrude out of the ribosome and fold. Ribosomal display is performed entirely in vitro, and it has two advantages over other selection technologies. First, the diversity of the libraries is not limited by the transformation efficiency of bacterial cells ($\sim 1x10^{11}$ to $1x10^{13}$), but only by the number of ribosomes and different mRNA molecules present in the test tube. Second, random mutations can be introduced easily after each selection round, as no cells must be transformed after any diversification step. In ribosomal display, the physical link between the genotype and the corresponding phenotype is accomplished by a complex consisting of mRNA, ribosome and protein.

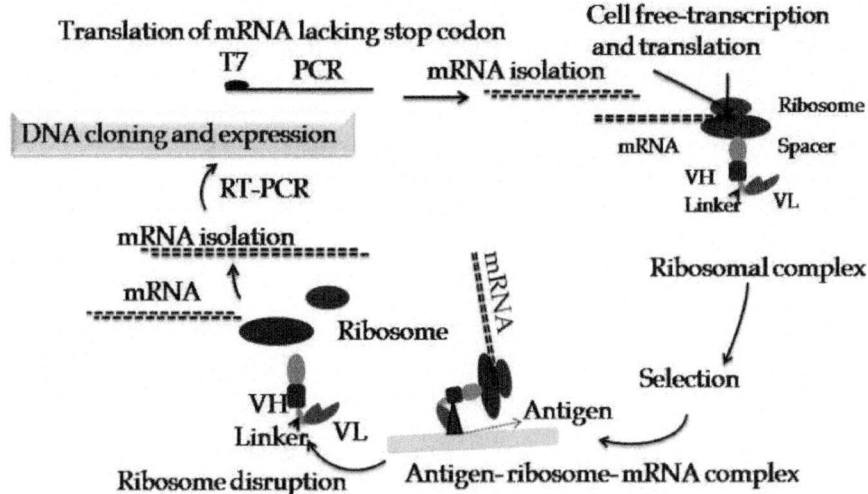

**Figure 3**. Schematic diagram for isolation of specific antibody fragment from ribosomal display

Ribosomal display has been used to isolate antibodies that bind to haptens with nanomolar affinities (Yau et al., 2003). A summary of the in vitro display systems available to researchers is shown in table 1.

## APPLICATIONS OF RABS AGAINST LOW MOLECULAR LIGANDS

A large number of rAbs have been used successfully to develop diagnostic kits, therapeutics and biosensors (Holliger et al., 2005; Huang et al., 2010; Kramer & Hock, 2003). The majority of the targets were large molecular weight analytes such as proteins and peptides. Prior to 1990, there were few reports of the isolation of rAbs against low molecular weight molecules (haptens) such as drugs of abuse, vitamins, hormones, metabolites, food toxins and environmental pollutants, including heavy metals and pesticides. Hapten-specific antibodies are necessary reagents for the development of immunoassays, immunosensor technologies (Charlton et al., 2001), and immunoaffinity chromatography purification columns (Sheedy & Hall, 2001). Commercial immunoassays for haptens such as small environmental contaminants still rely mostly on polyclonal antibodies rather than monoclonal or recombinant antibodies fragments (Sheedy et al., 2007). The complexity and costs associated with the production of anti-hapten antibodies by hybridoma technology and the preferential selection of antibodies that recognize the conjugated form of the haptens over antibodies that specifically recognize free haptens are two of the most important problems that have limited the development and application of antibodies that recognize haptens and other low molecules ligands. Moreover, some small molecular weight ligands will not trigger the animal immune system even when conjugated to a carrier protein, thereby making the production of antibodies against that such analytes very difficult.

**Table 1.** Comparing the main in vitro selection platforms for isolation of rAbs

| Name | Display | Library size | Main applications | Advantages | Disadvantages |
|---|---|---|---|---|---|
| Phage display | Monovalent Multivalent | $10^{10}$ to $10^{11}$ | Abs from natural & synthetic libraries; Affinity maturation & stability increase | Easy and versatile for large rAbs panels | Laborious to make large libraries; Not truly monovalent |
| Yeast surface display | Multivalent | $10^7$ | Abs from natural & synthetic libraries; Affinity maturation & stability increase | Rapid when used in combination with random mutagenesis | Small rAb panels, FACS expertise required |
| Ribosome display | Monovalent | $10^{12}$ to $10^{13}$ | Abs from natural & synthetic libraries; Affinity maturation & stability increase | Intrinsic mutagenesis, fastest of all systems | Small rAb panels, limited selection scope and technically sensitive |

In recent years, the production of recombinant antibodies to low molecular weight ligands has increased significantly, as shown in table 2. A single methyl or hydroxyl group can have a considerable effect on the biological properties of a steroid hormone. Similarly, protein phosphorylation, acetylation and sulfation, all of which are relatively simple posttranslational modifications in chemical terms, can dramatically affect signal transduction (Bikker et al., 2007; Hoffhines et al., 2006; Kehoe et al., 2006). Antibodies capable of discerning such relatively simple chemical modification are of great values in studying these effects. The display methods to tailor both affinity and specificity have generated antibodies capable of discerning minor difference between related small molecules far better than those obtained by immunization.

# IMPROVING THE SPECIFICITY AND AFFINITY OF RABS TO LOW MOLECULAR WEIGHT LIGANDS

Although recombinant antibody technology has been able to open the bottleneck in the isolation of antibodies against virtually any antigen, it remains difficult to obtain high affinity antibodies against small molecules using immune and naïve libraries. Various approaches have been utilized, including identifying the key binding residues, developing more effective procedures for selection of the most specific binders and avoiding interfacial effects that can compromise the yield and stability of rAbs.

**Table 2.** List of small molecule-specific recombinant antibodies

| Target Hapten | Ab format | Antibody library | *In vitro* display | Reference |
|---|---|---|---|---|
| Aflatoxin B1 | scFv | Naïve | Phage | (Moghaddam et al., 2001) |
| Digoxigenin | scFv | Naïve | Phage | (Dorsam et al., 1997) |
| Doxorubicin | scFv | Naïve | Phage | (Vaughan et al., 1996) |
| Estradiol | scFv | Naïve | Phage | (Dorsam et al., 1997) |
| Indole-3-acetic acid | VHH | Naïve | Phage | (Sheedy et al., 2006) |
| Fluorescein | scFv | Naïve | Phage | (Vaughan et al., 1996) |
| Phenyloxazolone | scFv | Naïve | Phage | (de Haard et al., 1999) |
| Picloram | VHH | Naïve | Ribosome | (Yau et al., 2003) |
| Progesterone | scFv | Naïve | Ribosome | (He et al., 1999) |
| Fumosinin B1 | scFv | Naïve | Phage | (Lauer et al., 2005) |
| Atrazine | scFv | Immune | Phage | (Li et al., 2000) |
| Azo-dye RR1 | VHH | Immune | Phage | (Spinelli et al., 2000) |
| Cortisol | scFv | Immune | Phage | (Chames & Baty, 1998) |
| Digoxin & analogues | scFv | Immune | Phage | (Short et al., 1995) |
| Isoproturon | scFv | Immune | Phage | (Li et al., 2000) |
| Mecoprop | scFv | Immune | Phage | (Li et al., 2000) |
| Simazine | scFv | Immune | Phage | (Li et al., 2000) |
| Triazine | scFv | Immune | Phage | (Kramer, 2002) |
| 4-Hydroxy-3-iodo-5-nitrophenol | scFv | Semi-synthetic | Phage | (van Wyngaardt et al., 2004) |
| Fluorescein | scFv | Semi-synthetic | Phage | (van Wyngaardt et al., 2004) |
| Microcystin LR | scFv | Semi-synthetic | Phage | (Strachan et al., 2002) |
| Phtalic acid | scFv | Semi-synthetic | Phage | (Strachan et al., 2002) |
| Trichlocarbon | VHH | Naïve | Phage | (Tabares-da Rosa et al., 2011) |
| 6-Monoacetylmorphine but not morphine | scFv | Naïve | Phage | (Moghaddam et al., 2003) |
| Metallic gold | Fv | Naïve | Phage | (Watanabe et al.,, 2008) |
| Anti-Aluminum | VHH | Semi-synthetic | Phage | (Hattori et al., 2010) |
| Anti-Cobalt | VHH | Semi-synthetic | Phage | (Hattori et al., 2010) |
| Anti-Uranium | scFv | Immune | Phage | (Zhu et al., 2011) |
| Domoic acid | scFv | Immune | Phage | (Shaw et al., 2008) |
| Azoxystrobin | VHH | Immune | Phage | (Makvandi-Nejad et al., 2011) |
| Methamidophos | scFv | Immune | Phage | (Li et al., 2006) |

There are, however, unique challenges to the development of antibodies that will perform well in assays for low molecular weight ligands. Antigen binding sites are generated by the cooperation between the variable domains of the HC and LC (VH/VL). The amino acids of FRs compose rigid scaffolds that position the amino acids in the CDRs in loops that extend outward from scaffold. These loops play important roles in making contact with the antigen. Hapten antigens have remained a great challenge for immunodiagnostics, because the hapten portion of the antigen often ends up almost buried inside the concave–shaped antigen binding pocket. The extended shape of this binding pocket then facilitates additional interactions between amino acid residues in binding site and portions of the hapten-protein conjugate present in the bridge between the hapten and the protein carrier. These additional interactions mean that the antibody often binds to the much more tightly to the hapten-protein conjugate than to the soluble hapten. In our laboratory, we have studied this phenomenon with 10 different anti-hapten antibodies. In this study, the antibody always bound more tightly to the protein conjugate than to the soluble antigen. The differences in affinity ranged from 1.5 to 1600 fold, depending upon the antibody being analyzed (Blake et al., 1996, Melton, 2010). Thus, when given a choice, anti-hapten antibodies almost always prefer binding to the hapten-protein conjugate, and additional soluble hapten is required to inhibit this interaction, thus reducing assay sensitivity. Selective panning and affinity maturation are methods available in recombinant technology for reducing selective binding of hapten antibodies to the hapten-protein conjugate.

## PANNING OPTIMIZATION

A variety of selection strategies have been reported for the isolation of high affinity rAbs against chelated metals and other haptens (Sheedy et al., 2007; Zhu et al., 2011). The most successful strategies employed first loose and then increasingly stringent panning conditions to enrich the population of phage antibodies as follow: (i) the concentration of coating antigen was gradually decreased during successive rounds of panning (Strachan et al., 2002; Zhu et al., 2011); (ii) soluble hapten was used to elute ligand-specific antibodies in place of the triethylamine more commonly used for elution; (iii) during the panning of immune scFv libraries, the conjugate carrier protein and/or other linker peptides were included for several intermediate incubation steps at high concentration and subsequently decreased to remove phage antibodies that bound to the protein conjugate rather than the soluble hapten; (iv) the phage antibodies were incubated with structural analogues of the hapten prior to incubation with the immobilized target hapten to eliminate phage antibodies with unwanted cross reactivities (Charlton et al., 2001; Zhu et al., 2011). Such panning optimization strategies have led to the isolation of antibodies with higher affinity and specificity and lower levels of cross-reactivity. For an example from the isolation of antibodies to metal-chelate complexes, such subtractive panning strategies were employed to isolate an antibody that bound tightly to uranium in complex with 2,9-dicarboxyl-1,10- phenanthroline, (DCP), but weakly to metal-free DCP. In successive rounds of panning, the phage antibody population was incubated with a high concentration of carrier protein (BSA) and increasing concentrations of soluble DCP in immunotubes coated with decreasing concentrations of a $UO_2^{2+}$-DCP-BSA conjugate as shown in table 3.

**Table 3.** General selection strategy for the isolation of scFvs that bind to a metal-loaded but not a metal-free chelator (Zhu et al., 2011).

| Round of Selection | Percent of maximum conjugate coated onto immunotubes | Percent of maximum metal-free chelator added to the phage binding buffer | Percent of maximum carrier protein added to the phage binding buffer |
|---|---|---|---|
| 1 | 100 | 10 | 100 |
| 2 | 100 | 10 | 100 |
| 3 | 10 | 20 | 100 |
| 4 | 10 | 20 | 100 |
| 5 | 1 | 100 | 100 |

## *In Vitro* Antibody Affinity Maturation

The affinity maturation procedure contains two stages: (a) making a modified antibody library with a larger diversity than the original library (b) selecting desired antibodies molecules from the library using the previously discussed in vitro display and panning methods. An antibody's affinity for its antigen is dependent on the identity and conformation of the amino acid sidechains in the CDRs of both the HC and LC. Improvement in the antigen-binding affinity can be attained using a number of strategies. The mostly common used are random mutagenesis, site-direct mutagenesis and chain shuffling. These processes are often referred to as in vitro affinity maturation, to distinguish the process from the affinity maturation that takes place in the animal. Although a considerable number of successful affinity maturation processes have been reported for antibodies against macromolecule antigens like proteins, affinity maturation for low weight molecules like haptens and metals is obviously more difficult, and consequently, only a limited number of successful studies have thus far been reported.

## *Random Mutagenesis (Error Prone PCR; E-p PCR)*

Random mutagenesis is the process that most closely mimics the in vivo process of somatic hypermutation. This process makes no assumptions as to which sites are the best to mutate in order to increase affinity, and it is also technically rather simple to execute. Error prone PCR uses low fidelity polymerization conditions to introduce a low level of point mutations randomly throughout a wide region of a target gene (e.g. , the entire VH and VL). Error prone PCR has been used to demonstrate the effect of mutation frequency on the affinity maturation of antibodies against both proteins and small ligands (Daugherty et al., 2000). When wild type antibodies to the hapten, diogoxin, were subjected to E-p PCR, the higher affinity clones isolated from libraries all contained aromatic residues substitutions in the antibody binding site. These resides were thought to be important for hydrophobic interaction with the planer aromatic structure of digoxin (Short et al., 1995). A disadvantage of E-p PCR is that surface-selection often enriches binders with increased tendency for dimerization, especially when using the scFv format. In addition, most of the mutants lacked detectable expression or lost antigen-binding affinity. A few mutants lost specificity and showed increased cross-reactivity to analogs (Fuji, 2004; Sheedy et al., 2007). Point mutation can cause profound effects on the binding affinity and specificity of an antibody for its small ligands. The affinity maturation processes reported for anti-hapten scFvs are listed in table 4.

## Site-Directed Mutagenesis

In site-directed mutagenesis, the investigator changes specific amino acid residues. Sitedirected mutagenesis is often used in combination with in silico modeling, crystallographic data, and ligand docking programs, which allow the investigator to hypothesize about the roles that individual binding site amino acid residues have in antigen binding. The CDRs of VH and VL are usually targeted for both haptens and protein antigens (Siegel, et al., 2008), and mutations in CDRs as opposed to within the framework residues generally contribute more to increases in affinity (Orcutt et al., 2011; Short et al., 2002). In one study, the most significant increase in affinity was correlated with mutations in the light chain CDR1 even though this CDR1 was not contacting the hapten directly (Valjakka et al., 2002). Hapten-specific antibodies whose affinity has been increased by site directed mutagenesis are listed in table 4.

**Table 4**. List of successful affinity maturations of anti-hapten antibodies

| Target hapten | Fold increase in affinity | Ab format/ *in vitro* display | Affinity maturation | Reference |
|---|---|---|---|---|
| phOx-GABA | 290 | scFv/Phage | Chain shuffling | (Marks et al., 1992) |
| Cortisol | 7.9 | scFv/Phage | Site-directed | (Chames et al., 1998) |
| Estradiol-17 β | 12 | Fab/Phage | Site-directed | (Kobayashi et al., 2010) |
| Fluorescein | 2600 | scFv/Yeast | Ep-PCR /DNA -shuffling | (Boder et al., 2000) |
| Testosterone | 35 | Fab /Phage | Site-directed | (Valjakka et al., 2002) |
| Tacrolimus | 15 | scFv/Yeast | Site-directed | (Siegel et al., 2008) |
| DOTA-chelate | 1000 | scFv/Yeast | Site-directed | (Orcutt et al., 2011) |

## Shuffling of Antibody Genes

Shuffling of antibody genes to create new antibody libraries can be accomplished in several ways: chain shuffling, DNA shuffling, and staggered extension processes.

### *Chain Shuffling*

In this procedure, one of the two chains (VH of VL) is fixed and combined with a repertoire of partner chains to yield a secondary library that can be searched for superior pairings against antigens. This approach takes advantage of "random" mutations that have been introduced into VH and VL germline genes in vivo. Phage display and yeast display are often used to facilitate the selection of improved binders from these secondary libraries (Lou et al., 2010; Marks, 2004; Persson et al., 2006). This procedure has also been used to increase the affinity of anti-hapten antibodies and the results are reviewed in table 4. Chain shuffling is only a suitable mutagenesis strategy when VH and VL sequences are available from immune libraries. Chain shuffling is, therefore, not useful with naïve libraries since heavy and light chains available in these libraries have not been exposed to the antigen of interest.

### *DNA Shuffling by Random Fragmentation and Reassembly*

DNA shuffling is based on repeated cycles of point mutagenesis, recombination and selection, which allows in vitro molecular evolution of protein (Stemmer, 1994). The process mimics somewhat the natural mechanism of molecular evolution (Ness et al., 1999). This shuffling technique involves the digestion of a large antibody gene with DNase I to create a pool of random DNA fragments. These fragments can then be reassembled into full-length genes by repeated cycles of annealing in the presence of DNA polymerase. DNA shuffling offers several advantages over more traditional mutagenesis strategies. It uses longer DNA sequences and also permits the selection of clones with mutations outside of the antibody binding site.

# MODIFICATION OF RABS FUSED WITH SIGNAL ENHANCER PROTEINS

Antibody engineering enables the preparation of fusion proteins combining scFvs and enzymes via the expression of a single scFv-enzyme fusion gene. Such recombinant scFvfusion proteins have been reported for numerous applications, including the detection of a plant virus (Griep et al., 1999), the human pathogen hantaviruses (Velappan et al., 2007), other protein targets such as Bacillus anthraces (Wang et al., 2006), cholera and ricin toxins (Swain et al., 2011), and the haptens morphine (Brennan et al., 2003) and 11-deoxycortisol (11-DC) (Kobayashi et al., 2006). These fusion proteins provide a much higher signal/noise ratio in the ELISA format than conventional enzyme-labeled antibodies because the fusion proteins can be obtained as a single molecule species having a 1:1 rAb/enzyme ratio, and thus are uncontaminated by unconjugated enzyme and rAb molecules. As an example, the sensitivity of a competitive immunoassay for 11-deoxycortisol was 10,000-fold higher when an scFv-alkaline phosphatase fusion protein replaced the standard enzyme-labeled secondary antibody (Kobayashi et al., 2006; Martin et al., 2006).

# BEYOND ANTIBODY FRAGMENTS (SCAFFOLD PROTEIN)

Conventional diagnostic immunoassays are limited to the analysis of a few hundred assays per day, whereas with antibody microarrays using individually addressable electrodes, thousands of assays can be run in parallel (Dill et al., 2004). Antibody fragments are providing valuable alternatives to full length mAbs for new biosensing devices because they provide small, stable, highly specific reagents against the target antigens. In addition, because the recombinant antibody is smaller than the intact IgG, the density of binding sites that can be immobilized on the surface of these sensors can be increased. The stability of surface-immobilized ligands is also crucial in immunoassay format. Therefore, a great deal of interest has been focused on simplifying the antibody scaffold, and molecular engineering has pushed the concepts of antibody miniaturization to develop more stable binders that are less sterically hindered when immobilized on surfaces.

To overcome the limitation of antibodies, the several alternative protein frameworks have been developed. Design of these protein frameworks, collectively called "scaffolds" or "scaffold proteins", usually involves the adaptation of structurally well-defined polypeptide frameworks by the introduction of novel functionality. The new functionality is added to those parts of the protein surface that are not considered important for protein folding or stability. The recent development of non-biological alternatives to antibodies, including both scaffold proteins and plastibodies, may create distinct opportunities for future improvements in immunoassay technology. This could be particularly relevant in applications where compatibility of the binding probe with organic solvents and the ability to withstand thermal and mechanical stress are required. Currently, there are more than 60 non-antibody scaffolds suggested as affinity ligands, primarily for therapeutic and diagnostic purposes (Binz et al., 2005a; Caravella & Lugovskoy, 2010; Lofblom et al.,

2010; Skerra, 2007). One of the current problems with bacterial expression of antibody fragments is that of disulfide bond formation, which occurs primarily in the periplasm of bacterial cells. Because of the intradomain disulfide bonds required for proper immunoglobulin folding, neither scFvs nor Fabs are compatible with intracellular expression and only very stable scFv fragments have been expressed in the cytoplasm of E. coli (Martineau & Betton, 1999; Ohage et al., 1999). The ideal scaffold therefore should be stable without disulfide bonds, expressed in high amounts in E. coli and compatible with current display techniques. The scaffold should contain loops or other structures on its surface that can be modified to form the binding site. This can be a natural binding site or created de novo. Randomizations made to the binding region should be able to generate binders with high specificity and affinity (Binz & Pluckthun, 2005b; Gebauer et al., 2009; Gronwall & Stahl, 2009; Kim et al., 2009; Lofblom et al., 2010). All of the scaffolds reported to date, including affibodies, anticalins, and designed ankyrin repeats (DARPin), can be engineered for interaction with analytes by mimicking the way the immune system shuffles sequences to create diversity in loop structures. The goal is to randomize the loops without affecting the overall structure and stability of protein. Thus, it is possible to engineer binding properties that are totally independent of their original biological function. An example of this strategy is the recently developed anticalin with picomolar affinity for DTPA-chelated lanthanides, especially YIII (Kim et al., 2009). This anticalin forms a tight noncovalent complex (with slow dissociation kinetics) under physiological conditions in the presence of the chelated metal ion and, after fusion with an appropriate targeting domain; it may provide an ideal tool for applications in 'pretargeting' radioimmunotherapy. Notably, the only established non-Ig scaffold that intrinsically provides pockets and thus allows tight and specific complexation of small molecules is the one of the lipocalins.

## CONCLUSION

Immunoassay techniques provide simple, powerful and inexpensive methods for the measurement of small ligands. However, the progress of the development of new immunoassays and related immunotechnologies is still limited by the availability of antibodies with the desired affinities and specificities for given applications. Advances in molecular biology have led to the ability to synthesize antibodies in vitro, completely without the use of animals. Recombinant molecular technology that can generate variability, combined with high-throughput screening methodologies, can be used to produce engineered antibody-like molecules and novel antibody-mimic domains on scaffold proteins. The rAbs fused with other functional proteins can enhance the sensitivity of antibody-based assays and reduce the cost and labor involved in chemically synthesizing conjugates. Antibody engineering had already matured into a technology available to the general scientific community. Further advances will lead to better binding proteins that will permit the development of novel, high-throughput sensing systems for low molecular weight ligands.

## ACKNOWLEDGMENTS

The authors acknowledge funding from the Office of Science, Department of Energy (DESC0004959) and from the USPHS, NIEHS (U19-ES020677).

## REFERENCES

1.    Barbas, C. F., 3rd, Bain, J. D., Hoekstra, D. M., & Lerner, R. A. (1992). Semisynthetic combinatorial antibody libraries: a chemical solution to the diversity problem. Proc Natl Acad Sci U S A, 89(10), 4457-4461.

2.    Bikker, F. J., Mars-Groenendijk, R. H., Noort, D., Fidder, A., & van der Schans, G. P. (2007). Detection of sulfur mustard adducts in human callus by phage antibodies. Chem Biol Drug Des, 69(5), 314-320.

3.    Binz, H. K., Amstutz, P., Kohl, A., Stumpp, M. T., Briand, C., Forrer, P., Grutter, M. G., & Pluckthun, A. (2004). High-affinity binders selected from designed ankyrin repeat protein libraries. Nat Biotechnol, 22(5), 575-582.

4.    Binz, H. K., Amstutz, P., & Pluckthun, A. (2005a). Engineering novel binding proteins from nonimmunoglobulin domains. Nat Biotechnol, 23(10), 1257-1268.

5.    Binz, H. K., & Pluckthun, A. (2005b). Engineered proteins as specific binding reagents. Curr Opin Biotechnol, 16(4), 459-469.

6.    Blake, D. A., Chakrabarti, P., Khosraviani, M., Hatcher, F. M., Westhoff, C. M., Goebel, P., Wylie, D. E., & Blake, R. C., 2nd (1996). Metal binding properties of a monoclonal antibody directed toward metal-chelate complexes. J Biol Chem, 271(44), 27677- 27685.

7.    Boder, E. T., Midelfort, K. S., & Wittrup, K. D. (2000). Directed evolution of antibody fragments with monovalent femtomolar antigen-binding affinity. Proc Natl Acad Sci U S A, 97(20), 10701-10705.

8.    Boder, E. T., & Wittrup, K. D. (2000). Yeast surface display for directed evolution of protein expression, affinity, and stability. Methods Enzymol, 328, 430-444.

9.  Bradbury, A. R., & Marks, J. D. (2004). Antibodies from phage antibody libraries. J Immunol Methods, 290(1-2), 29-49.

10. Brennan, J., Dillon, P., & O'Kennedy, R. (2003). Production, purification and characterization of genetically derived scFv and bifunctional antibody fragments capable of detecting illicit drug residues. J Chromatogr B Analyt Technol Biomed Life Sci, 786(1- 2), 327-342.

11. Caravella, J., & Lugovskoy, A. (2010). Design of next-generation protein therapeutics. Curr Opin Chem Biol, 14(4), 520-528.

12. Chames, P., & Baty, D. (1998). Engineering of an anti-steroid antibody: amino acid substitutions change antibody fine specificity from cortisol to estradiol. Clin Chem Lab Med, 36(6), 355-359.

13. Chames, P., Coulon, S., & Baty, D. (1998). Improving the affinity and the fine specificity of an anti-cortisol antibody by parsimonious mutagenesis and phage display. J Immunol, 161(10), 5421-5429.

14. Chao, G., Lau, W. L., Hackel, B. J., Sazinsky, S. L., Lippow, S. M., & Wittrup, K. D. (2006). Isolating and engineering human antibodies using yeast surface display. Nat Protoc, 1(2), 755-768.

15. Charlton, K., Harris, W. J., & Porter, A. J. (2001). The isolation of super-sensitive anti-hapten antibodies from combinatorial antibody libraries derived from sheep. Biosens Bioelectron, 16(9-12), 639-646.

16. Daugherty, P. S., Chen, G., Iverson, B. L., & Georgiou, G. (2000). Quantitative analysis of the effect of the mutation frequency on the affinity maturation of single chain Fv antibodies. Proc Natl Acad Sci U S A, 97(5), 2029-2034.

17. de Haard, H. J., van Neer, N., Reurs, A., Hufton, S. E., Roovers, R. C., Henderikx, P., de Bruine, A. P., Arends, J. W., & Hoogenboom, H. R. (1999). A large non-immunized human Fab fragment phage library that permits rapid isolation and kinetic analysis of high affinity antibodies. J Biol Chem, 274(26), 18218-18230.

18. Dill, K., Montgomery, D. D., Ghindilis, A. L., & Schwarzkopf, K. R. (2004). Immunoassays and sequence-specific DNA detection on a microchip using enzyme amplified electrochemical detection. J Biochem Biophys Methods, 59(2), 181-187.

19. Dorsam, H., Rohrbach, P., Kurschner, T., Kipriyanov, S., Renner, S., Braunagel, M., Welschof, M., & Little, M. (1997). Antibodies to steroids from a small human naive IgM library. FEBS Lett, 414(1), 7-13.

20. Dreier, B., & Pluckthun, A. (2011). Ribosome display: a technology for selecting and evolving proteins from large libraries. Methods Mol Biol, 687, 283-306.

21.  Feldhaus, M. J., Siegel, R. W., Opresko, L. K., Coleman, J. R., Feldhaus, J. M., Yeung, Y. A., Cochran, J. R., Heinzelman, P., Colby, D., Swers, J., Graff, C., Wiley, H. S., & Wittrup, K. D. (2003). Flow-cytometric isolation of human antibodies from a nonimmune Saccharomyces cerevisiae surface display library. Nat Biotechnol, 21(2), 163-170.

22.  Fujii, I. (2004). Antibody affinity maturation by random mutagenesis. Methods Mol Biol, 248, 345-359.

23.  Gebauer, M., & Skerra, A. (2009). Engineered protein scaffolds as next-generation antibody therapeutics. Curr Opin Chem Biol, 13(3), 245-255.

24.  Griep, R. A., van Twisk, C., van der Wolf, J. M., & Schots, A. (1999). Fluobodies: green fluorescent single-chain Fv fusion proteins. J Immunol Methods, 230(1-2), 121-130.

25.  Gronwall, C., & Stahl, S. (2009). Engineered affinity proteins--generation and applications. J Biotechnol, 140(3-4), 254-269.

26.  Hanes, J., & Pluckthun, A. (1997). In vitro selection and evolution of functional proteins by using ribosome display. Proc Natl Acad Sci U S A, 94(10), 4937-4942.

27.  Hattori, T., Umetsu, M., Nakanishi, T., Togashi, T., Yokoo, N., Abe, H., Ohara, S., Adschiri, T., & Kumagai, I. (2010). High affinity anti-inorganic material antibody generation by integrating graft and evolution technologies: potential of antibodies as biointerface molecules. J Biol Chem, 285(10), 7784-7793.

28.  He, M., Menges, M., Groves, M. A., Corps, E., Liu, H., Bruggemann, M., & Taussig, M. J. (1999). Selection of a human anti-progesterone antibody fragment from a transgenic mouse library by ARM ribosome display. J Immunol Methods, 231(1-2), 105-117.

29.  Hoffhines, A. J., Damoc, E., Bridges, K. G., Leary, J. A., & Moore, K. L. (2006). Detection and purification of tyrosine-sulfated proteins using a novel anti-sulfotyrosine monoclonal antibody. J Biol Chem, 281(49), 37877-37887.

30.  Holliger, P., & Hudson, P. J. (2005). Engineered antibody fragments and the rise of single domains. Nat Biotechnol, 23(9), 1126-1136.

31.  Hoogenboom, H. R. (2005). Selecting and screening recombinant antibody libraries. Nat Biotechnol, 23(9), 1105-1116.

32.  Hoogenboom, H. R., & Winter, G. (1992). By-passing immunisation. Human antibodies from synthetic repertoires of germline VH gene segments rearranged in vitro. J Mol Biol, 227(2), 381-388.

33.  Huang, L., Muyldermans, S., & Saerens, D. (2010). Nanobodies(R): proficient tools in diagnostics. Expert Rev Mol Diagn, 10(6), 777-785.

34. Huse, W. D., Sastry, L., Iverson, S. A., Kang, A. S., Alting-Mees, M., Burton, D. R., Benkovic, S. J., & Lerner, R. A. (1989). Generation of a large combinatorial library of the immunoglobulin repertoire in phage lambda. Science, 246(4935), 1275-1281.

35. Hust, M., & Dubel, S. (2005). Phage display vectors for the in vitro generation of human antibody fragments. Methods Mol Biol, 295, 71-96.

36. Kehoe, J. W., Velappan, N., Walbolt, M., Rasmussen, J., King, D., Lou, J., Knopp, K., Pavlik, P., Marks, J. D., Bertozzi, C. R., & Bradbury, A. R. (2006). Using phage display to select antibodies recognizing post-translational modifications independently of sequence context. Mol Cell Proteomics, 5(12), 2350-2363.

37. Kim, H. J., Eichinger, A., & Skerra, A. (2009). High-affinity recognition of lanthanide(III) chelate complexes by a reprogrammed human lipocalin 2. J Am Chem Soc, 131(10) , 3565-3576.

38. Kobayashi, N., Iwakami, K., Kotoshiba, S., Niwa, T., Kato, Y., Mano, N., & Goto, J. (2006). Immunoenzymometric assay for a small molecule,11-deoxycortisol, with attomolerange sensitivity employing an scFv-enzyme fusion protein and anti-idiotype antibodies. Anal Chem, 78(7), 2244-2253.

39. Kobayashi, N., Oyama, H., Kato, Y., Goto, J., Soderlind, E., & Borrebaeck, C. A. (2010). Twostep in vitro antibody affinity maturation enables estradiol-17beta assays with more than 10-fold higher sensitivity. Anal Chem, 82(3), 1027-1038.

40. Kohler, G., & Milstein, C. (1975). Continuous cultures of fused cells secreting antibody of predefined specificity. Nature, 256(5517), 495-497.

41. Kontermann, R. E. (2010). Alternative antibody formats. Curr Opin Mol Ther, 12(2), 176-183.

42. Kramer, K. (2002). Synthesis of a group-selective antibody library against haptens. J Immunol Methods, 266(1-2), 209-220.

43. Kramer, K., & Hock, B. (2003). Recombinant antibodies for environmental analysis. Anal Bioanal Chem, 377(3), 417-426.

44. Lauer, B., Ottleben, I., Jacobsen, H. J., & Reinard, T. (2005). Production of a single-chain variable fragment antibody against fumonisin B1. J Agric Food Chem, 53(4), 899-904.

45. Li, T., Zhang, Q., Liu, Y., Chen, D., Hu, B., Blake, D.A., & Liu, F. (2006) Production of recombinant antibodies against methamidophos from a phage-display library of a hyperimmunized mouse. J Agric Food Chem, 54(24), 9085-9091.

46. Li, Y., Cockburn, W., Kilpatrick, J. B., & Whitelam, G. C. (2000). High affinity ScFvs from a single rabbit immunized with multiple haptens. Biochem Biophys Res Commun, 268(2), 398-404.

47. Lofblom, J., Feldwisch, J., Tolmachev, V., Carlsson, J., Stahl, S., & Frejd, F. Y. (2010). Affibody molecules: engineered proteins for therapeutic, diagnostic and biotechnological applications. FEBS Lett, 584(12), 2670-2680.

48. Lou, J., Geren, I., Garcia-Rodriguez, C., Forsyth, C. M., Wen, W., Knopp, K., Brown, J., Smith, T., Smith, L. A., & Marks, J. D. (2010). Affinity maturation of human botulinum neurotoxin antibodies by light chain shuffling via yeast mating. Protein Eng Des Sel, 23(4), 311-319.

49. Makvandi-Nejad, S., Fjallman, T., Arbabi-Ghahroudi, M., Mackenzie, C. R., & Hall, J. C. (2011). Selection and expression of recombinant single domain antibodies from a hyper-immunized library against the hapten azoxystrobin. J Immunol Methods. in press.

50. Marks, J. D. (2004). Antibody affinity maturation by chain shuffling. Methods Mol Biol, 248, 327-343.

51. Marks, J. D., & Bradbury, A. (2004). PCR cloning of human immunoglobulin genes. Methods Mol Biol, 248, 117-134.

52. Marks, J. D., Griffiths, A. D., Malmqvist, M., Clackson, T. P., Bye, J. M., & Winter, G. (1992). By-passing immunization: building high affinity human antibodies by chain shuffling. Biotechnology (N Y), 10(7), 779-783.

53. Marks, J. D., Hoogenboom, H. R., Bonnert, T. P., McCafferty, J., Griffiths, A. D., & Winter, G. (1991). By-passing immunization. Human antibodies from V-gene libraries displayed on phage. J Mol Biol, 222(3), 581-597.

54. Martin, C. D., Rojas, G., Mitchell, J. N., Vincent, K. J., Wu, J., McCafferty, J., & Schofield, D. J. (2006). A simple vector system to improve performance and utilisation of recombinant antibodies. BMC Biotechnol, 6, 46.

55. Martineau, P., & Betton, J. M. (1999). In vitro folding and thermodynamic stability of an antibody fragment selected in vivo for high expression levels in Escherichia coli cytoplasm. J Mol Biol, 292(4), 921-929.

56. Melton, S. J. (2010) Ph.D. Thesis, Biomedical Sciences Graduate Program, Tulane University School of Medicine, New Orleans, LA, USA.

57. Moghaddam, A., Borgen, T., Stacy, J., Kausmally, L., Simonsen, B., Marvik, O. J., Brekke, O. H., & Braunagel, M. (2003). Identification of scFv antibody fragments that specifically recognise the heroin metabolite

6-monoacetylmorphine but not morphine. J Immunol Methods, 280(1-2), 139-155.

58. Moghaddam, A., Lobersli, I., Gebhardt, K., Braunagel, M., & Marvik, O. J. (2001). Selection and characterisation of recombinant single-chain antibodies to the hapten Aflatoxin-B1 from naive recombinant antibody libraries. J Immunol Methods, 254(1- 2), 169-181.

59. Ness, J. E., Welch, M., Giver, L., Bueno, M., Cherry, J. R., Borchert, T. V., Stemmer, W. P., & Minshull, J. (1999). DNA shuffling of subgenomic sequences of subtilisin. Nat Biotechnol, 17(9), 893-896.

60. Ohage, E. C., Wirtz, P., Barnikow, J., & Steipe, B. (1999). Intrabody construction and expression. II. A synthetic catalytic Fv fragment. J Mol Biol, 291(5), 1129-1134.

61. Orcutt, K. D., Slusarczyk, A. L., Cieslewicz, M., Ruiz-Yi, B., Bhushan, K. R., Frangioni, J. V., & Wittrup, K. D. (2011). Engineering an antibody with picomolar affinity to DOTA chelates of multiple radionuclides for pretargeted radioimmunotherapy and imaging. Nucl Med Biol, 38(2), 223-233.

62. Persson, H., Lantto, J., & Ohlin, M. (2006). A focused antibody library for improved hapten recognition. J Mol Biol, 357(2), 607-620.

63. Sblattero, D., & Bradbury, A. (2000). Exploiting recombination in single bacteria to make large phage antibody libraries. Nat Biotechnol, 18(1), 75-80.

64. Schoonbroodt, S., Frans, N., DeSouza, M., Eren, R., Priel, S., Brosh, N., Ben-Porath, J., Zauberman, A., Ilan, E., Dagan, S., Cohen, E. H., Hoogenboom, H. R., Ladner, R. C., & Hoet, R. M. (2005). Oligonucleotide-assisted cleavage and ligation: a novel directional DNA cloning technology to capture cDNAs. Application in the construction of a human immune antibody phage-display library. Nucleic Acids Res, 33(9), e81.

65. Shaw, I., O'Reilly, A., Charleton, M., & Kane, M. (2008). Development of a high-affinity antidomoic acid sheep scFv and its use in detection of the toxin in shellfish. Anal Chem, 80(9), 3205-3212.

66. Sheedy, C., & Hall, J. C. (2001). Immunoaffinity purification of chlorimuron-ethyl from soil extracts prior to quantitation by enzyme-linked immunosorbent assay. J Agric Food Chem, 49(3), 1151-1157.

67. Sheedy, C., MacKenzie, C. R., & Hall, J. C. (2007). Isolation and affinity maturation of hapten-specific antibodies. Biotechnol Adv, 25(4), 333-352.

68. Sheedy, C., Yau, K. Y., Hirama, T., MacKenzie, C. R., & Hall, J. C. (2006). Selection , characterization, and CDR shuffling of naive llama single-domain antibodies selected against auxin and their cross-reactivity with auxinic herbicides from four chemical families. J Agric Food Chem, 54(10), 3668-3678.

69. Short, M. K., Jeffrey, P. D., Kwong, R. F., & Margolies, M. N. (1995). Contribution of antibody heavy chain CDR1 to digoxin binding analyzed by random mutagenesis of phage-displayed Fab 26-10. J Biol Chem, 270(48), 28541-28550.

70. Short, M. K., Krykbaev, R. A., Jeffrey, P. D., & Margolies, M. N. (2002). Complementary combining site contact residue mutations of the anti-digoxin Fab 26-10 permit high affinity wild-type binding. J Biol Chem, 277(19), 16365-16370.

71. Sidhu, S. S., Weiss, G. A., & Wells, J. A. (2000). High copy display of large proteins on phage for functional selections. J Mol Biol, 296(2), 487-495.

72. Siegel, R. W., Baugher, W., Rahn, T., Drengler, S., & Tyner, J. (2008). Affinity maturation of tacrolimus antibody for improved immunoassay performance. Clin Chem, 54(6), 1008-1017.

73. Skerra, A. (2007). Alternative non-antibody scaffolds for molecular recognition. Curr Opin Biotechnol, 18(4), 295-304.

74. Spinelli, S., Frenken, L. G., Hermans, P., Verrips, T., Brown, K., Tegoni, M., & Cambillau, C. (2000). Camelid heavy-chain variable domains provide efficient combining sites to haptens. Biochemistry, 39(6), 1217-1222.

75. Stemmer, W. P. (1994). Rapid evolution of a protein in vitro by DNA shuffling. Nature, 370(6488), 389-391.

76. Strachan, G., McElhiney, J., Drever, M. R., McIntosh, F., Lawton, L. A., & Porter, A. J. (2002). Rapid selection of anti-hapten antibodies isolated from synthetic and semisynthetic antibody phage display libraries expressed in Escherichia coli. FEMS Microbiol Lett, 210(2), 257-261.

77. Swain, M. D., Anderson, G. P., Serrano-Gonzalez, J., Liu, J. L., Zabetakis, D., & Goldman, E. R. (2011). Immunodiagnostic reagents using llama single domain antibody-alkaline phosphatase fusion proteins. Anal Biochem, 417(2), 188-194.

78. Tabares-da Rosa, S., Rossotti, M., Carleiza, C., Carrion, F., Pritsch, O., Ahn, K. C., Last, J. A., Hammock, B. D., & Gonzalez-Sapienza, G. (2011). Competitive selection from single domain antibody libraries

allows isolation of high-affinity antihapten antibodies that are not favored in the llama immune response. Anal Chem, 83(18):7213-7220

79.  Valjakka, J., Hemminki, A., Niemi, S., Soderlund, H., Takkinen, K., & Rouvinen, J. (2002). Crystal structure of an in vitro affinity- and specificity-matured anti-testosterone Fab in complex with testosterone. Improved affinity results from small structural changes within the variable domains. J Biol Chem, 277(46), 44021-44027.

80.  van Wyngaardt, W., Malatji, T., Mashau, C., Fehrsen, J., Jordaan, F., Miltiadou, D., & du Plessis, D. H. (2004). A large semi-synthetic single-chain Fv phage display library based on chicken immunoglobulin genes. BMC Biotechnol, 4, 6.

81.  Vaughan, T. J., Williams, A. J., Pritchard, K., Osbourn, J. K., Pope, A. R., Earnshaw, J. C., McCafferty, J., Hodits, R. A., Wilton, J., & Johnson, K. S. (1996). Human antibodies with sub-nanomolar affinities isolated from a large non-immunized phage display library. Nat Biotechnol, 14(3), 309-314.

82.  Velappan, N., Martinez, J. S., Valero, R., Chasteen, L., Ponce, L., Bondu-Hawkins, V., Kelly, C., Pavlik, P., Hjelle, B., & Bradbury, A. R. (2007). Selection and characterization of scFv antibodies against the Sin Nombre hantavirus nucleocapsid protein. J Immunol Methods, 321(1-2), 60-69.

83.  Wang, S. H., Zhang, J. B., Zhang, Z. P., Zhou, Y. F., Yang, R. F., Chen, J., Guo, Y. C., You, F., & Zhang, X. E. (2006). Construction of single chain variable fragment (ScFv) and BiscFv-alkaline phosphatase fusion protein for detection of Bacillus anthracis. Anal Chem,78(4), 997-1004.

84.  Watanabe, H., Nakanishi, T., Umetsu, M., & Kumagai, I. (2008). Human anti-gold antibodies: biofunctionalization of gold nanoparticles and surfaces with anti-gold antibodies. J Biol Chem, 283(51), 36031-36038.

85.  Wesolowski, J., Alzogaray, V., Reyelt, J., Unger, M., Juarez, K., Urrutia, M., Cauerhff, A., Danquah, W., Rissiek, B., Scheuplein, F., Schwarz, N., Adriouch, S., Boyer, O., Seman, M., Licea, A., Serreze, D. V., Goldbaum, F. A., Haag, F., & Koch-Nolte, F. (2009).Single domain antibodies: promising experimental and therapeutic tools in infection and immunity. Med Microbiol Immunol, 198(3), 157-174.

86.  Yau, K. Y., Groves, M. A., Li, S., Sheedy, C., Lee, H., Tanha, J., MacKenzie, C. R., Jermutus, L., & Hall, J. C. (2003). Selection of hapten-specific single-domain antibodies from a non-immunized llama ribosome display library. J Immunol Methods, 281(1-2), 161- 175.

87.  Zahnd, C., Amstutz, P., & Pluckthun, A. (2007). Ribosome display: selecting and evolving proteins in vitro that specifically bind to a target. Nat Methods, 4(3), 269-279.

88.  Zhu, X., Kriegel, A. M., Boustany, C. A., & Blake, D. A. (2011). Single-chain variable fragment (scFv) antibodies optimized for environmental analysis of uranium. Anal Chem, 83(10), 3717-3724.

# Chapter 6

# PARALLEL IN VIVO DNA ASSEMBLY BY RECOMBINATION: EXPERIMENTAL DEMONSTRATION AND THEORETICAL APPROACHES

Zhenyu Shi, Anthony G. Wedd, Sally L. Gras

School of Chemistry, University of Melbourne, Parkville, Victoria, Australia, Bio21 Molecular Science and Biotechnology Institute, Parkville, Victoria, Australia

## ABSTRACT

The development of synthetic biology requires rapid batch construction of large gene networks from combinations of smaller units. Despite the availability of computational predictions for well-characterized enzymes, the optimization of most synthetic biology projects requires combinational constructions and tests. A new building-brick-style parallel DNA assembly framework for simple and flexible batch construction is presented here. It is based on robust recombination steps and allows a variety of DNA assembly techniques to be organized for complex constructions (with or without scars). The assembly of five DNA fragments into a host genome was performed as an experimental demonstration.

## INTRODUCTION

Recombinant DNA technology was the first technique proposed for targeted manipulation of DNA and has had a wide-ranging and lasting impact on many fields [1]. However, its limitations include the appearance of operational restriction sites inside the DNA sequence of interest and the challenges of *in* Genetic engineering of long DNA fragments. As a result, a defined set of rules has evolved for assembly from standardized units (bricks). The BioBrick system was one of the first attempts to standardize assembly using recombinant DNA technology [2] and has been applied in a number of synthetic biology projects [3], [4]. Improvements allowed the issues of junction scars in translation to be avoided (BglBricks; [5]) and permitted ligation of multiple DNA fragments in a single step [6].

Recently, *in vitro* recombination techniques such as sequence and ligation independent cloning (SLIC) have been developed [7] and applied to prepare small fragments as part of the synthesis of bacterial genomes [8], [9]. It is possible to apply SLIC to the building brick strategy, but the relatively long boundaries favour pre-designed seamless assembly as SLIC is sequence independent. The efficiency of the method is similar to that of recombinant DNA technology. Hence the longer boundary is a disadvantage for small standard unit systems. An improved SLIC approach (the Gibson isothermal method) has been commercialized [10], [11]. The In-Fusion [12] and uracil-specific excision reagent (USER) [13] techniques are based on *in vitro* single-strand annealing. They employ relatively short overhangs of 15 to 30 bp, and can be modified into *in vitro* building brick systems [14].

Site-specific recombination methods have been applied widely to *in vitro* cloning. The Invitrogen Gateway system [15]–[18] uses Lambda phage attachment sequence recombination and is one of the most popular approaches as it allows multiple sites to be accessed [19]. The Cre-loxP [20]–[23] and FLP-frt [24]–[26] systems are also very efficient. However, they are less favoured for *in vitro* cloning as they do not have the binary recombination features of the attachment sequence system (such as the attachment sequence recombination: attB + attP = attL + attR) but remain the same after recombination. The Integron system has a similar binary mechanism to that of the Lambda phage attachment system and has been applied to the assembly of multiple genes [27].

All the above strategies are *in vitro* technologies. They rely on both high reaction and high transformation efficiencies. Standardized building brick systems aim to simplify the construction process, both for design and experiments. Since complex designs can be managed via a software platform, the reduction of experimental complexity is an important issue. One possibility is to carry out the assembly *in vivo* to avoid all *in vitro* operations. *In vivo* recombination, such as homologous recombination in *Bacillis subtillis* [28], [29] and in yeast [8], [9], [30]–[32], are more popular for the parallel assembly of large fragments. Although *in vivo* site-specific recombination is more efficient than *in vivo* homologous recombination, it has rarely been applied to DNA assembly except for some iterative systems[33]–[35] and individual integrations [36]. Lambda red-based recombinations have been applied widely in *E. coli* for gene knockout [37]–[40], for the generation of mutations [41] and for the integration of fragments into plasmids or genomes [35], [42]–[45]. The full length prophage protein recE was also reported to be able to assemble multiple fragments *in vivo* [46]. It is apparent that these extremely efficient *in vivo* recombination methods have potential in parallel DNA assembly.

A novel theoretical framework is proposed here in order to both explore the potential of parallel DNA assembly by *in vivo* recombination and to develop a standard engineering interface. It proves to be an efficient system for assembly of operons into small and medium-sized gene systems. High recombination efficiency was demonstrated by the experimental assembly of five genes.

# MATERIALS AND METHODS

## *In Silico* **Simulation Environment**

The program Vexcutor (available free from http://www.synthenome.com/) was applied to simulate and manage all *in silico* steps in the assembly process [47]. Vexcutor is a software platform that simulates molecular cloning and genetic engineering. The designs and assembly steps of all theories presented here were simulated by Vexcutor. The simulation confirms the designs are workable and provides step-by-step detailed information for the assembly procedure.

## **Plasmid and DNA materials**

### *Strains and Plasmids*

Bacterial strains and plasmids used in the experimental demonstration are listed in Tables 1 and 2.

**Table 1:** Bacterial strains

| Strain | Important Phenotypes | Description | Source |
|---|---|---|---|
| E. coli Trans1T1 | | Initial host for assembly. Later found to be problematical due to a mutation in the HK022 attB core sequence that causes low recombination efficiency between its resulting attL and the native HK022 attR. | Transgen |
| E. coli DH5α | recA endA | Host for plasmid construction. | Transgen |
| E. coli BW25142 | ΔuidA4::pir-116 | Host for R6Kγ plasmid construction. | [36] |

**Table 2:** Plasmids

| Plasmid | Description | Source |
|---|---|---|
| pINT-ts | pINT-ts expresses λ phage *int* when induced by heat. | [36] |
| pAH57 | pAH57 expresses λ phage *int* and xis when induced by heat. | [36] |
| pAH63 | pAH63 contains λ phage *attP*. | [36] |
| pAH69 | pAH69 expresses HK022 phage *int* when induced by heat. | [36] |
| pAH70 | pAH70 contains HK022 phage *attP*. | [36] |
| pAH129 | pAH129 expresses Phi80 phage *int* and *xis* when induced by heat. | [36] |
| pAH83 | pAH83 expresses HK022 phage *int* and *xis* when induced by heat. | [36] |
| pCMR | pCMR expresses the HK022-λ chimera *int* and xis when induced by heat. It specifically catalyzes HK022 attL and attR recombination. | This work |
| pCMP | pCMR expresses the HK022-λ chimera *int* and *xis*, as well as the *pir* gene. | This work |
| pAHP | pCMR expresses the HK022 *int*, as well as the *pir* gene. | This work |
| pUnitR | The assembly unit vector, HK022 attR version. | This work |
| pUnitP | The assembly unit vector, HK022 attP version. | This work |
| TARGET | The vector for converting wild-type *E. coli* to the assembly host. | This work |
| pUnitExR | The extraction vector, HK022 attR version. | This work |
| pUnitExP | The extraction vector, HK022 attP version. | This work |
| pUnitRHP | The assembly unit vector HK022 attR with T7 promoter and optimized mRNA structure. | This work |
| pUnitPHP | The assembly unit vector, HK022 attP with T7 promoter and optimized mRNA structure. | This work |
| pUnitRHPluxA | The library vector for luxA, HK022 attR version. | This work |
| pUnitRHPluxB | The library vector for luxB, HK022 attR version. | This work |
| pUnitRHPluxC | The library vector for luxC, HK022 attR version. | This work |
| pUnitRHPluxD | The library vector for luxD, HK022 attR version. | This work |
| pUnitRHPluxE | The library vector for luxE, HK022 attR version. | This work |
| pUnitPHPluxA | The library vector for luxA, HK022 attP version. | This work |
| pUnitPHPluxB | The library vector for luxB, HK022 attP version | This work |
| pUnitPHPluxC | The library vector for luxC, HK022 attP version | This work |
| pUnitPHPluxD | The library vector for luxD, HK022 attP version | This work |
| pUnitPHPluxE | The library vector for luxE, HK022 attP version | This work |
| pUnitExRCD | The extracted vector with assembled luxC and luxD. | This work |
| pUnitExPABE | The extracted vector with assembled luxA, luxB and luxE. | This work |
| pSB406 | The template plasmid for cloning the lux genes. | [51] |

## *Reagents and Bacterial Hosts*

Enzymes and antibiotics used in the experimental demonstration.

## *DNA Sequences for* in Silico *Simulation*

All DNA sequences for simulation were obtained from GenBank and BioBricks.

A schematic diagram of the method tested experimentally using the Single-Selective-Marker Recombination Assembly System (SRAS) for parallel DNA assembly developed here is given in Figure 1. The next section describes the preparation of the experimental tools required for SRAS, followed by a description of methods involved in the experimental demonstration of SRAS.

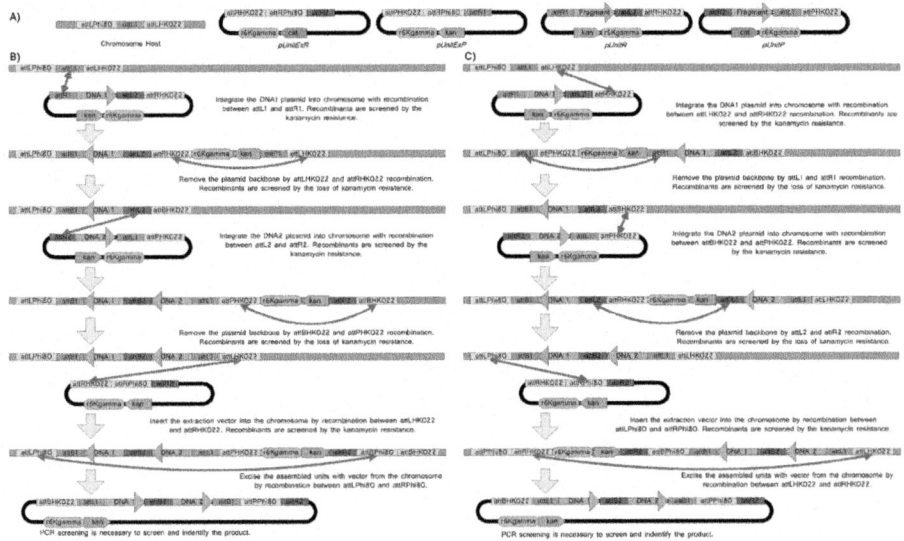

**Figure 1:** Assembly process details of SRAS. The schematic design of the Single-Selective-Marker Recombination Assembly System (SRAS) is shown here (A), consisting of two unit vectors pUnitR, pUnitP, the excision vectors pUnitExR and pUnitExP and the E.coli host chromosome. One method of I/O assembly is shown here for the assembly of two DNA fragments DNA1 and DNA 2 (B): integration occurs via reactive ends first and excision via topology breakers. A second method of I/O assembly is shown here for the assembly of two DNA fragments DNA 1 and DNA2 (C): integration occurs via topology breakers first and excision via reactive ends. The screening procedures and results are identical to B part of this figure. attLPhi80 and attPPhi80 stand for the Phi80 phage attL and attR, respectively. attLHK022, attRHK022, attB-HK022 and attPHK022 stand for the HK022 phage attL, attR, attB and attP respectively. The chloroamphenicol and kanamycin resistance genes are designated cat and kan respectively. The cat gene is not employed in the schematic above but could be used as a second single selective marker in parallel rounds of DNA assembly.

## Preparation of the Experimental Tools for Experimental Demonstration of SRAS

A number of tools are required for the experimental demonstration of SRAS. These include the two unit plasmids (pUnitP and pUnitR), the two extraction plasmids (pUnitExP and pUnitExR), the TARGET plasmid within the host *Escherichia coli* (*E. coli*) and helper plasmids that assist with integration and excision of DNA. These tools are shown in Figure 2. The construction of these tools, integration of the TARGET plasmid and modification of unit plasmids to incorporate the T7 promoter and terminator sequences is described below.

## A) Unit Vectors

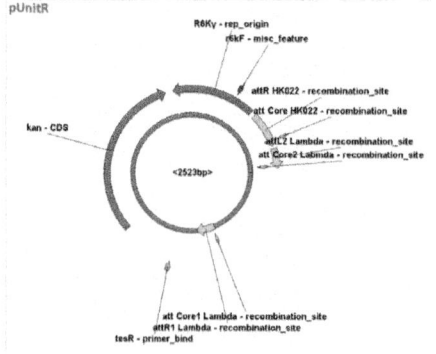

**2 DNA Extraction**
2523bp C;

**3 DNA Extraction**
2620bp C;

## B) Extraction Vectors

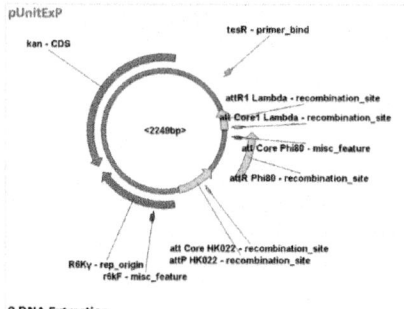

**0 DNA Extraction**
2249bp C;

**1 DNA Extraction**
2152bp C;

## C) Assembly Helper Vectors

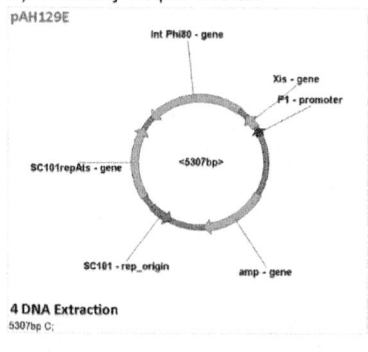

**4 DNA Extraction**
5307bp C;

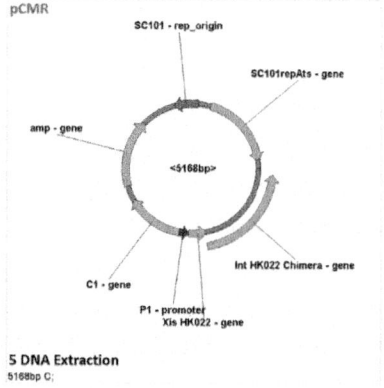

**5 DNA Extraction**
5168bp C;

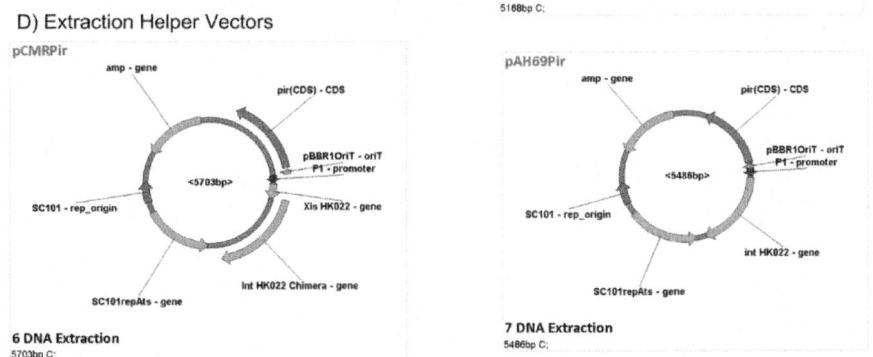

**Figure 2:** The experimental materials constructed for the demonstration of the Single-Selective-Marker Recombination Assembly System (SRAS). (A) The pUnitP and pUnitR unit vectors for constructing fragment libraries. (B) The pUnitExP and pUnitExR extraction vectors. (C) The assembly and helper Vectors that integrate the library vectors pUnitP and pUnitR into the host cell chromosome. (D) The extraction helper vectors. The helper vector contains a *pir* gene that allows the replication of the extracted vectors. (E) The host strain containing the TARGET sequence for SRAS.

## Construction of Unit Plasmids, Extraction Plasmids and TARGET Plasmid

A series of short DNA components including the kanamycin-resistant operon and R6Kgamma from pKD13, the chloramphenicol-resistant operon (consisting of the *cat* gene and its promoter) derived from pKD3, the synthesized lambda attL1, attL2, attR1 and attR2, the HK022 attP and attR cloned from CRIM plasmid pAH70 and its integration product, the phi80 attL and attR cloned from the integration product of the CRIM plasmid pAH153, and a cloning site region from pBHR68 [48] were assembled into the two unit plasmids (pUnitP and pUnitR), the two extraction plasmids (pUnitExP and pUnitExR) and TARGET plasmid by recombinant DNA technology (Figure 3). The detailed procedures for the construction of these plasmids and sequences can be found in the corresponding Vexcutor file (vxt file; Construction of the Unit and Extraction Plasmids.vxt), which shows the individual steps involved in the construction of each of these plasmids and the sequences *in silico*.

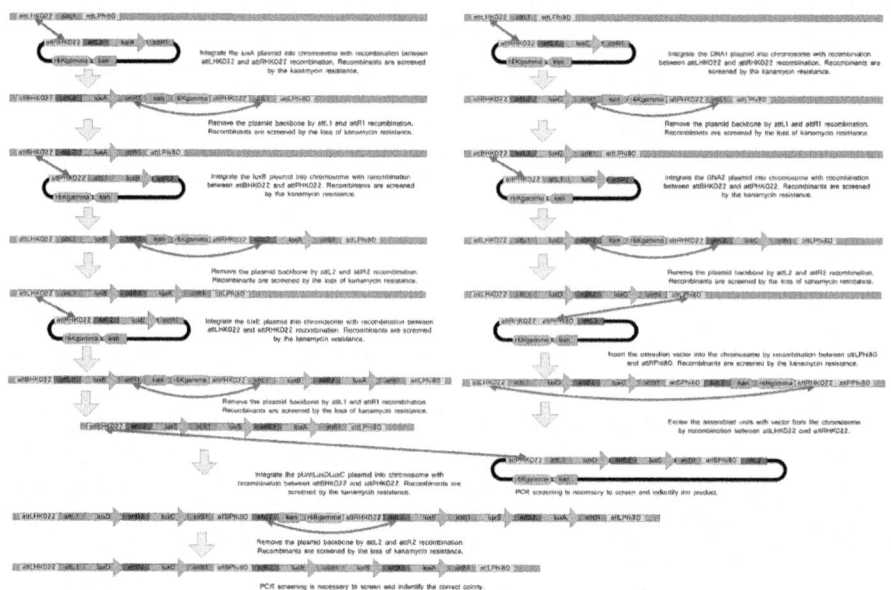

**Figure 3:** SRAS. Schematic of the Single-Selective-Marker Recombination Assembly System (SRAS) shown in Figure 1, as applied to the experimental assembly of Lux genes (A): Lux A, LuxB, LuxC, LuxD and LuxE. The screening procedures and results are identical toFigure 1. attLPhi80 and attPPhi80 stand for the Phi80 phage attL and attR, respectively. attLHK022, attRHK022, attBHK022 and attPHK022 stand for the HK022 phage attL, attR, attB and attP respectively. The chloroamphenicol and kanamycin resistance genes are designated cat and kan respectively.

## Construction of the Helper Plasmid

The helper plasmid, pCMR, for integration was constructed by fusing the first 330 codon of HK022 Int (pAH83) with the codons from the Lambda Int (pAH57), because this hybrid Int gene can specifically catalyze the LxR reaction of HK022 attL and attR [49], [50]. Plasmids pCMRPir and pAH69pir were constructed by inserting the pir gene into pCMR and a CRIM helper plasmid pAH69. pAH129E was constructed by destroying the c1 repressor gene in the CRIM helper plasmid, pAH129. Construction of the Helper Plasmids.vxt, which can be accessed as described above.

## Integration of the TARGET Plasmid

The TARGET plasmid was integrated into the *Escherichia coli* (*E. coli*) TOP10 strain by the HK022 BxP recombination; the plasmid backbone was then replaced with lambda red recombination with pKD13 to act as a template for polymerase chain reaction (PCR); finally the pKD13 kan gene was removed by pCP20 resulting in the HOST strain with the desired chromosomal sequence shown schematically in Figure 1 A. Details of the preparation of the HOST strain and sequences are provided in the second Vexcutor. Construction of Host Strain.vxt.

## Modification of the Punit Vectors

A synthesized T7 promoter and terminator region was inserted into the pUnitP and pUnitR plasmids. Add T7 Promoter and Terminator to the Unit Vectors.vxt. Note that the T7 sequence is omitted from Figure 1A, as it is not a necessary part of the assembly unit.

Together, the four plasmids and host chromosome shown in Figure 1A form the basic experimental tools units required to apply the SRAS.

# EXPERIMENTAL DEMONSTRATION OF THE SRAS

## Construction of the Punit Vectors Punitluxa-E

To demonstrate the application of SRAS, the five genes in the lux Operon (sequence luxC, luxD, luxA, luxB, luxE) in pSB406 [51] were individually subcloned into the pUnitP and pUnitR plasmids to form the five assembly units for DNA assembly named pUnitLuxA, pUnitLuxB, pUnitLuxC, pUnitLuxD and pUnitLuxE respectively. Each gene was cloned by PCR and Kanamycin resistance used to confirm the integration of each gene within the pUnit vector.

## Construction of the Host Strains Hostluxdluxc and Hostluxeluxbluxa

To perform the DNA assembly from these initial five pUnit vectors, the LuxC and luxD genes were first sequentially assembled into the E. coli chromosome generating the strain HOSTLuxDLuxC, as shown to the right of Figure 3. Integration of LuxC involved recombination using the attL1 and attR1 sites, followed by the removal of the pUnit backbone by recombination between attLHK022 and attRK022. Colonies were screened for Kanamycin resistance or loss of Kanamycin resistance in these two steps, as shown in Figure 3. For example, colonies in which LuxC had integrated into the host were selected on media containing Kanamycin. In the next step, following excision of the pUnit backbone, the HOSTLuxC strains were selected by first plating bacteria on media without Kanamycin to form colonies; then 12 colonies were transferred to media with Kanamycin to select those colonies that were Kanamycin sensitive. The original live colonies on Kanamycin free media were then used for subsequent rounds of integration and assembly, which followed the same sequential steps. Lux D was integrated using recombination between attL2 and attR2 and the backbone excised using recombination between attBHK022 and attPHK022, as shown in Figure 3.

The luxA, luxB and luxE genes were next sequentially assembled into the E. coli host chromosome generating strain HOSTLuxELuxBLuxA. These steps, involved the same stages described above and shown in Figure 3.

## Excision of the Luxdluxc and Luxeluxbluxa Sequences

The DNA sequences HOSTLuxDLuxC and HOSTLuxELuxBLuxA were extracted from the host chromosome using recombination between the sites attLPhi80 and attRPhi80, as shown inFigures 1 and 3. These new plasmids were named pUnitLuxDLuxC and pUnitLuxELuxBLuxA respectively. PCR was used to verify the sequence of the DNA between the reactive site ends for each of the five genes. The identity of these sequences was confirmed by DNA sequencing and compared to the sequence of the individual genes cloned in the pUnit vectors pUnitLuxA-E. The sequence of the primers and sequenced region is given in the seventh Vexcutor file Sequencing results of the extracted plasmids.vxt. To view the sequences select any 'Compare DNA box', select view DNA and open the DNA tab. The sequence of each gene is shown in one of the two sub tabs; the region defined by the primers is selected by clicking on the schematic on the right.

## Construction of the Hostluxdluxcluxeluxbluxa Strain

In the final step, the luxDC genes in pUnitLuxDLuxC were integrated into the chromosome of HOSTLuxELuxBLuxA strain, using the recombination between attP HK022 and attB HK022, resulting in the strain HOSTLuxDLuxCLuxELuxBLuxA (Figure 3). The integration within the host chromosome was then confirmed by PCR. Recombination Process.vxt. Follow the prompts given above and open box '103 PCR Verification' to view the sequences. The PCR primers amplify the DNA between attR region at the end of the LuxE gene in the HOSTLuxELuxBLuxA strain and the attL region of the LuxC gene from the integrated pUnitLuxDLuxC. The length of the PCR product was also determined by electrophoresis.

# RESULTS

## Concepts

Effective DNA assembly requires specific and efficient reactions between any two specified ends of given *DNA fragments*. If unrelated sequences are to be excluded from assembly products, each fragment needs two tightly flanking *reactive ends* (recombination sites). A DNA fragment with two reactive ends is defined as an *assembly unit*. The assembly process for linear assembly units is simply the reaction of reactive ends. The internal sequence between the two reactive ends will not influence the assembly product. Two units in linear carriers can be assembled by one reaction between their reactive ends (see Figure 4). However, most of the stable replicable DNA molecules in the widely-used laboratory bacterial host *E. coli* are circular.

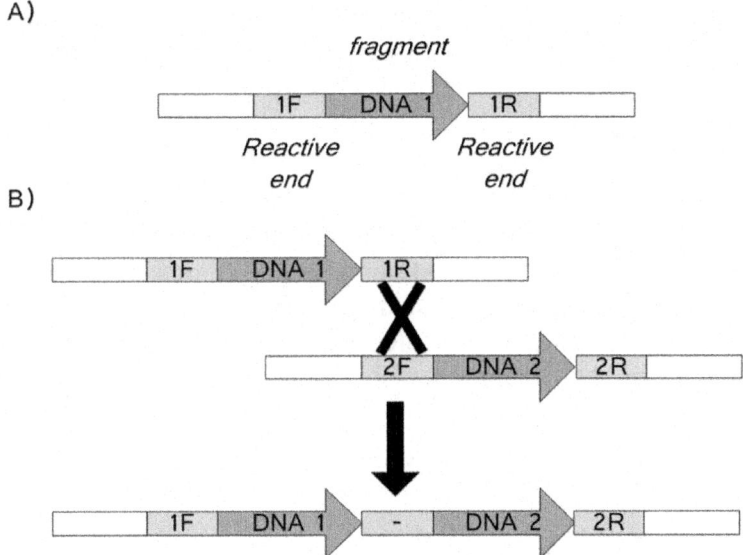

**Figure 4:** Assembly units and linear assembly. 1F, 1R, 2F and 2R are the reactive ends. (A) The conformation of a typical fragment for assembly; (B) A typical assembly reaction between two linear fragments DNA1 and DNA2. The designation '-' represents the scar left after reaction of 1R and 2F.

An assembly unit in a circular carrier (plasmid, phagemid, chromosome) needs an additional recombination site as a *topology breaker* to ensure that combination between circular conformation plasmids is identical to that between the equivalent linear forms (Figure 5). Two linear DNA molecules exchange parts during recombination, while two circular DNA molecules insert and fuse with each other, i.e., circular DNA molecules will keep fusing if there are no topology breakers.

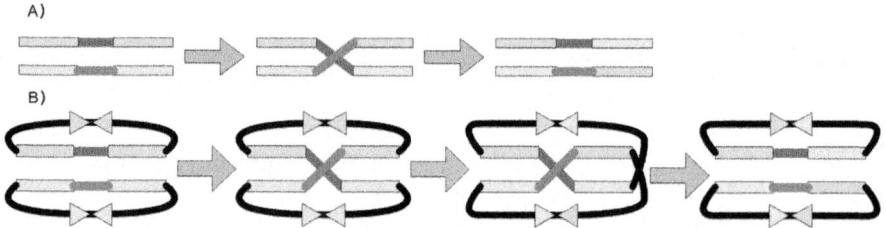

**Figure 5:** The need for topology breakers for circular plasmids. Exchange is shown occurring in the recombination of two linear fragments (shown in red and green) (A). Exchange is also shown for two circular plasmids, which requires an additional recombination (B). Topology breaker sites are shown as triangles.

*Selective markers*, which are usually constitutively expressed antibiotic resistant genes, can simplify the screening process. When the assembled fragments need to be recovered from host vectors or genomes, extra *extraction sites* (which are also recombination sites) may also be required. Therefore, DNA fragments, reactive ends, topology breakers and selective markers are the necessary elements in an assembly; extraction sites are optional. In each assembly cycle, positive clones can be screened by the use of selective markers.

There are three rules for the assembly strategy proposed here:- active ends must become inactive after the assembly reaction; topology breakers must be reactive throughout the assembly process; assembled products must have the same structure as the assembly units to allow another round of assembly to proceed.

Rules 1 and 3 are similar to the concepts developed for iGEM BioBricks [2].

## Theory for a Single-Selective-Marker Recombination Assembly System (SRAS)

Following assembly, the selective marker needs to be removed. The iterative part of this cycle is similar to that of a previous report [34], except that the reactive ends are changed to *attL* and *attR* from the Phi80 or HK022 phage attachment sequence recombination system to satisfy Rule 1. Additional extraction vectors that enable further assembly are also required. The appropriate design of extraction vectors allows the system to obey Rule 2. Briefly, there are two ways to achieve this goal:- an assembly unit can be integrated into the host vector or genome by recombination at the reactive ends; the vector backbone is then removed by a second recombination at the topology breakers; an assembly unit can be integrated into the host vector or genome by recombination at the topology breakers; the vector backbone is then removed by a second recombination at the reactive ends.

The design of the assembly units developed here and schematic diagrams of the assembly steps taken to assemble DNA fragments are shown in Figure 1. There is one chromosomal host, two extraction vectors (pUnitExR and pUnitExP) and two unit vectors (pUnitR and pUnitP). Parts B and C of this figure illustrate two alternative assembly procedures for this system, where in each case two DNA fragments are assembled and then excised. Notably, the chromosomal host works as a *de facto* exclusive single-copy selective marker to ensure that all grown colonies are recombinants. As the whole assembly process features integration and excision plus extraction, it is referred to as an In/Out-Extract strategy.

## An Experimental Demonstration of In/Out-Extract SRAS

The In/Out-Extract SRAS system described in Figure 1 was selected for an experimental demonstration. Before assembly could begin, the individual pUnit vectors (pUnitP and pUnitR), the two extraction plasmids (pUnitExP and pUnitExR) and the HOST strain containing the TARGET sequence were constructed from readily available biological components. These materials, shown in Figure 1 and 2, are the minimum set of components required for the assembly of any DNA sequence from its individual genes via the In/Out-Extract SRAS system. The preparation of these materials and the subsequent assembly steps described below were tracked *in silico* using the Vexcutor program, which was designed to manage complex assembly pathways and sequences in processes such as SRAS.

The five genes in the lux Operon (luxC, luxD, luxA, luxB, luxE) were selected for the experimental demonstration of the In/Out-Extract SRAS system, as they represent a small well characterized set of prokaryotic genes that have been widely applied as a reporter system[52]. Each of the five genes were assembled into a pUnit vector, as shown in Figure 2 A, ready for the assembly process. Each gene is flanked by a T7 terminator and promoter allowing construction of the gene sequence in any order. The arrangement of a promoter with each gene potentially allows for different promoters to be integrated allowing altered levels of gene expression in later developments of this system. The sequence LuxDLuxCLuxELuxBLuxA was chosen for experimental demonstration here to differentiate from the naturally occurring order in the lux operon. As reported recently, realignment of such natural operons can result in higher performance [53].

The luxC and luxD genes were sequentially integrated into the E. coli host chromosome generating the strain HOSTLuxDLuxC, following the scheme shown in Figure 3 where the integration occurs via reactive ends (e.g. attL1 and attR1) and excision between topology breakers (e.g. attRHK022 and attLHK022). Successful colonies were selected at these steps using Kanamycin resistance or sensitivity respectively. The luxE, luxB and luxA genes were assembled sequentially in the same manner; lux A followed by lux B and then lux E, forming a second E. coli host strain HOSTLuxELuxBLuxA (see Figure 3). At this point, the two assembled sequences HOSTLuxDLuxC and HOSTLuxELuxBLuxA were extracted from the host chromosome using the recombination sites attLPhi80 and attRPhi80 generating the plasmids named pUnitLuxDLuxC and pUnitLuxELuxBLuxA respectively.

The identity of the assembled DNA sequences was confirmed by PCR using the reactive site ends flanking each lux gene. The sequencing primers were ATF, ATR, BTF, BTR, CTF, CTR, DTF, DTR, ETF and ETR, corresponding to the reactive end sites attB1, attB2, and attBPhi80 flanking the lux D, C, E, B and A genes. The sequence obtained by PCR was found to match that obtained for each gene in each individual pUnit vector (Figure 2A). For example, Box 24 Compare DNA (see Vexcutor file 7 for raw sequence data) shows the sequence alignment of the LuxE gene in the assembled pUnitLuxELuxBLuxA and the original pUnitRHPLuxE starting vector. The amplified sequence is CTACATCAATAAAACTTAA for both plasmids. This PCR verification step is suggested for all assembly schemes presented here, as this step confirms DNA assembly has occurred and eliminates the possibility of incorporation of unwanted by-products that could complicate subsequent rounds of assembly.

Having confirmed the successful integration and initial assembly steps of the two and three gene sequences, the smaller of the two DNA sequences containing the luxD and luxC genes in pUnitLuxDLuxC was integrated into the chromosome of HOSTLuxELuxBLuxA strain in a single integration step (using recombination between attP HK022 and attB HK022), producing the final host strain named HOSTLuxDLuxCLuxELuxBLuxA (Figure 3). This final assembly was confirmed by PCR of the joining sequence DNA between LuxDLuxC and LuxEluxBluxA in the host genome, using primers for their reactive ends, specifically the aTTR of LuxE and the aTTL of LuxC. The length of the PCR sequence was determined to be 1.6–1.7 Kbp by gel electrophoresis, consistent with the theoretical prediction of 1682 bp.

The PCR product successfully demonstrates the assembly of the five individual gene sequences from the five original pUnit vectors. A total of six integration steps were required to assemble the five DNA sequences using a total of six pUnit vectors, with several steps occurring in parallel. This parallel approach will be a particular advantage where larger gene assemblies are required. Notably, hundreds of colonies appeared following integration at each stage of the process indicating the robust nature of this *E.coli*-based assembly process.

## DISCUSSION

A theoretical framework is proposed for the design of an *in vivo* parallel assembly system. The components required for the In/Out-Extract SRAS approach were developed experimentally and the assembly of a set of five genes from the lux Operon (luxC, luxD, luxA, luxB, luxE) was realized. The only *in vitro* operation required, following the construction of the starting materials containing the individual genes to be assembled, is the extraction of plasmids and confirmation of sequence identity by PCR. The experimental demonstration performed here provides a proof of concept of the In/Out-Extract SRAS system using a small well-characterized operon. Theoretically, the absence of *in vitro* operations permits the assembly of large fragments via this technique with an assembly turnover cycle as short as 1–2 days for 2 fragments ($2^N$ fragments in N cycles or N to 2N days, or $O(\log(N))$) in terms of time complexity for computer science).

An inherent component of synthetic biology is the synthesis of chromosomes and genomes. Methods such as the Gibson isothermal assembly and yeast recombination have been developed for parallel DNA assembly and chromosome fabrication [8]–[11], [54]. In addition, building-brick-based methods are still popular for the construction and optimization of small and medium-sized genetic and metabolic systems [2], [4], [5], [14], [19], [27]. In metabolic engineering projects where most components need to be optimized, traditional recombinant DNA methods and building brick methods are still favoured [5], [55]–[57]. The flexibility of metabolic networks can be probed by elementary mode analysis [58] and the metabolic distribution can be analyzed by flux analysis based on chromatographic and mass spectrometric techniques [59]–[63]. However, the performance of genetic systems must be simulated currently with simplified models [64]–[67]. In addition, prediction of protein function based on known domains, patterns and motifs presently cannot predict the exact activity of an uncharacterized protein [68]–[72]. Consequently, screening of a series of enzymes that have similar activities or screening of series of different alignments and combinations of enzymes must be performed for a sound optimization study.

The advantage of the building brick method for such optimization is that the built-in adapters used for alignment and combination can be used in any order with no additional subcloning. This potentially results in lower costs for both time and labour when larger assemblies are considered. Furthermore, the native DNA fragments can be of any size, so the building brick systems have the second advantage of low sensitivity to unit size. Lastly, a lower dependency on *in vitro* operations constitutes a third advantage.

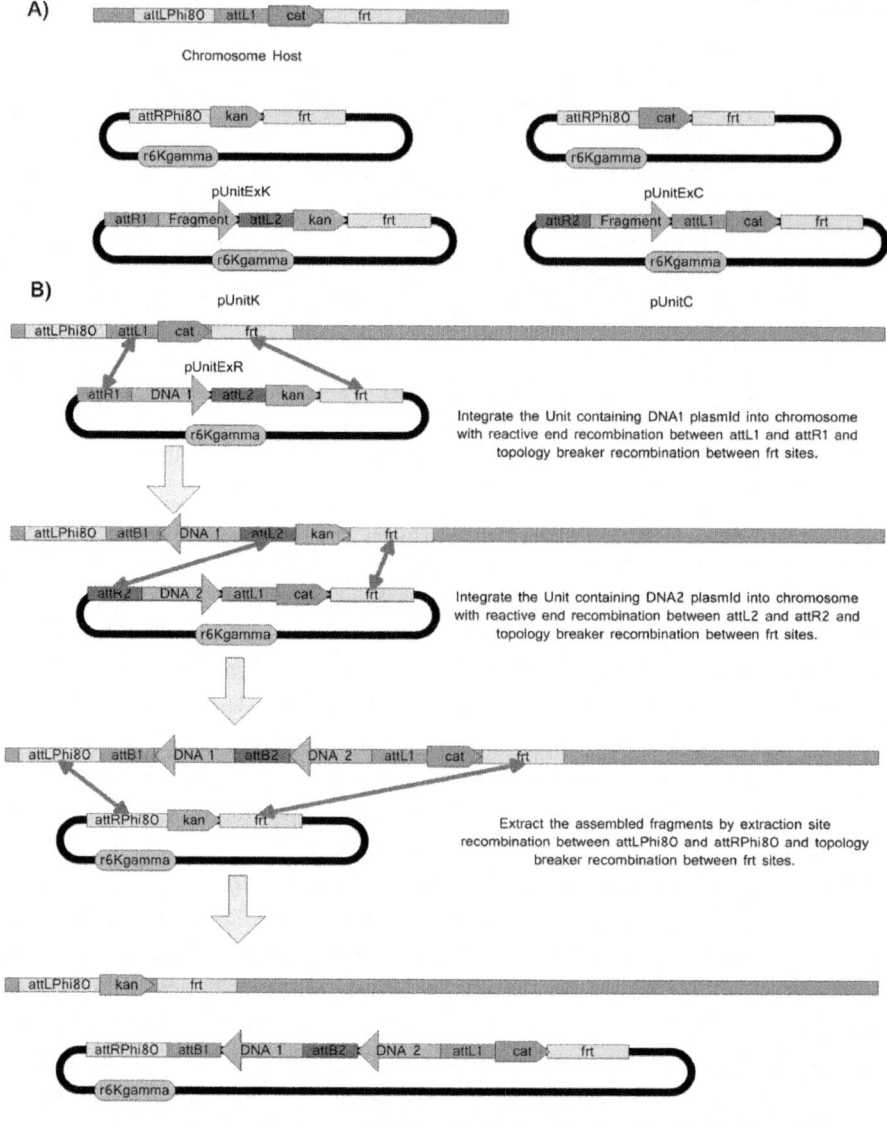

**Figure 6:** DRAS. The design of the DRAS system is shown comprising chromosome host, pUnitExK, pUnitExC, pUnitK and pUnitC (A). The Swap/Ex-

tract assembly process is shown for two fragments DNA1 and DNA2 (B). The chloramphenicol, gentamycin and kanamycin resistance genes are designated cat, gen and kan, respectively. R6Kgamma is the R6Kγ conditional replication site when PI protein is available in the cell. Frt is the frt recombination site. Mutated lambda phage attachment sequences are designated as attL1, attL2, attR1 and attR2.

The SRAS approach demonstrated here could be further extended by the use of a Double-Selective Marker Recombination system (DRAS), a Triple-Selective-Marker Recombination Assembly System (TRAS), a linear plasmid approach to double or triple selective marker systems (linear DRAS and linear TRAS) or a seamless linear Bi-Swap TRAS. These variations on the parallel method of DNA assembly are discussed in turn here.

When the system contains more than one selective marker, the In/Out-Extract strategy can still be implemented with minor modifications. The In/Out steps can be simplified by the swap of the two selective markers. Extraction vectors are also required. A design of this method is shown in Figure 6. It is designated as a Swap-Extract strategy. Briefly, recombination at the reactive ends and topology breaking occur concurrently and successful recombinants are screened out by substitution of the antibiotic resistant unit. The choice of topology breakers will dramatically affect the complexity of the system. Since topology breakers need to remain reactive after recombination (to obey Rule 2), frt from yeast and loxP from the P1 Phage are better options than attachment sites that feature binary recombination. In addition, the chromosomal host here has the same exclusive selective effect as occurs for the SRAS.

When three or more selective markers are available, both the In/Out-Extract and Swap/Extract strategies can be implemented. Moreover, the selective marker swap strategy can be applied between unit plasmids so that integration into the host is not necessary and the extraction step is not required. This is designated a bidirectional swap (Bi-Swap) strategy. A typical design employing the Bi-Swap strategy for the directional assembly of two DNA fragments is shown inFigure 7.

A)

B)

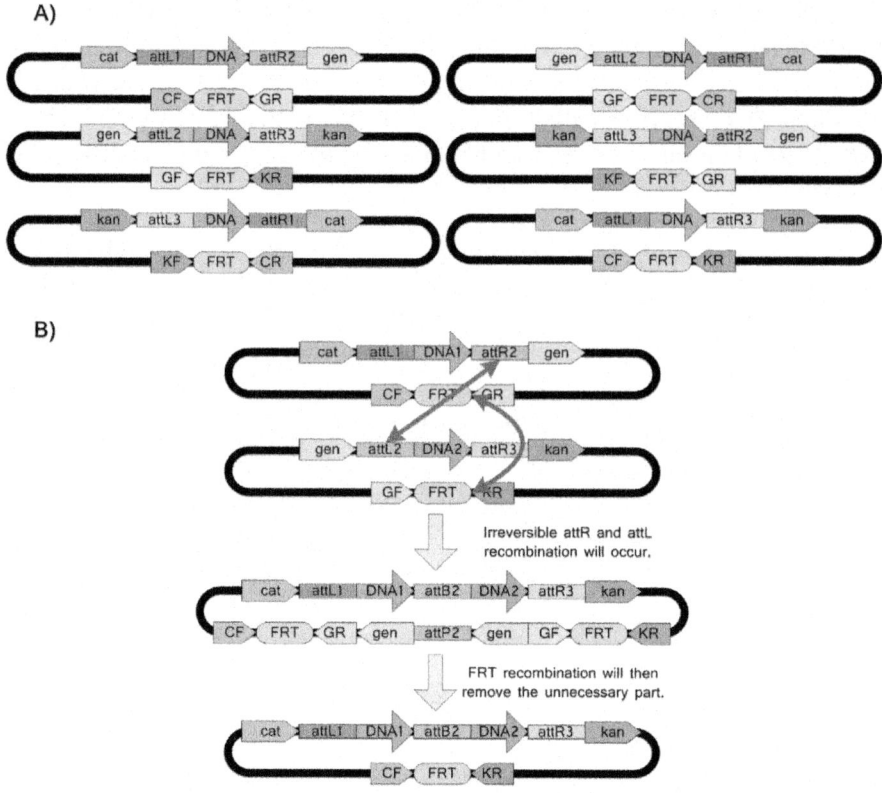

Irreversible attR and attL
recombination will occur.

FRT recombination will then
remove the unnecessary part.

PCR verification by CF and KR is necessary to rule out by-products.

**Figure 7:** Bi-Swap TRAS. The design of Bi-Swap TRAS containing the vectors C-DNA-G, G-DNA-K, K-DNA-C, G-DNA-C, K-DNA-G and C-DNA-K (A), where CF, CR, GR, GF, KR and KF are verification primer sites. The chloramphenicol, genta-mycin and kanamycin resistance genes are designated cat, gen and kan, respective-ly. In order to screen a recombination product with a new combination of antibiotic resistance, at least three antibiotic resistance markers must be used. Since the DNA sequence is directional and two antibiotic resistance markers must be different to per-form screening, the total number of Unit plasmids is 6 (3×2). In this case, an assembly Unit vector has one Unit vector that can swap and extend its left arm or one Unit vector that can swap and extend its right arm. For example, if C-DNA1-G is the beginning unit, K-DNA2-C can be used to extend the left arm of C-DNA1-G and G-DNA2-K can be used to extend the right arm of C-DNA1-G. The assembly process of the fragments DNA1 and DNA2 (B). G-DNA2-K is used to swap and extend the right arm of C-DNA1-G. Because this is a circular plasmid system, fusion intermediates could form after recombination (but will contain all three antibiotic resistant markers). Therefore, PCR and counter-selection are considered necessary to indentify the correct products.

In *E. coli* and many other prokaryotes, there are also linear plasmids such as the N15 phage plasmids [73]. In the case of linear plasmids or genomes, the linear ends are ideal topology breakers because no recombination at the topology breakers is required (although telomerase can be considered as a kind of recombinase). Therefore, it is possible to use a single recombination strategy to finish the whole assembly process. Figure 8 shows a possible design for a linear Bi-Swap strategy where five DNA fragments are assembled in order. Two more advanced Bi-Swap BioBrick. These approaches are an extension on those presented in Figure 7 and 8 that employ recombination sites as the reactive ends that drive assembly.

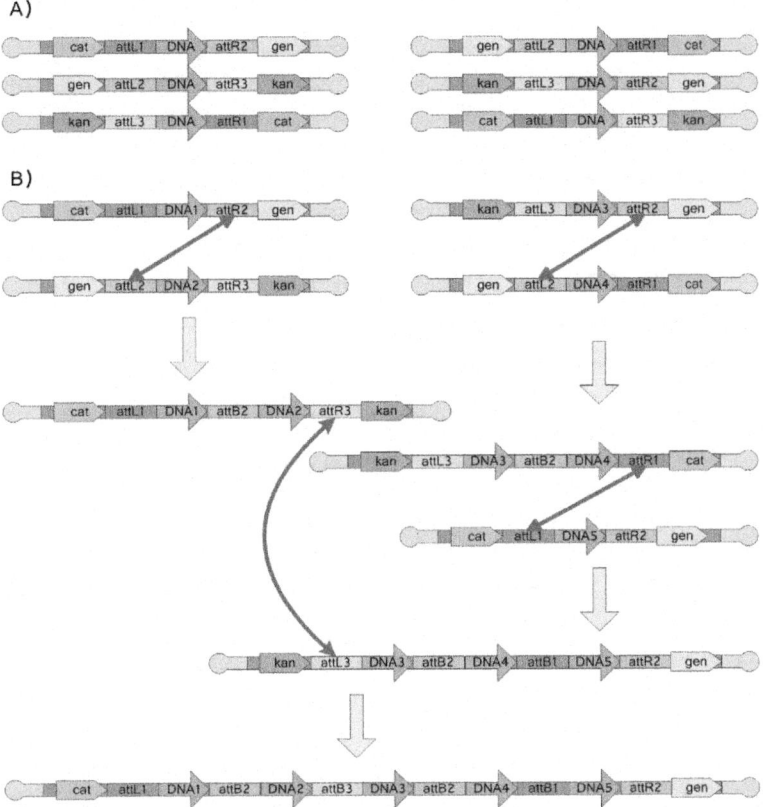

**Figure 8:** Linear Bi-Swap TRAS. The schematic design of linear Bi-Swap TRAS containing the unit vectors C-DNA-G, G-DNA-K, K-DNA-C, G-DNA-C, K-DNA-G and C-DNA-K (A). The yellow round hairpin ends represent N15 plasmids. The chloramphenicol, gentamycin and kanamycin resistance genes are designated cat, gen and kan, respectively. The assembly process for five DNA fragments DNA1, DNA2, DNA3, DNA4 and DNA5 (B). Because this is a linear plasmid system, theoretically no fusion intermediate could form after recombination. PCR and counter-selection may be nec-

essary to indentify the correct products to avoid the false positive case where two unit plasmids exist in the same cell.

A final variation for TRAS, is a seamless model based on the MAGIC [74], Landing Pad [42] or recE [46] recombination systems. An example of a seamless linear Bi-Swap TRAS is shown for two DNA fragments in Figure 9, where the homing endonucleases expressed by the host strain are used to make site-specific double-strand breaks in the DNA. An alternative strategy is to express the required homing endonucleases on the assembly unit plasmids as each plasmid will be cut by the product of the other plasmid when both plasmids are present in the cell.

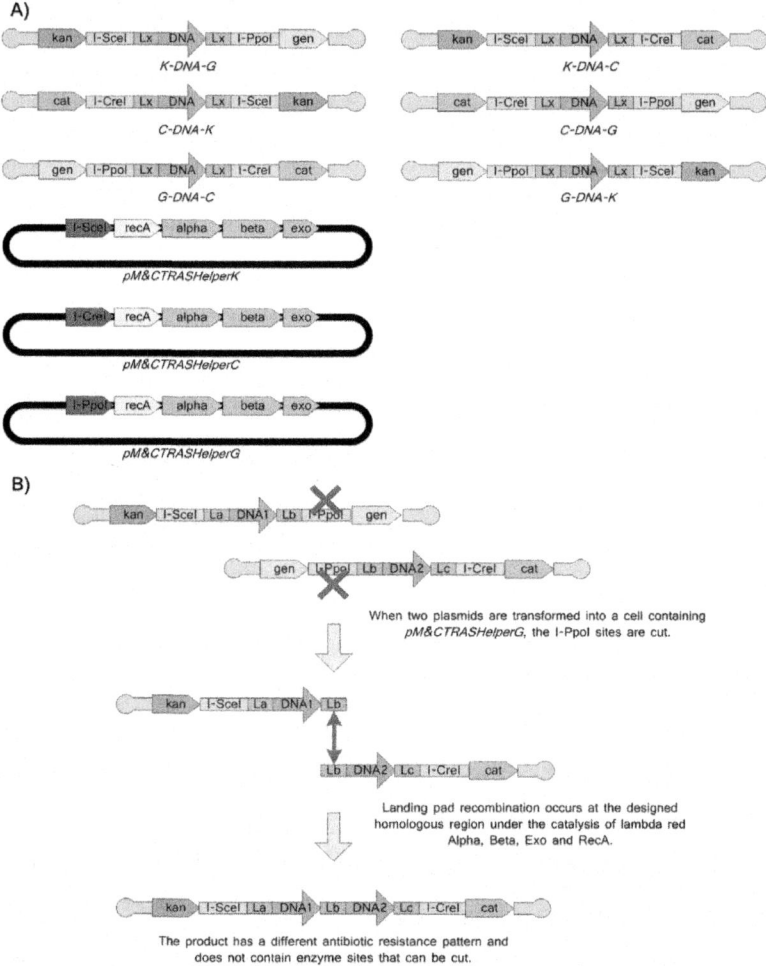

**Figure 9:** Seamless assembly system M&C (meet-and-cut) Linear-TRAS. A Design of a seamless assembly system M&C (meet-and-cut) Linear-TRAS containing the unit

vectors of *C-DNA-G*, *G-DNA-K*, *K-DNA-C*, *G-DNA-C*, *K-DNA-G* and *C-DNA-K*(A), where the chloramphenicol, gentamycin and kanamycin resistance genes are designated cat, gen and kan, respectively. This system also includes the helper plasmids *pM&CTRASHelperK*, *pM&CTRASHelperC* and *pM&CTRASHelperG* for removing the kan, cat and gen markers respectively. I-CreI, I-SceI and I-PpoI in yellow rectangles indicate the homing endonuclease sites. Lx, La, Lb and Lc stand for the landing pad. I-CreI, I-SceI and I-PpoI in red triangle-ended rectangles indicate the genes encoding for the homing endonucleases. The alpha, beta and exo stand for the lambda red phage alpha, beta and exo genes. The recA stands for the E. coli recA gene. The assembly of two fragments *K-DNA1-G* and *G-DNA2-C* under the catalysis of *pM&CTRASHelperG* (B). Similar to the system described in Figure 5, PCR and counter-selection may be necessary for indentifying the correct products to avoid the false positive case where two unit plasmids exist in the same cell.

A further modification could be the addition of a bacterial conjugation component [75] to the TRAS vectors, allowing the assembly process to be performed by mixing two bacterial strains to facilitate conjugation. A third helper strain could be incorporated in this modification to provide this conjugation ability.

*In vivo* recombination based building brick methods presented here have the potential to be logistically simpler than other *in vitro* methods where the assembly of large numbers of fragments is required, due to the assembly of $2^N$ fragments in N cycles and shorter turnover time, as they require only plasmid transformation steps and fewer *in vitro* manipulations. The number of steps and theoretical time taken to assemble 32 fragments with different methods including the SRAS method, the Bi-Swap TRAS method (with and without conjugation), the Gibson isothermal assembly method and Recombinant DNA (or BioBrick) method is presented in Table 3. The SRAS method does not present any immediate advantages over the established methods in terms of time and number of steps, due to the use of time consuming counter selection and separate integration and excision steps. These shortcomings are overcome in both modified TRAS approaches. These differ in their need for *in vitro* steps; the Bi-Swap TRAS with conjugation is performed *in vivo*, whereas the Bi-Swap TRAS requires plasmid purification. Both TRAS approaches are quicker than the Gibson Isothermal Assembly and BioBrick methods, which rely on *in vitro* operations. The Gibson isothermal assembly method also assembles 4 fragments in each turnover cycle, while the SRAS, TRAS and Recombinant DNA method assemble 2.

**Table 3:** Number of operation steps of different strategies for assembling 32 fragments

| Operations | SRAS | Bi-Swap TRAS with Conjugation | Bi-Swap TRAS | Gibson Isothermal Assembly | Recombinant DNA (BioBrick) |
|---|---|---|---|---|---|
| Incubation (liquid) | 185 | 63 | 63 | 41 | 63 |
| Plasmid Extraction | 48 | 0 | 63 | 41 | 63 |
| Digestion | 0 | 0 | 0 | 41 | 63 |
| DNA Extraction | 0 | 0 | 0 | 41 | 63 |
| Ligation | 0 | 0 | 0 | 0 | 31 |
| Gibson Assembly | 0 | 0 | 0 | 21 | 0 |
| Conjugation | 0 | 31 | 0 | 0 | 0 |
| Electroporation | 170 | 0 | 31 | 21 | 31 |
| PCR Verification | 77 | 31 | 31 | 21 | 31 |
| Incubation (plate) | 294 | 31 | 31 | 21 | 31 |
| Number of in vivo manipulations | 295 | 0 | 94 | 165 | 251 |
| Estimated Period | 24 days | 5 days | 5 days | 8 days | 10 days |

doi:10.1371/journal.pone.0056854.t003

Previous studies have shown that that the rate of *in vivo* recombination is near 100% for the integration of a single copy of CRIM plasmid into the *E. coli* genome [36]. This integration consists of two molecular steps: the transformation and the *in vivo* integration. In our study, we also found that the number of colonies transformed with the six Lux pUnit vectors for *in vivo* integration in the SRAS assembly was similar to that observed for the transformation of pUnit into the *E. coli* S17-1 *pir* host in which the *pUnit* plasmids can replicate, i.e., the *in vivo* integration step did not reduce cell viability, implying that the *in vivo* integration efficiency is near to 100%. In the theoretical case of Swap or Bi-Swap approaches (Figures 6 and 8), the by-products caused by the topology breaker sites (frt or loxP) can also be ruled out by counter screening for lost antibiotic resistance. Thus each of the proposed methods is expected to exhibit a high recombination efficiency – consistent with the experimental observations made here. PCR screening of the assembled genes did not reveal any artifacts from the SRAS method.

In contrast, most *in vitro* operations are less than 100% efficient as circular by-products can form, as can unpredictable mismatched products [76]. In addition, *in vitro* operations usually require demanding conditions to obtain optimum efficiency. It is apparent that the proposed *in vivo* assembly strategies have the potential of significantly increased efficiencies.

One advantage of the Bi-Swap strategy (Figures 6 and 7) is that this approach allows additional units to be added to any unit vector and vectors can be joined to any other unit. This Bi-swap model could also be useful for screening by adding antibiotic-resistant transcriptional units between the EcoRI and XbaI restriction enzyme sites and between the SpeI and PstI restriction enzyme sites. In addition, the topology breaker concepts from the new DNA assembly theory can also be applied to *in vitro* methods.

The SRAS and similar methods described here are expected to leave attachment *attB* scars with the sequence ACAAGTTTGTACAAAAAAGCAGGCT for attB1 Lambda, ACCACTTTGTACAAGAAAGCTGGGT for attB2 Lambda and AACCTTTTTCACCTAAAGTGCACC for attB HK022 between the assembled genes at a frequency of one scar per round of integration, as shown in Figures 1 and 2. These scars might prove to be an issue for a system containing many *attB* copies following the assembly of large numbers of DNA fragments because of the potential for homologous recombination. A similar issue could also occur for highly similar DNA fragments. However, due to the high recombination efficiency, the Biswap synthesis outlined here is ideal for constructing small genetic systems containing less than 20 genes. For larger systems where homologous recombination between attB scars may be a concern, the Landing pad Bi-Swap TRAS system.

In addition, in the case where different domains of one protein are assembled into a single reading frame or two or more genes assembled into one operon, attB scars would likely affect the open-reading frame and may affect the operon structure. The strength of the SRAS and similar methods therefore lies in making combinations and alignments of genes and the Landing Pad Bi-Swap method would be a better choice in cases where sequences have multiple domains. The approaches presented here could all prove useful, however, for difficult to combine sequences.

While an assessment of mRNA and protein expression levels of the five assembled genes from the lux Operon was beyond the scope of the present study, it is anticipated that the availability of methods for rapid DNA assembly, such as those presented here, will stimulate the study of the complementary areas of gene expression and metabolic regulation. An understanding of these fields will be essential to testing the function of the assembled DNA and integrating new pathways, such as metabolic networks or artificial chromosomes into the desired host cell.

In summary, this study has proposed and demonstrated a new framework for the design of DNA building bricks and their *in vivo* or *in vitro* assembly. The proposed system, when optimized as described in the modified TRAS and other approaches described, has the potential to reduce the labor and time associated with DNA assembly.

# REFERENCES

1.    Lobban PE, Kaiser AD (1973) Enzymatic end-to end joining of DNA molecules. J Mol Biol 78: 453–471. doi: 10.1016/0022-2836(73)90468-3

2.    Knight TJ (2003) Idempotent Vector Design for Standard Assembly of Biobricks. DSpace.

3.    Norville JE, Derda R, Gupta S, Drinkwater KA, Belcher AM, et al. (2010) Introduction of customized inserts for s-treamlined assembly and optimization of BioBrick synthetic genetic circuits. J Biol Eng 4: 17. doi: 10.1186/1754-1611-4-17

4.    Shetty RP, Endy D, Knight TF Jr (2008) Engineering BioBrick vectors from BioBrick parts. J Biol Eng 2: 5. doi: 10.1186/1754-1611-2-5

5.    Anderson JC, Dueber JE, Leguia M, Wu GC, Goler JA, et al. (2010) BglBricks: A flexible standard for biological part assembly. J Biol Eng 4: 1. doi: 10.1186/1754-1611-4-1

6.    .Shetty R, Lizarazo M, Rettberg R, Knight TF (2011) Assembly of BioBrick standard biological parts using three antibiotic assembly. Methods Enzymol 498: 311–326. doi: 10.1016/b978-0-12-385120-8.00013-9

7.    .Li MZ, Elledge SJ (2007) Harnessing homologous recombination in vitro to generate recombinant DNA via SLIC. Nat Methods 4: 251–256. doi: 10.1038/nmeth1010

8.    Gibson DG, Benders GA, Andrews-Pfannkoch C, Denisova EA, Baden-Tillson H, et al. (2008) Complete chemical synthesis, assembly, and cloning of a Mycoplasma genitalium genome. Science 319: 1215–1220. doi: 10.1126/science.1151721

9.    Gibson DG, Benders GA, Axelrod KC, Zaveri J, Algire MA, et al. (2008) One-step assembly in yeast of 25 overlapping DNA fragments to form a complete synthetic Mycoplasma genitalium genome. Proc Natl Acad Sci U S A 105: 20404–20409. doi: 10.1073/pnas.0811011106

10.   Gibson DG, Smith HO, Hutchison CA 3rd, Venter JC, Merryman C (2010) Chemical synthesis of the mouse mitochondrial genome. Nat Methods 7: 901–903. doi: 10.1038/nmeth.1515

11.   .Gibson DG, Young L, Chuang RY, Venter JC, Hutchison CA 3rd, et al (2009) Enzymatic assembly of DNA molecules up to several hundred kilobases. Nat Methods 6: 343–345. doi: 10.1038/nmeth.1318

12.   Clonetech In-Fusion Cloning System. http://bioinfoclontechcom/infusion/.

13.   Bitinaite J, Rubino M, Varma KH, Schildkraut I, Vaisvila R, et al. (2007) USER friendly DNA engineering and cloning method by uracil excision. Nucleic Acids Res 35: 1992–2002. doi: 10.1093/nar/gkm041

14. Sleight SC, Bartley BA, Lieviant JA, Sauro HM (2010) In-Fusion BioBrick assembly and re-engineering. Nucleic Acids Res 38: 2624–2636. doi: 10.1093/nar/gkq179

15. Hartley JL, Brasch MA (1999) Recombinational cloning using engineered recombination sites. United States Patent US5888732.

16. Esposito D, Garvey LA, Chakiath CS (2009) Gateway cloning for protein expression. Methods Mol Biol 498: 31–54. doi: 10.1007/978-1-59745-196-3_3

17. Fu C, Wehr DR, Edwards J, Hauge B (2008) Rapid one-step recombinational cloning. Nucleic Acids Res 36: e54. doi: 10.1093/nar/gkn167

18. .Aslanidis C, de Jong PJ (1990) Ligation-independent cloning of PCR products (LIC-PCR). Nucleic Acids Res 18: 6069–6074. doi: 10.1093/nar/18.20.6069

19. Petersen LK, Stowers RS (2011) A Gateway MultiSite recombination cloning toolkit. PLoS One 6: e24531. doi: 10.1371/journal.pone.0024531

20. Van Duyne GD (2001) A structural view of cre-loxp site-specific recombination. Annu Rev Biophys Biomol Struct 30: 87–104. doi: 10.1146/annurev.biophys.30.1.87

21. .Kuhn R, Torres RM (2002) Cre/loxP recombination system and gene targeting. Methods Mol Biol 180: 175–204. doi: 10.1385/1-59259-178-7:175

22. Chen QJ, Xie M, Ma XX, Dong L, Chen J, et al. (2010) MISSA is a highly efficient in vivo DNA assembly method for plant multiple-gene transformation. Plant Physiol 153: 41–51. doi: 10.1104/pp.109.152249

23. .Ma L, Dong J, Jin Y, Chen M, Shen X, et al. (2011) RMDAP: a versatile, ready-to-use toolbox for multigene genetic transformation. PLoS One 6: e19883. doi: 10.1371/journal.pone.0019883

24. Schweizer HP (2003) Applications of the Saccharomyces cerevisiae Flp-FRT system in bacterial genetics. J Mol Microbiol Biotechnol 5: 67–77. doi: 10.1159/000069976

25. .Pan G, Luetke K, Sadowski PD (1993) Mechanism of cleavage and ligation by FLP recombinase: classification of mutations in FLP protein by in vitro complementation analysis. Mol Cell Biol 13: 3167–3175.

26. .Senecoff JF, Bruckner RC, Cox MM (1985) The FLP recombinase of the yeast 2-micron plasmid: characterization of its recombination site. Proc Natl Acad Sci U S A 82: 7270–7274. doi: 10.1073/pnas.82.21.7270

27.  .Zhang L, Zhao G, Ding X (2011) Tandem assembly of the epothilone biosynthetic gene cluster by in vitro site-specific recombination. Sci Rep 1: 141. doi: 10.1038/srep00141

28.  .Itaya M (1999) Genetic transfer of large DNA inserts to designated loci of the Bacillus subtilis 168 genome. J Bacteriol 181: 1045–1048.

29.  Itaya M, Fujita K, Kuroki A, Tsuge K (2008) Bottom-up genome assembly using the Bacillus subtilis genome vector. Nat Methods 5: 41–43. doi: 10.1038/nmeth1143

30.  Kouprina N, Leem SH, Solomon G, Ly A, Koriabine M, et al. (2003) Segments missing from the draft human genome sequence can be isolated by transformation-associated recombination cloning in yeast. EMBO Rep 4: 257–262. doi: 10.1038/sj.embor.embor766

31.  .Leem SH, Noskov VN, Park JE, Kim SI, Larionov V, et al. (2003) Optimum conditions for selective isolation of genes from complex genomes by transformation-associated recombination cloning. Nucleic Acids Res 31: e29. doi: 10.1093/nar/gng029

32.  .Noskov VN, Kouprina N, Leem SH, Ouspenski I, Barrett JC, et al. (2003) A general cloning system to selectively isolate any eukaryotic or prokaryotic genomic region in yeast. BMC Genomics 4: 16. doi: 10.1186/1471-2164-4-16

33.  Bennett GN (2003) Recombination assembly of large DNA fragments. United States Patent 7267984.

34.  Lin L, Liu YG, Xu X, Li B (2003) Efficient linking and transfer of multiple genes by a multigene assembly and transformation vector system. Proc Natl Acad Sci U S A 100: 5962–5967. doi: 10.1073/pnas.0931425100

35.  .Dafhnis-Calas F, Xu Z, Haines S, Malla SK, Smith MC, et al. (2005) Iterative in vivo assembly of large and complex transgenes by combining the activities of phiC31 integrase and Cre recombinase. Nucleic Acids Res 33: e189. doi: 10.1093/nar/gni192

36.  Haldimann A, Wanner BL (2001) Conditional-replication, integration, excision, and retrieval plasmid-host systems for gene structure-function studies of bacteria. J Bacteriol 183: 6384–6393. doi: 10.1128/jb.183.21.6384-6393.2001

37.  .Datsenko KA, Wanner BL (2000) One-step inactivation of chromosomal genes in Escherichia coli K-12 using PCR products. Proc Natl Acad Sci U S A 97: 6640–6645. doi: 10.1073/pnas.120163297

38.  Lesic B, Rahme LG (2008) Use of the lambda Red recombinase system to rapidly generate mutants in Pseudomonas aeruginosa. BMC Mol Biol 9: 20. doi: 10.1186/1471-2199-9-20

39. .Husseiny MI, Hensel M (2005) Rapid method for the construction of Salmonella enterica Serovar Typhimurium vaccine carrier strains. Infect Immun 73: 1598–1605. doi: 10.1128/iai.73.3.1598-1605.2005

40. Posfai G, Plunkett G 3rd, Feher T, Frisch D, Keil GM, et al (2006) Emergent properties of reduced-genome Escherichia coli. Science 312: 1044–1046. doi: 10.1126/science.1126439

41. .Ellis HM, Yu D, DiTizio T, Court DL (2001) High efficiency mutagenesis, repair, and engineering of chromosomal DNA using single-stranded oligonucleotides. Proc Natl Acad Sci U S A 98: 6742–6746. doi: 10.1073/pnas.121164898

42. Kuhlman TE, Cox EC (2010) Site-specific chromosomal integration of large synthetic constructs. Nucleic Acids Res 38: e92. doi: 10.1093/nar/gkp1193

43. Chiang CJ, Chen PT, Chao YP (2008) Replicon-free and markerless methods for genomic insertion of DNAs in phage attachment sites and controlled expression of chromosomal genes in Escherichia coli. Biotechnol Bioeng 101: 985–995. doi: 10.1002/bit.21976

44. Minaeva NI, Gak ER, Zimenkov DV, Skorokhodova AY, Biryukova IV, et al. (2008) Dual-In/Out strategy for genes integration into bacterial chromosome: a novel approach to step-by-step construction of plasmid-less marker-less recombinant E. coli strains with predesigned genome structure. BMC Biotechnol 8: 63. doi: 10.1186/1472-6750-8-63

45. .Sharan SK, Thomason LC, Kuznetsov SG, Court DL (2009) Recombineering: a homologous recombination-based method of genetic engineering. Nat Protoc 4: 206–223. doi: 10.1038/nprot.2008.227

46. .Fu J, Bian X, Hu S, Wang H, Huang F, et al.. (2012) Full-length RecE enhances linear-linear homologous recombination and facilitates direct cloning for bioprospecting. Nat Biotechnol.

47. Shi Z, Wedd AG, Gras SL (2012) Vexcutor: An All-in-one General Cloning Simulation Software Platform. Submitting freely available from http://www.synthenome.com/.

48. Spiekermann P, Rehm BH, Kalscheuer R, Baumeister D, Steinbuchel A (1999) A sensitive, viable-colony staining method using Nile red for direct screening of bacteria that accumulate polyhydroxyalkanoic acids and other lipid storage compounds. Arch Microbiol 171: 73–80. doi: 10.1007/s002030050681

49. .Dorgai L, Yagil E, Weisberg RA (1995) Identifying determinants of recombination specificity: construction and characterization of mutant

bacteriophage integrases. J Mol Biol 252: 178–188. doi: 10.1006/jmbi.1995.0486

50.   .Yagil E, Dorgai L, Weisberg RA (1995) Identifying determinants of recombination specificity: construction and characterization of chimeric bacteriophage integrases. J Mol Biol 252: 163–177. doi: 10.1006/jmbi.1995.0485

51.   .Winson MK, Swift S, Fish L, Throup JP, Jorgensen F, et al. (1998) Construction and analysis of luxCDABE-based plasmid sensors for investigating N-acyl homoserine lactone-mediated quorum sensing. FEMS Microbiol Lett 163: 185–192. doi: 10.1111/j.1574-6968.1998.tb13044.x

52.   Winson MK, Swift S, Hill PJ, Sims CM, Griesmayr G, et al. (1998) Engineering the luxCDABE genes from Photorhabdus luminescens to provide a bioluminescent reporter for constitutive and promoter probe plasmids and mini-Tn5 constructs. FEMS Microbiol Lett 163: 193–202. doi: 10.1111/j.1574-6968.1998.tb13045.x

53.   Gahan CG (2012) The bacterial lux reporter system: applications in bacterial localisation studies. Curr Gene Ther 12: 12–19. doi: 10.2174/156652312799789244

54.   .Ellis T, Adie T, Baldwin GS (2011) DNA assembly for synthetic biology: from parts to pathways and beyond. Integr Biol (Camb) 3: 109–118. doi: 10.1039/c0ib00070a

55.   Kirby J, Keasling JD (2008) Metabolic engineering of microorganisms for isoprenoid production. Nat Prod Rep 25: 656–661. doi: 10.1039/b802939c

56.   .Steen EJ, Chan R, Prasad N, Myers S, Petzold CJ, et al. (2008) Metabolic engineering of Saccharomyces cerevisiae for the production of n-butanol. Microb Cell Fact 7: 36. doi: 10.1186/1475-2859-7-36

57.   Jang YS, Lee J, Malaviya A, Seung do Y, Cho JH, et al. (2012) Butanol production from renewable biomass: rediscovery of metabolic pathways and metabolic engineering. Biotechnol J 7: 186–198. doi: 10.1002/biot.201100059

58.   Trinh CT, Wlaschin A, Srienc F (2009) Elementary mode analysis: a useful metabolic pathway analysis tool for characterizing cellular metabolism. Appl Microbiol Biotechnol 81: 813–826. doi: 10.1007/s00253-008-1770-1

59.   Noh K, Wiechert W (2011) The benefits of being transient: isotope-based metabolic flux analysis at the short time scale. Appl Microbiol Biotechnol 91: 1247–1265. doi: 10.1007/s00253-011-3390-4

60. Zamboni N (2011) 13C metabolic flux analysis in complex systems. Curr Opin Biotechnol 22: 103–108. doi: 10.1016/j.copbio.2010.08.009

61. Niklas J, Schneider K, Heinzle E (2010) Metabolic flux analysis in eukaryotes. Curr Opin Biotechnol 21: 63–69. doi: 10.1016/j.copbio.2010.01.011

62. Schwender J (2008) Metabolic flux analysis as a tool in metabolic engineering of plants. Curr Opin Biotechnol 19: 131–137. doi: 10.1016/j.copbio.2008.02.006

**63.** .Wiechert W (2002) An introduction to 13C metabolic flux analysis. Genet Eng (N Y) 24: 215–238. doi: 10.1007/978-1-4615-0721-5_10

64. JA (2004) BioJADE: A Design and Simulation Tool for Synthetic Biological Systems. AITR 2004–003.

65. Villalobos A, Ness JE, Gustafsson C, Minshull J, Govindarajan S (2006) Gene Designer: a synthetic biology tool for constructing artificial DNA segments. BMC Bioinformatics 7: 285. doi: 10.1186/1471-2105-7-285

**66.** .Czar MJ, Cai Y, Peccoud J (2009) Writing DNA with GenoCAD. Nucleic Acids Res 37: W40–47. doi: 10.1093/nar/gkp361

**67.** .Mirschel S, Steinmetz K, Rempel M, Ginkel M, Gilles ED (2009) PROMOT: modular modeling for systems biology. Bioinformatics 25: 687–689. doi: 10.1093/bioinformatics/btp029

68. Sael L, Chitale M, Kihara D (2012) Structure- and sequence-based function prediction for non-homologous proteins. J Struct Funct Genomics 13: 111–123. doi: 10.1007/s10969-012-9126-6

69. Li B, Kihara D (2012) Protein docking prediction using predicted protein-protein interface. BMC Bioinformatics 13: 7. doi: 10.1186/1471-2105-13-7

70. Fleishman SJ, Whitehead TA, Strauch EM, Corn JE, Qin S, et al. (2011) Community-wide assessment of protein-interface modeling suggests improvements to design methodology. J Mol Biol 414: 289–302.

71. Trost B, Kusalik A (2011) Computational prediction of eukaryotic phosphorylation sites. Bioinformatics 27: 2927–2935. doi: 10.1093/bioinformatics/btr525

72. Chitale M, Kihara D (2011) Computational protein function prediction: Framework and challenges.

73. Ravin NV (2011) N15: the linear phage-plasmid. Plasmid 65: 102–109. doi: 10.1016/j.plasmid.2010.12.004

74. Li MZ, Elledge SJ (2005) MAGIC, an in vivo genetic method for the rapid construction of recombinant DNA molecules. Nat Genet 37: 311–319. doi: 10.1038/ng1505

75. Griffiths AJ, Miller JH, Suzuki DT, Lewontin RC, Gelbart WM (2000) An Introduction to Genetic Analysis 7th ed.

76. Engler C, Gruetzner R, Kandzia R, Marillonnet S (2009) Golden gate shuffling: a one-pot DNA shuffling method based on type IIs restriction enzymes. PLoS One 4: e5553. doi: 10.1371/journal.pone.0005553

# Chapter 7

## A PHF8 HOMOLOG IN C. ELEGANS PROMOTES DNA REPAIR VIA HOMOLOGOUS RECOMBINATION

Changrim Lee, Seokbong Hong, Min Hye Lee, Hyeon-Sook Koo

Department of Biochemistry, College of Life Science & Biotechnology, Yonsei University, Seoul, Republic of Korea

## ABSTRACT

PHF8 is a JmjC domain-containing histone demethylase, defects in which are associated with X-linked mental retardation. In this study, we examined the roles of two PHF8 homologs, JMJD-1.1 and JMJD-1.2, in the model organism *C. elegans* in response to DNA damage. A deletion mutation in either of the genes led to hypersensitivity to interstrand DNA crosslinks (ICLs), while only mutation of *jmjd-1.1* resulted in hypersensitivity to double-strand DNA breaks (DSBs). In response to ICLs, JMJD-1.1 did not affect the focus formation of FCD-2, a homolog of FANCD2, a key protein in the Fanconi anemia pathway. However, the dynamic behavior of RPA-1 and RAD-51 was affected by the mutation: the accumulations of both proteins at ICLs appeared normal, but their subsequent disappearance was retarded, suggesting that later steps of homologous recombination were defective. Similar changes in the dynamic behavior of RPA-1 and RAD-51 were seen in response to DSBs, supporting a role of JMJD-1.1 in homologous recombination. Such a role was also supported by our finding that the hypersensitivity of *jmjd-1.1* worms to ICLs was rescued by knockdown of *lig-4*, a homolog of Ligase 4 active in nonhomologous end-joining. The hypersensitivity of *jmjd-1.1* worms to ICLs was increased by *rad-54* knockdown, suggesting that JMJD-1.1 acts in parallel with RAD-54 in modulating chromatin structure. Indeed, the level of histone H3 Lys9 tri-methylation, a marker of heterochromatin, was higher in *jmjd-1.1* cells than in wild-type cells. We conclude that the histone demethylase JMJD-1.1 influences homologous recombination either by relaxing heterochromatin structure or by indirectly regulating the expression of multiple genes affecting DNA repair.

## INTRODUCTION

Histones H3 and H4 are methylated on several amino acid residues, and this methylation can proceed up to the triple level at individual amino acids, usually lysines or less frequently arginines. Histone methylation patterns differ greatly between active and inactive genes, suggesting that histone methylation has a great impact on chromatin structure [1]. A well-known example is the histone H3 Lys9 tri-methylation and Lys4 hypomethylation in heterochromatin [2]. DNA damage signaling as well as gene expression, is influenced by histone methylation, as in the case of 53BP1 binding to chromatin containing methylated H3K79 or H4K20 [3,4]. The DNA damage checkpoint protein MDC1 is demethylated in response to double-strand DNA breaks, and this allows its ubiquitination [5]. The methylation of the tumor suppressor p53 promotes its association with 53BP1 leading to transcriptional activation and apoptosis [6].

Histones methylated on lysines are demethylated by two classes of enzymes, LSD and JmjC demethylases, with very different reaction mechanisms [1,7,8]. The LSD (lysine-specific demethylase) family has only two members in mammals and removes methyl groups by oxidizing amines using FAD and oxygen. The other family, JmjC (with Jumonji C domains), has more than twenty members, and requires $Fe^{2+}$ and $\alpha$-ketoglutarate for catalysis. A JmjC demethylase, PHF8, targets histone H3 mono- and di-methyl Lys9 (H3K9me1 and H3K9me2), and its mutation is associated with X-linked mental retardation (XLMR) [9]. In the model organism *C. elegans*, there are two close homologs of PHF8, JMJD-1.1 and JMJD-1.2, encoded by F43G6.6 and F29B9.2 open reading frames, respectively. Interestingly, JMJD-1.2 is expressed in neurons and is involved in the movement of worms, which agrees with the fact that its human homolog PHF8 is associated with XLMR. JMJD-1.2 targets H3K9me2 like PHF8, as well as H3 di-methyl Lys27 (H3K27me2) [9], and induces the expression of target genes by binding to H3K4me3 via a PHD domain [10].

In this work, we investigated whether and how the two PHF8 homologs in *C. elegans*, JMJD-1.1 and JMJD-1.2, affect worm survival in response to DNA damage. JMJD-1.2 demethylates H3K9me2 and H3K27me2, and its closest homolog JMJD-1.1 is likely to have a similar specificity for particular histone residues. It is known that the di- and tri-methylation of H3K9 and H3K27 have been associated with inactive genes and heterochromatin [1]. Therefore, it seemed very likely that the two PHF8 homologs influence responses to DNA damage by modulating chromatin structure. Indeed, we show here that the two proteins, especially JMJD-1.1, influence DNA repair in response to interstrand DNA crosslinks and double-strand DNA breaks.

# RESULTS

## JMJD-1.1 is required for Resistance to Interstrand DNA Crosslinks and Double-strand DNA Breaks in C. Elegans

JMJD-1.1 and JMJD-1.2 proteins are very similar in amino acid sequences to human PHF8, which is a histone demethylase associated with X-linked mental retardation. There is more than 35% identity between PHF8 and *C. elegans* homologs over a 380 amino acid stretch and 55% in the JmjC domain (Fig 1). JMJD-1.1, in particular, has 60% similarity to JMJD-1.2 over the entire length of the polypeptide, suggesting that the two *C. elegans* proteins are probably paralogs. We obtained *jmjd-1.1(tm3980)* and *jmjd-1.2(tm3713)* deletion mutant worms generated by the National Bioresource Project (Japan) and outcrossed them several times with wild-type N2 strain to remove any background mutations. The two mutants have deletions of more than 500 nucleotides containing exons, and are very likely to be null mutants. In the case of *jmjd-1.1(tm3980)*, two successive exons are deleted, and the most likely transcript is the second exon joined to the fifth exon. Actually, the transcript was found to encode a prematurely terminated polypeptide of 135 amino acids due to a frameshift, as verified by reverse transcription followed by a gene-specific polymerase chain reaction. In addition, the level of the transcript in mutant worms amounted to only 20% of that in wild-type N2, as measured by quantitative polymerase chain reaction after reverse transcription. Therefore, the greatly reduced mRNA level and the premature termination eliminating the JmjC domain suggest that the *jmjd-1.1(tm3980)* allele is completely null.

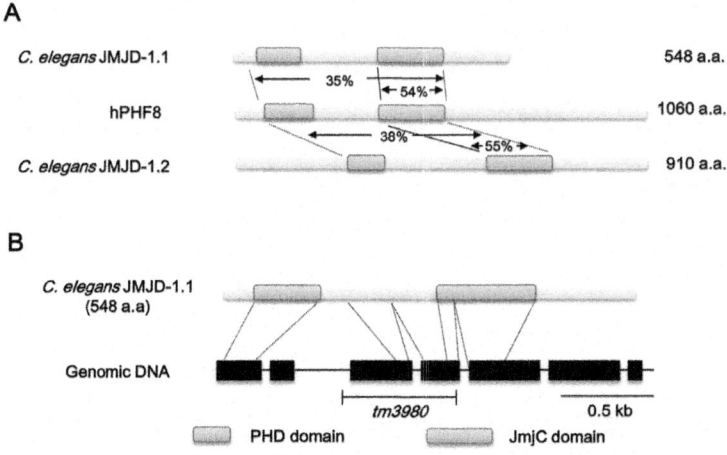

**Figure 1:** Schematic representation of human PHF8 together with its *C. elegans* homologs, and a simplified gene structure of *C. elegans jmjd-1.1*.

(A) Comparison of the amino acid sequences of human histone demethylase PHF8 and its *C. elegans* homologs. The alignment of conserved domains and similarity scores were obtained using NCBI Protein Blast. (B) The structure of *C. elegans jmjd-1.1*(F43G6.6), from exon 1 to 7 (black boxes), and the deletion site in the allele *tm3980*.

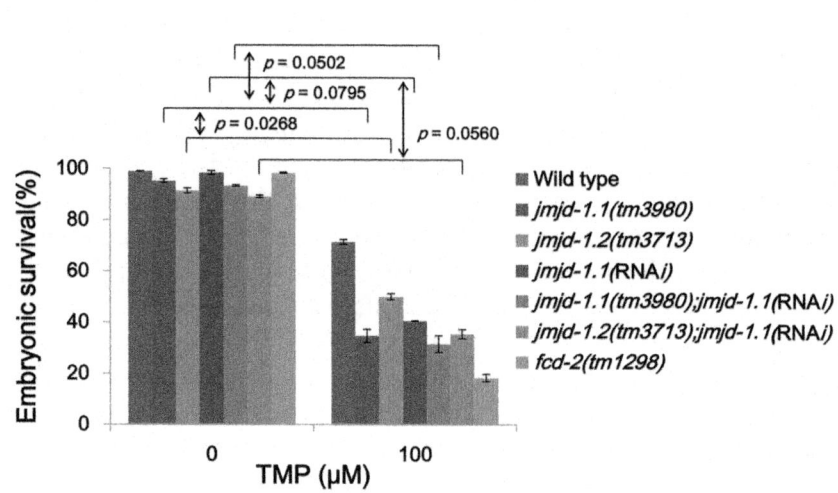

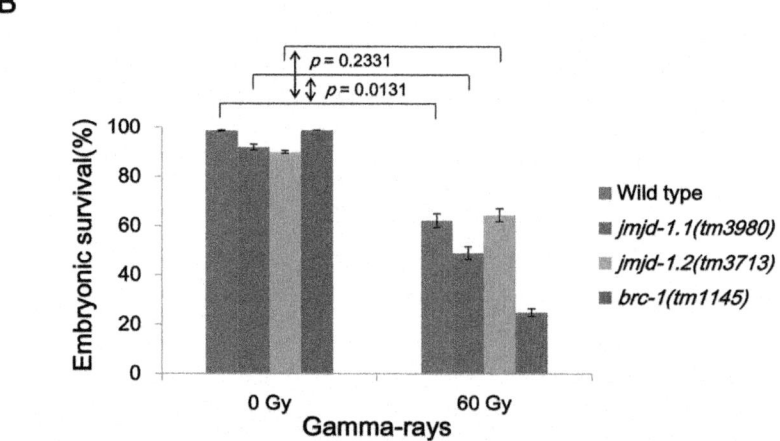

**Figure 2:** *jmjd-1.1* mutant worms are hypersensitive to ICLs and DSBs.

In order to probe the roles of the histone demethylases in DNA damage responses, the *jmjd-1.1(tm3980)* and *jmjd-1.2(tm3713)* mutants were subjected to various types of DNA damage at the L4 larval stage (Fig 2). In *C. elegans* hermaphrodite gonads, proliferating germ cells undergo promeiotic development to produce oocytes and sperms, and after self-fertilization, early-embryos are laid by the worms. Embryos that developed from treated germ cells in the proliferating region of the gonad were scored for hatching to L1 larvae. Besides the two *jmjd-1*mutants, two other mutants were also used as positive controls for different types of DNA damage, along with wild-type N2 as a negative control. FCD-2 (a *C. elegans* FANCD2 homolog) is a key component of the Fanconi anemia (FA) pathway, which participates in the repair of interstrand crosslinks (ICLs) [11]. Therefore, *fcd-2(tm1298)* mutant worms were used as a positive control for hypersensitivity to ICLs. BRC-1 (a *C. elegans* BRCA1 homolog) has an indispensable role in promoting end-resection in the initial stage of homologous recombination (HR) [12], and the mutant *brc-1(tm1145)* was used as a positive control for hypersensitivity to DSBs.

(A) L4 stage worms were collected and incubated with trimethyl psoralen (TMP, 100 μM) for 40 min and exposed to UVA light (150 J/m$^2$). (B) L4 stage worms were exposed to γ-rays at 60 Gy. In every experiment, eggs were collected between 24 and 40 h post treatment, and their survival was scored 24 h later. Mutations for *fcd-2* and *brc-1* (*C.elegans* FANCD2 and BRCA1 homologs, respectively) were used as positive controls for the corresponding type of DNA damage. Error bars indicate SEM. *p* values were obtained by calculating the difference (Δ) in embryonic survival between 0 and 100 TMP for each strain, and comparing the Δ values in the strains by two-way ANOVA.

When *jmjd-1.1(tm3980)* or *jmjd-1.2(tm3713)* worms were exposed to ICLs induced by photo-activated psoralen, both mutants were hypersensitive compared to the wild-type (Fig 2A). The*jmjd-1.1(tm3980)* mutant showed a higher sensitivity to ICLs than *jmjd-1.2(tm3713)* (*p* = 0.0268, two-way ANOVA test), implying a more important role of JMJD-1.1 in repair of ICLs. Since the two mutants were both hypersensitive to ICLs, we tested whether the two corresponding proteins had overlapping functions. For this purpose, we reduced *jmjd-1.1* expression by RNA interference (RNA*i*) in wild-type, *jmjd-1.1*, and *jmjd-1.2* strains. Knockdown of *jmjd-1.1* in the wild-type background decreased embryonic survival after ICL treatment as effectively as the*jmjd-1.1(tm3980)* mutation, indicating almost complete inhibition of JMJD-1.1 expression (*p* = 0.0795, *jmjd-1.1*(RNA*i*) vs. *jmjd-1.1*(*tm3980*)). As anticipated, knockdown of *jmjd-1.1* in *jmjd-1.1(tm3980)* worms did not further decrease resistance to ICL treatment (*p* = 0.0502, *jmjd-1.1(tm3980);jmjd-1.1*(RNA*i*) vs. *jmjd-1.1*(*tm3980*); *p*≥0.05 was treated as not significantly different). This result showed that the knockdown was specific for *jmjd-1.1* and also that the hypersensitivity of *jmjd-1.1(tm3980)* worms to ICLs did not result from cryptic background mutations. The double deficiency strain *jmjd-1.2(tm3713);jmjd-

*1.1(RNAi)* was not significantly different from *jmjd-1.1(RNAi)* in embryonic survival ($p = 0.0560$, *jmjd-1.2(tm3713);jmjd-1.1(RNAi)* vs. *jmjd-1.1(RNAi)*), pointing to some functional overlap between JMJD-1.1 and JMJD-1.2. A double deficiency strain was also generated by crossing the two single mutants. However, unexpectedly the double mutant was less sensitive to ICLs than either of the single mutant, unlike the doubly-deficient strains generated by combining one mutation with knockdown of the other gene ($p = 0.0058$, *jmjd-1.1(tm3980)* vs. *jmjd-1.1(tm3980);jmjd-1.2(tm3713)*; $p = 0.0333$, *jmjd-1.2(tm3713)* vs. *jmjd-1.1(tm3980);jmjd-1.2(tm3713)*). A change that compensates for the absence of the two enzyme activities may have occurred in the double mutant during its maintenance over multiple generations.

Only *jmjd-1.1(tm3980)* worms showed some hypersensitivity to double-strand DNA breaks (DSBs) induced by γ-rays ($p = 0.0131$, wild-type N2 vs. *jmjd-1.1(tm3980)*; $p = 0.2331$, N2 vs.*jmjd-1.2(tm3713)* at 60 Gy) (Fig 2B). In addition, embryonic survival after UVC treatment was not affected in either of the mutants (data not shown). Since *jmjd-1.1* mutant was more sensitive to ICLs than *jmjd-1.2*, and only it was hypersensitive to γ-rays, we focused on the role of JMJD-1.1 in response to DNA damage, especially toward ICLs, in further work.

## JMJD-1.1 is a Downstream Effector of the Fanconi Anemia Pathway during ICL Repair

In mammalian cells, the FA pathway participates in the initial stage of ICL repair, which involves the recognition of ICLs and their conversion to DSBs. FANCD2 is a key player in the FA pathway, shunting the DSB intermediate to the HR pathway [13–15]. The accumulation of nuclear foci of FCD-2, a FANCD2 homolog in *C. elegans*, at ICL sites is a good indicator of successful activation of the FA pathway [16]. We found that the extent of FCD-2 focus formation after ICL treatment in the proliferating mitotic germ cells of *jmjd-1.1(tm3980)* and wild-type worms was not significantly different (Fig 3, $p>0.09$ at all the time points, Student's *t*-test). In addition, the immuno-signal declined gradually to similar extents in the two strains until 24 h. These results indicate that JMJD-1.1 does not affect FCD-2 activation and has a role downstream of the FA pathway in dealing with ICLs.

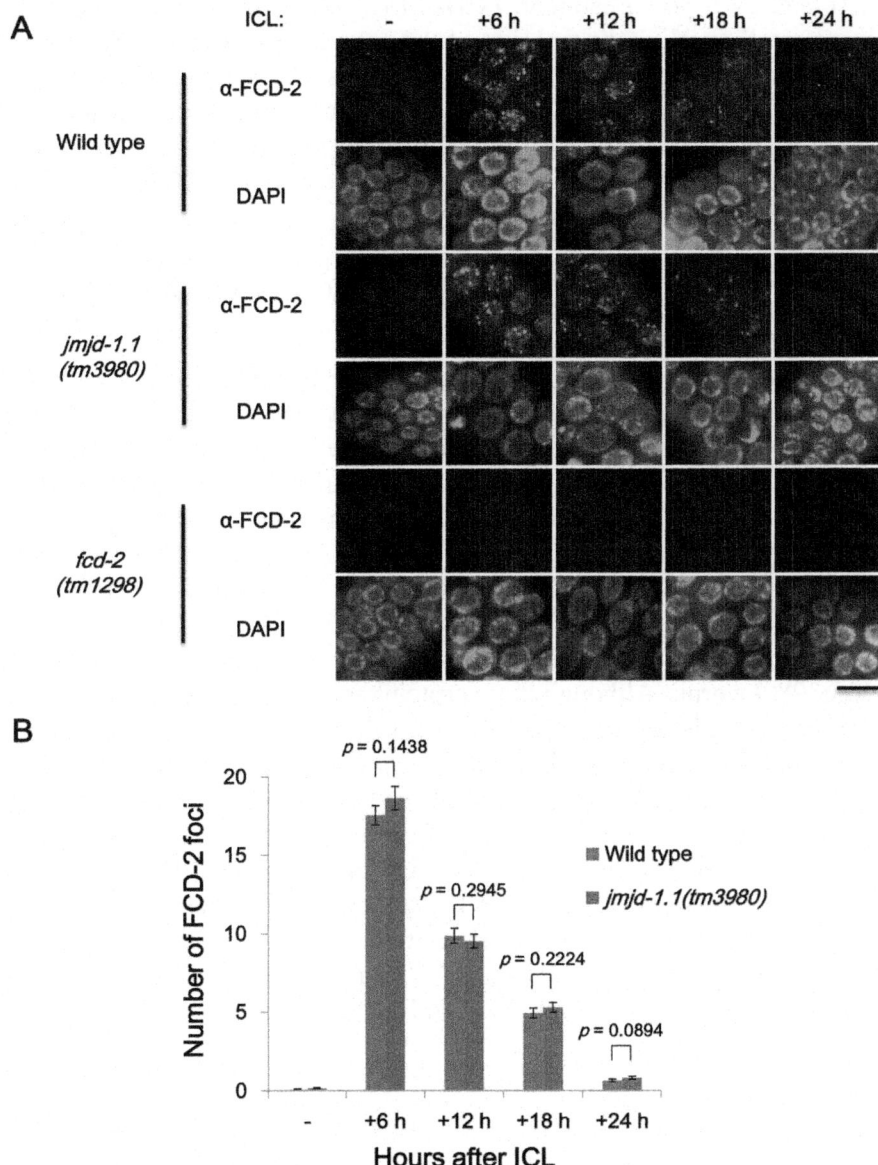

**Figure 3:** FCD-2 focus formation is unaffected upon ICL induction in mitotic germ cells of *jmjd-1.1* worms.

L4 stage worms of wild type, *jmjd-1.1(tm3980)*, and *fcd-2(tm1298)* were exposed to photoactivatable trimethyl psoralen (TMP, 200 μM) for 40 min and then to UVA light (150 J/m$^2$). (A) Gonads were dissected, fixed, and immuno-stained with antibody against FCD-2 (*C. elegans* FANCD2 homolog) at 6, 12, 18 and 24 h post treatment. FCD-2 foci were observed in germ cells in the mitotically proliferating region of the gonad, and were not present in the negative control strain, *fcd-2(tm1298)*. Scale bar, 10 μm. (B) The focal plane with the maximum number of FCD-2 foci was chosen for each nucleus of the germ cells, and the numbers of FCD-2 foci per focal plane were compared in wild-type and *jmjd-1.1* worms at various time points (n = 100 for each bar). *p* values were obtained by Student's *t*-test.

## Prominent role of JMJD-1.1 in Homologous Recombination during ICL and DSB Repairs

ICLs are converted to DSBs in the FA pathway as a result of DNA cleavages by endonucleases such as XPF, MUS81, SLX1, and FAN1 [17,18]. In order for DSBs to be repaired by HR, extensive end resection of DSBs needs to take place; then RPA binds to the exposed single-strand DNA, and RAD51 replaces RPA prior to DNA strand invasion [14]. Mammalian cells defective in HR exhibit prolonged accumulation (or slow dissociation) of RPA and RAD51 proteins at DNA lesions [19,20]. Therefore, we investigated the accumulation and dissociation kinetics of RPA-1 and RAD-51 (*C. elegans* RPA70 and RAD51 homologs, respectively) in mitotic germ cells of wild-type and *jmjd-1.1(tm3980)* worms. Although these proteins accumulated normally at 6 h after ICL treatment (also at 9 h for RAD-51), striking defects in HR were revealed by the subsequent sluggish dissociation of these proteins in the mutant worms (Fig 4): accumulation of RPA-1 and RAD-51 were still observed in the mutant worms at 18 h post treatment, but they disappeared almost completely in wild-type worms. Similar results were observed for the repair of DSBs induced by γ-rays. The formation of RPA-1 and RAD-51 foci in mitotic germ cells was normal in the mutant, but their dissociation was slowed down. In summary, JMJD-1.1 is needed for repair of both ICLs and DSBs via HR, specifically in the step following DSB end resection and subsequent RPA-1/RAD-51 loading.

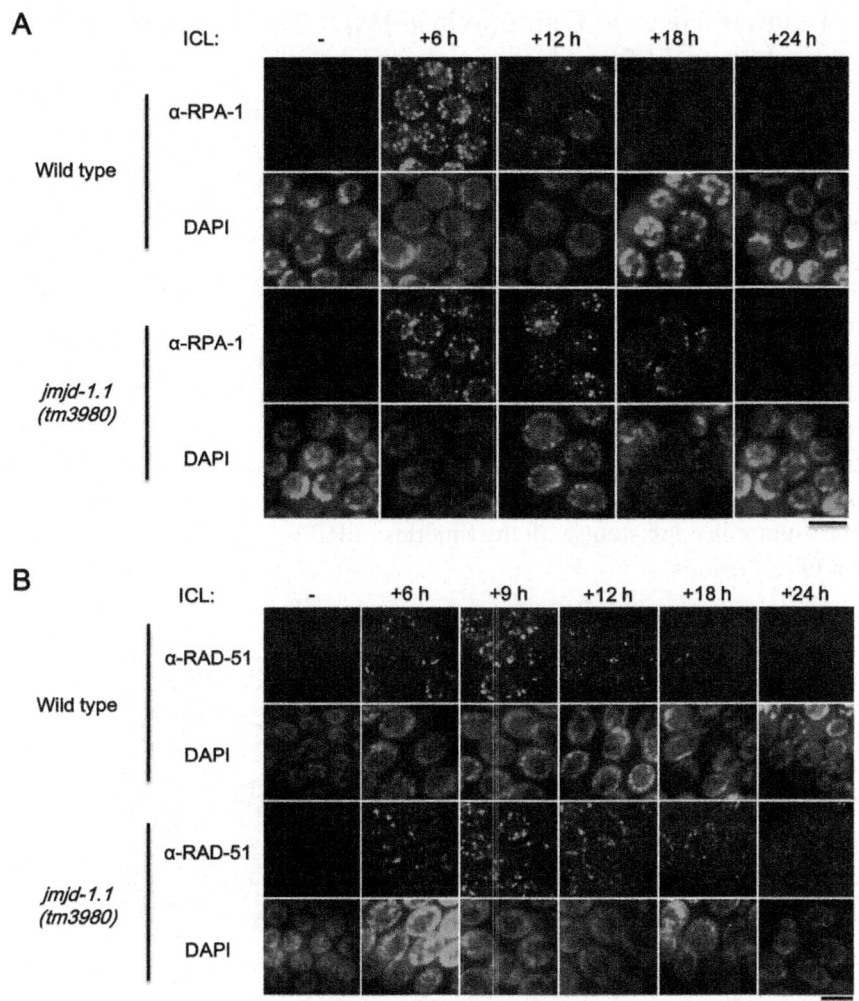

**Figure 4:** Slow dissipation of RPA-1 and RAD-51 foci induced by ICLs in mitotic germ cells of *jmjd-1.1* mutant worms.

Prolonged accumulations of (A) RPA-1 (*C. elegans* RPA70 homolog) and (B) RAD-51 (*C. elegans* RAD51 homolog) foci upon ICL formation in the nuclei of mitotic germ cells of *jmjd-1.1* worms. L4 stage worms were exposed to trimethyl psoralen as in Fig 3, and gonads were immuno-stained with the indicated antibodies at 6, 12, 18 and 24 h post treatment (with an additional time point at 9 h in (B)). Scale bar, 10 μm.

## The Putative Histone Demethylase JMJD-1.1 Modulates Heterochromatin Structure

Spatiotemporal regulation of chromatin structure is critical for the DNA repair machinery to perform its work efficiently [21,22]. We therefore examined the level of histone H3 tri-methyl Lys9 (H3K9me3), a well-established heterochromatin marker, in mitotic germ cells after ICL or DSB formation. The immuno-signal for H3K9me3 became weaker after DNA damage induction, but it recovered gradually with the progression of DNA repair in both wild-type and mutant cells. However, the amount of heterochromatin in the mitotic germ cells of mutant worms was higher than in wild-type worms both before and after the induction of ICLs and DSBs. The differences in H3K9me3 level were also evident in a western analysis of worm extracts (Fig 5B), where the signal for H3K9me3 in the *jmjd-1.1* mutant was about 1.5 fold that in the wild type both before and after ICL treatment. The higher level of heterochromatin in the mutant is consistent with the kinetics of RPA-1 and RAD-51 dissociation from DNA lesions.

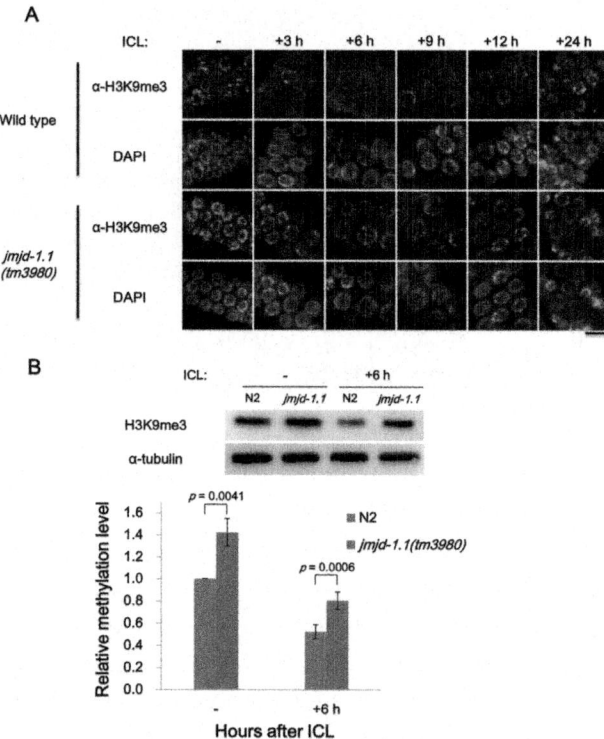

**Figure 5:** Retarded relaxation of heterochromatin structure upon ICL formation in mitotic germ cells of *jmjd-1.1* mutant worms.

L4 stage worms were collected and treated with trimethyl psoralen as in Fig 3. (A) The gonads were immuno-stained with antibody against histone H3K9me3 as an indicator of heterochromatin at 3, 6, 9, 12 and 24 h post treatment. Scale bar, 10 μm. (B) After DNA damage induction, worms were grown for a further 6 h, and extracts were prepared. Proteins were separated on a 12% SDS-polyacrylamide gel and transferred to a nitrocellulose membrane. After probing the membrane for histone H3K9me3 and α-tubulin, band intensities were measured and plotted in the bar graph. Each bar represents an average of 9 measurements, obtained in 3 independent experiments each with 3 technical repetitions of the gel electrophoresis. $p$ values were obtained by Student's $t$-test.

Wild-type worms nearly completed repair of DNA damage within 12 h, as can be seen from the fact that most of RPA-1/RAD-51 had dissociated from the DNA lesions by 12 h post treatment. In contrast, DNA damage in the mutant worms was still under repair for up to 18 h post treatment, as shown by the presence of by non-detached RPA-1/RAD-51. The higher level of heterochromatin, both before and after DNA damage, is likely to be one of the main factors inhibiting DNA repair in the mutant worms. Based on a previous study of the dual demethylase activity of JMJD-1.2 on H3K9me2 and H3K27me2 [9], we carried out immuno-staining to assess any changes in the levels of these two histone signatures in mitotic germ cells. Contrary to our expectation, neither differed between wild-type and *jmjd-1.1* worms, suggesting that JMJD-1.1 targets other signatures such as H3K9me3.

## JMJD-1.1 Affects the Same Step in the HR Pathway as RAD-54

We further addressed the epistatic relationship between JMJD-1.1 and genes of the two main DSB repair pathways, HR and NHEJ (nonhomologous end-joining). For this analysis, we depleted *C. elegans rad-54* or *lig-4* (*C. elegans* RAD54 or LIG4 homologs, respectively). RAD54 and LIG4 participate in HR and NHEJ, respectively, in mammalian cells, and their conservation in *C. elegans* provides excellent ways of turning on and off each of these DSB repair pathways. In several reports, hypersensitivities to DNA damage caused by the absence of key HR factors were reversed by depletion or mutation of crucial components of NHEJ [13,23]. Thus in mammalian systems, phenotypes due to *BRCA1* mutation are rescued by mutation of *53BP1* which inhibits the end-resection of HR and promotes NHEJ [24,25]. Likewise in *C. elegans*, the hypersensitivity of *BRCA1* worms to DSBs was rescued by a *LIG4* mutation, and that of *RAD54* knockdown worms was reversed by *53BP1* or *LIG4* mutations [26]. Similarly, the hypersensitivity of *FANCD2* worms and Fanconi anemia patient cells to ICLs was rescued by inhibiting NHEJ [13].

Therefore, we examined the effect of depleting either *lig-4* or*rad-54* in *jmjd-1.1* mutants. Since *jmjd-1.1(tm3980)* worms are impaired in HR, we expected that the embryonic lethality caused by ICLs might be reversed by depleting *lig-4*. Moreover, if the main role of JMJD-1.1 were in HR, *rad-54* depletion in the mutant background might not have any additional effects on overall embryonic viability. As expected, embryonic lethality due to the *jmjd-1.1* mutation was fully rescued by *lig-4* knockdown, supporting the view that JMJD-1.1 is involved in HR during ICL repair (Fig 6A, $p = 0.0038$, *jmjd-1.1(tm3980);lig-4(RNAi)* vs.*jmjd-1.1(tm3980)*). However, *rad-54* depletion in the *jmjd-1.1* mutant background led to massive lethality: death induced by ICLs was 27±2% in the wild type, 56±5% in *jmjd-1.1(tm3980)*, 38±4% in *rad-54(RNAi)*, and 76±4% in *jmjd-1.1(tm3980);rad-54(RNAi)*. This suggested that JMJD-1.1 plays a role in parallel to RAD-54 in HR (Fig 6B, $p = 0.0177$, *jmjd-1.1(tm3980);rad-54(RNAi)* vs. *jmjd-1.1(tm3980)*; $p = 0.0019$, *jmjd-1.1(tm3980);rad-54(RNAi)* vs. *rad-54(RNAi)*). One possible explanation may be that JMJD-1.1 and RAD-54 act at the same step of HR, but have different effects on chromatin structure. In agreement with this view, we have found that knockdown of *rad-54* or use of a hypomorphic mutation of *rad-54* slows down the disappearance of RAD-51 foci after γ-irradiation (data not shown), as does the *jmjd-1.1*mutation.

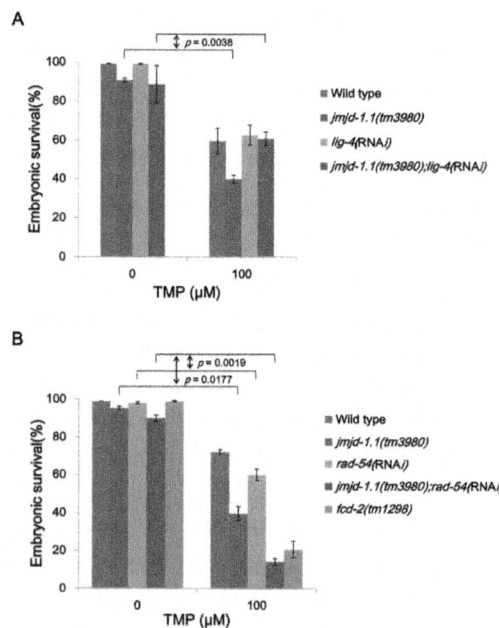

**Figure 6:** Epistasis test demonstrating a role of JMJD-1.1 in HR, in parallel with RAD-54, in response to ICLs.

(A) Wild-type and *jmjd-1.1(tm39809)* worms were fed with *E. coli* cells expressing double-strand RNA for (A) *lig-4* and (B) *rad-54* from the L1 stage. At the L4 stage, worms were subjected to trimethyl psoralen as in Fig 2. Eggs were collected between 24 and 40 h post treatment, and their survival was scored after 24 h. *p* values were obtained by calculating the differences (Δ) in embryonic survival between 0 and 100 TMP for each strain and comparing the Δ values between strains by two-way ANOVA.

## DISCUSSION

A recent report describing the similarities between human histone demethylase PHF8 and its *C.elegans* homolog JMJD-1.2, and the physiological importance of these enzymes led us to examine the roles of the JmjC domain-containing histone demethylases in response to DNA damage [9]. *C. elegans* has two homologs of human PHF8, the JmjC domain-containing proteins JMJD-1.1 and JMJD-1.2. PHF8 and the two *C. elegans* homologs share two conserved domains: the PHD-finger recognizes and selects target substrate specific for histone demethylase, while JmjC is the catalytic domain (Fig 1) [9]. In this study, we have linked the putative histone demethylase JMJD-1.1 to DNA damage responses, and this in turn led to understanding its activity in terms of chromatin structure. *jmjd-1.1(tm3980)* worms exhibit hypersensitivity to ICLs and DSBs (Fig 2), but not to pyrimidine dimers induced by UVC radiation (data not shown). The exceptionally severe effect on embryonic survival after ICL treatment demonstrated that JMJD-1.1 is dedicated more to ICL repair than to DSB repair. However, interestingly, *jmjd-1.2(tm3713)* worms showed a milder degree of hypersensitivity to ICLs than *jmjd-1.1(tm3980)* worms and no hypersensitivity to DSBs (Fig 2).

The *jmjd-1.1* mutation did not affect FCD-2 (FANCD2 homolog) focus formation in response to ICLs, indicating that the initial FA pathway converting ICLs to DSBs was not affected by the mutation. Therefore, the HR pathway acting subsequently to FA pathway was suspected to be defective in the mutant. Indeed, the disappearance of RAD-51 foci not their accumulation was retarded (Fig 4B). The slow disappearance of RAD-51 foci was also observed in *jmjd-1.1* after γ-ray treatment, reflecting defects in HR progression. In addition to RAD-51 foci, RPA-1 foci, which usually form at earlier times than those of RAD-51, also disappeared slowly in the mutant. Thus, JMJD-1.1 appears to function in HR during ICL and DSB repair, at a step after RAD-51 loading onto single-strand DNA. In addition to JMJD-1.1 and JMJD-1.2, a few *C.elegans* histone demethylases have been reported to be important for genome stability previously. A H3K4 demethylase SPR-5 is required for normal meiotic homologous recombination and for resistance to double-strand DNA breaks

[27,28]. JMJD-2 demethylates H3K9 and H3K36, and its depletion leads to increased apoptosis and RAD-51 focus formation in pachytene-stage germ cells [29].

Immunostaining of *C. elegans* germ cells against the heterochromatin signature, histone H3K9me3, strongly supported the notion that the difference in the dissociation kinetics of RPA-1 and RAD-51 between wild-type and *jmjd-1.1* worms derived from different chromatin relaxation levels (Fig 5): the immuno-signal was slightly but significantly more intense in *jmjd-1.1* germ cells both before and after treatment. This observation shows that JMJD-1.1 acts downstream of the extensive DSB end resection step by modulating chromatin structure.

Nevertheless, accompanying changes in the transcriptome, possibly involving transcripts of DNA repair proteins, may have also contributed to the ICL-hypersensitivity of the *jmjd-1.1*mutant.

Because of the retarded relaxation of heterochromatin structure in *jmjd-1.1* cells, our next question was which histone signature is modulated by JMJD-1.1. Since JMJD-1.2 has enzymatic activity on histone H3K9me2 and H3K27me2, we thought that JMJD-1.1 might also target these substrates. However, we were not able to distinguish the levels of these two signatures between wild-type and *jmjd-1.1* worms after ICL formation. Interestingly, a homolog of PHF8 and JMJD-1, EPE1, is an antisilencing factor in *S. pombe*, as it regulates the extent of heterochromatin domains [30]. The protein is recruited to heterochromatin protein 1 (HP1) bound to methylated histone H3K9 and selectively enriched at heterochromatin boundaries [31].

In an epistasis test, the sensitivity of *jmjd-1.1* worms to ICLs reverted to the level of wild-type worms after *lig-4* knockdown. This is very similar to the observation that phenotypes of a mutant of *fcd-2*, which promotes HR on DNA intermediates derived from ICLs, was rescued by a *lig-4* mutation [13]. This similarity between *jmjd-1.1* and *fcd-2* in the genetic interaction with*lig-4*, agrees well with our argument that JMJD-1.1 functions in HR. This argument was supported by our results showing the effect of JMJD-1.1 on the disappearance of RAD-51 foci after ICL and DSB formations. In contrast to our expectation based on the role of JMJD-1.1 in HR, *rad-54* RNA*i* synergized with the absence of JMJD-1.1 after ICL treatment. This revealed that JMJD-1.1 in its effect on HR is not fully epistatic to RAD-54. One possible explanation could be that both of JMJD-1.1 and RAD-54 regulates HR by chromatin structure modulation. RAD54 is a SNF2/SWI2 family protein with ATPase activity and is thought to be involved in RAD51 loading and its subsequent dissociation, as well as in branch migration in yeasts and mammals [32]. The RAD54 homolog in *C. elegans* affects RAD-51 dissociation (data not shown) as does JMJD-1.1, suggesting that both proteins probably regulate the same step of HR but via different mechanisms. However, another possibility is that JMJD-1.1 also influences a minor DSB repair pathway such as single-strand annealing (SSA) and alternative end-joining (Alt-EJ) as well as HR, thereby showing synergism with RAD-54.

Our research has generated new insights into the role of JMJD-1.1, a putative histone demethylase, in terms of DNA damage responses. Its novelty resides in the fact that *C. elegans*JMJD-1.1 and JMJD-1.2 are very similar in terms of amino acid sequence, but have both distinct and overlapping roles in DNA damage responses. Another interesting finding is that JMJD-1.1 affects homologous recombination in response to ICLs and DSBs, probably by modulating chromatin structure in parallel with RAD-54. It will be important to determine which histone proteins JMJD-1.1 acts on to relax chromatin structure and also which repair gene transcripts, if any, it regulates.

# MATERIALS AND METHODS

## Strains

The standard wild-type strain, Bristol N2, was obtained from the *Caenorhabditis* Genetics Center (Minneapolis, MN, USA). The mutants *jmjd-1.1(tm3980)*, *jmjd-1.2(tm3713)*, *fcd-2(tm1298)*, and *brc-1(tm1145)* were generated as part of the National Bioresource Project (Japan). The *jmjd-1.1(tm3980)* and *jmjd-1.2(tm3713)* strains were outcrossed with N2 males 7 and 5 times, respectively, to remove background mutations, while the *fcd-2(tm1298)* and *brc-1(tm1145)* strains were outcrossed 6 and 2 times, respectively. The double mutant *jmjd-1.1(tm3980);jmjd-1.2(tm3713)* was generated by crossing the cognate single mutants. *C.elegans* strains were maintained at 20°C on nematode growth medium (NGM) plates seeded with *E. coli* OP50-I cells.

## Bacteria-mediated RNA*i*

RNA*i* was performed by feeding *C. elegans* with *E. coli* HT115(DE3) cells expressing double-stranded RNA (dsRNA) for a given target gene. *E. coli* transformants expressing dsRNA for*rad-54* (W06D4.6), *lig-4* (C07H6.1), *jmjd-1.1* (F43G6.6), or *jmjd-1.2* (F29B9.2) gene were obtained from the *C. elegans* RNAi v1.1 Feeding Library (Open Biosystems). Cells expressing each dsRNA were first streaked on LB plates containing 50 µg/mL ampicillin, and a single colony was grown at 37°C to an $OD_{600nm}$ of 1.5 in LB medium containing 50 µg/mL ampicillin and 50 µg/mL tetracycline. The cells were then seeded onto solidified NGM plates containing 50 µg/mL ampicillin, 50 µg/mL tetracycline and 1 mM isopropyl β-D-1-thiogalactoside (IPTG), and were used the following day. L1 stage worms were placed on the plates and consumed the cells expressing dsRNA.

## Embryonic Survival Assay

To induce interstrand DNA crosslinks, L4 stage worms were incubated with 100 µM TMP (4,5,8-trimethylpsoralen, Sigma–Aldrich) for 40 min in a dark chamber. The worms were transferred onto NGM plates and exposed to 150 J/m$^2$ UVA (UVL-28 EL, UVP). For double-strand break formation, L4 stage worms were irradiated with 60 Gy of γ-rays from a $^{137}$Cs source (IBL 437C, CIS Biointernational). Eggs were collected between 24 and 40 h after induction of DNA damage, and their viability into larvae was scored after 24 h.

## Antibodies

*C. elegans* FCD-2, RPA-1, and RAD-51 primary antibodies had been previously raised in rats [16], and were used at 1:50, 1:100, and 1:100 dilutions, respectively. The anti-histone primary antibodies against H3 tri-methyl Lys9 and di-methyl Lys9 were purchased from Abcam, and used at 1:50 and 1:20 dilutions in immunostaining, respectively.

## Immunostaining

Worms were placed on glass slides covered with 1x PTw, and their gonads were extracted with an ethanol-cleansed blade. They were fixed in 3% paraformaldehyde (0.1 M K$_2$HPO$_4$, pH 7.2) for 15 min followed by addition of 1 ml 100% methanol (-20°C). They were then washed five times with 1 ml 1x PTw, and left in 200 µl of blocking solution (goat serum mixed with an equal volume of 1x PTw) for 1 h at room temperature. After blocking, the gonads were incubated with the indicated primary antibody at 4°C overnight. They were then washed five times with 1x PTw and incubated with FITC-conjugated goat secondary antibody (Roche, 1:1000 dilution in blocking solution) matching the host (mouse, rat, or rabbit) of the primary antibody at room temperature for 1 h. After three washes with 1x PTw, the gonads were stained with 1 µg/mL DAPI for 15 min followed by 2 washes with 1x PTw. Lastly, anti-fade solution was added and the gonads were observed under a fluorescence microscope (DMR HC, Leica).

## Western Analysis

L4 stage worms were treated with 200 µM photoactivated TMP for 40 min followed by 150 J/m$^2$UVA. The worms were grown further for 6 h and boiled in reducing SDS sample buffer. Proteins were separated by 12% SDS-PAGE and transferred onto a nitrocellulose membrane. Rabbit polyclonal antibody against H3K9me3 (1:500 dilution), mouse monoclonal antibody against H3K9me2 (1:500), and anti-α-tubulin monoclonal mouse antibody (1:5,000) were used as primary antibodies, followed by anti-rabbit or anti-mouse HRP antibodies (Santa Cruz Biotechnology) as secondary antibodies. Electrochemical luminescence assays were performed using WESTSAVEUp (AbFRONTIER). Luminescence signals were detected with a LAS-3000 imaging system (Fujifilm), and the band intensities were measured using a Quantity One version 4.6.6 (Bio-Rad).

## ACKNOWLEDGMENTS

The *C. elegans* N2 worms were obtained from the *Caenorhabditis* Genetics Center (Minneapolis, MN, USA). The mutants *jmjd-1.1(tm3980)*, *jmjd-1.2(tm3713)*, *fcd-2(tm1298)*, and*brc-1(tm1145)* that have been generated by the National Bioresource Project (Japan), were obtained from Dr. Shohei Mitani (Tokyo Women's Medical University School of Medicine).

## AUTHOR CONTRIBUTIONS

Conceived and designed the experiments: HSK CL. Performed the experiments: CL SH MHL. Analyzed the data: CL SH HSK. Contributed reagents/materials/analysis tools: HSK. Wrote the paper: CL HSK.

## REFERENCES

1.  Kooistra SM, Helin K. Molecular mechanisms and potential functions of histone demethylases. Nat Rev Mol Cell Biol. 2012; 13: 297–311. doi: 10.1038/nrm3327. pmid:22473470

2.  Li F, Huarte M, Zaratiegui M, Vaughn MW, Shi Y, Martienssen R, et al. Lid2 is required for coordinating H3K4 and H3K9 methylation of heterochromatin and euchromatin. Cell 2008; 135: 272–283. doi: 10.1016/j.cell.2008.08.036. pmid:18957202

3.  Mallette FA, Mattiroli F, Cui G, Young LC, Hendzel MJ, Mer G, et al. RNF8- and RNF168-dependent degradation of KDM4A/JMJD2A triggers 53BP1 recruitment to DNA damage sites. EMBO J. 2012; 31: 1865–1878. doi: 10.1038/emboj.2012.47. pmid:22373579

4.  Wakeman TP, Wang Q, Feng J, Wang XF. Bat3 facilitates H3K79 dimethylation by DOT1L and promotes DNA damage-induced 53BP1 foci at G1/G2 cell-cycle phases. EMBO J. 2012; 31: 2169–2181. doi: 10.1038/emboj.2012.50. pmid:22373577

5.   Watanabe S, Watanabe K, Akimov V, Bartkova J, Blagoev B, Lukas J, et al. JMJD1C demethylates MDC1 to regulate the RNF8 and BRCA1-mediated chromatin response to DNA breaks. Nat Struct Mol Biol. 2013; 20: 1425–1433. doi: 10.1038/nsmb.2702. pmid:24240613

6.   Huang J, Sengupta R, Espejo AB, Lee MG, Dorsey JA, Richter M, et al. p53 is regulated by the lysine demethylase LSD1. Nature 2007; 449: 105–108. pmid:17805299 doi: 10.1038/nature06092

7.   Pedersen MT, Helin K. Histone demethylases in development and disease. Trends Cell Biol. 2010; 20: 662–671. doi: 10.1016/j.tcb.2010.08.011. pmid:20863703

8.   Hoffmann I, Roatsch M, Schmitt ML, Carlino L, Pippel M, Sippl W, et al. The role of histone demethylases in cancer therapy. Mol Oncol. 2012; 6: 683–703. doi: 10.1016/j.molonc.2012.07.004. pmid:22902149

9.   Kleine-Kohlbrecher D, Christensen J, Vandamme J, Abarrategui I, Bak M, Tommerup N, et al. A functional link between the histone demethylase PHF8 and the transcription factor ZNF711 in X-linked mental retardation. Mol Cell 2010; 38: 165–178. doi: 10.1016/j.molcel.2010.03.002. pmid:20346720

10.  Lin H, Wang Y, Wang Y, Tian F, Pu P, Yu Y, et al. Coordinated regulation of active and repressive histone methylations by a dual-specificity histone demethylase ceKDM7A from *Caenorhabditis elegans*. Cell Res. 2010; 20: 899–907. doi: 10.1038/cr.2010.84. pmid:20567262

11.  Lee KY, Yang I, Park JE, Baek OR, Chung KY, Koo HS. Developmental stage- and DNA damage-specific functions of *C. elegans* FANCD2. Biochem Biophys Res Commun. 2007; 352: 479–485. pmid:17126808 doi: 10.1016/j.bbrc.2006.11.039

12.  Boulton SJ, Martin JS, Polanowska J, Hill DE, Gartner A, Vidal M. BRCA1/BARD1 orthologs required for DNA repair in *Caenorhabditis elegans*. Curr Biol. 2004; 14: 33–39. pmid:14711411 doi: 10.1016/j.cub.2003.11.029

13.  Adamo A, Collis SJ, Adelman CA, Silva N, Horejsi Z, Ward JD, et al. Preventing nonhomologous end joining suppresses DNA repair defects of Fanconi anemia. Mol Cell 2010; 39: 25–35. doi: 10.1016/j.molcel.2010.06.026. pmid:20598602

14.  Ciccia A, Elledge SJ. The DNA damage response: making it safe to play with knives. Mol Cell 2010; 40: 179–204. doi: 10.1016/j.molcel.2010.09.019. pmid:20965415

15. Deans AJ, West SC. DNA interstrand crosslink repair and cancer. Nat Rev Cancer 2011; 11: 467–480. doi: 10.1038/nrc3088. pmid:21701511

16. Lee KY, Chung KY, Koo HS. The involvement of FANCM, FANCI, and checkpoint proteins in the interstrand DNA crosslink repair pathway is conserved in *C. elegans*. DNA Repair 2010; 9: 374–382. doi: 10.1016/j. dnarep.2009.12.018. pmid:20075016

17. Stoepker C, Hain K, Schuster B, Hilhorst-Hofstee Y, Rooimans MA, Steltenpool J, et al. SLX4, a coordinator of structure-specific endonucleases, is mutated in a new Fanconi anemia subtype. Nat Genet. 2011; 43:138–141. doi: 10.1038/ng.751. pmid:21240277

18. Kottemann MC, Smogorzewska A. Fanconi anaemia and the repair of Watson and Crick DNA crosslinks. Nature 2013; 493: 356–363. doi: 10.1038/nature11863. pmid:23325218

19. Godthelp BC, Artwert F, Joenje H, Zdzienicka MZ. Impaired DNA damage-induced nuclear Rad51 foci formation uniquely characterizes Fanconi anemia group D1. Oncogene 2002; 21: 5002–5005. pmid:12118380 doi: 10.1038/sj.onc.1205656

20. Long DT, Räschle M, Joukov V, Walter JC. Mechanism of RAD51-dependent DNA interstrand cross-link repair. Science 2011; 333: 84–87. doi: 10.1126/science.1204258. pmid:21719678

21. Chiolo I, Minoda A, Colmenares SU, Polyzos A, Costes SV, Karpen GH. Double-strand breaks in heterochromatin move outside of a dynamic HP1a domain to complete recombinational repair. Cell 2011; 144: 732–744. doi: 10.1016/j.cell.2011.02.012. pmid:21353298

22. Gospodinov A, Herceg Z. Chromatin structure in double strand break repair. DNA Repair 2013; 12: 800–810. doi: 10.1016/j.dnarep.2013.07.006. pmid:23919923

23. Pace P, Mosedale G, Hodskinson MR, Rosado IV, Sivasubramaniam M, Patel KJ. Ku70 corrupts DNA repair in the absence of the Fanconi anemia pathway. Science 2010; 329: 219–223. doi: 10.1126/science.1192277. pmid:20538911

24. Cao L, Xu X, Bunting SF, Liu J, Wang RH, Cao LL, et al. A selective requirement for 53BP1 in the biological response to genomic instability induced by Brca1 deficiency. Mol Cell 2009; 35: 234–241. doi: 10.1016/j. molcel.2009.06.037

25. Bunting SF, Callén E, Wong N, Chen HT, Polato F, Gunn A, et al. 53BP1 inhibits homologous recombination in Brca1-deficient cells by blocking resection of DNA breaks. Cell 2010; 141: 243–254. doi: 10.1016/j. cell.2010.03.012. pmid:20362325

26. Ryu JS, Kang SJ, Koo HS. The 53BP1 homolog in *C. elegans* influences DNA repair and promotes apoptosis in response to ionizing radiation. PLoS One 2013; 8: e64028. doi: 10.1371/journal.pone.0064028. pmid:23667696

27. Katz DJ, Edwards TM, Reinke V, Kelly WG. A *C. elegans* LSD1 demethylase contributes to germline immortality by reprogramming epigenetic memory. Cell 2009; 137: 308–320. doi: 10.1016/j.cell.2009.02.015. pmid:19379696

28. Nottke AC, Beese-Sims SE, Pantalena LF, Reinke V, Shi Y, Colaiácovo MP. SPR-5 is a histone H3K4 demethylase with a role in meiotic double-strand break repair. Proc Natl Acad Sci USA 2011; 108: 12805–12810. doi: 10.1073/pnas.1102298108. pmid:21768382

29. Whetstine JR, Nottke A, Lan F, Huarte M, Smolikov S, Chen Z, et al. Reversal of histone lysine trimethylation by the JMJD2 family of histone demethylases. Cell 2006; 125: 467–481. pmid:16603238 doi: 10.1016/j.cell.2006.03.028

30. Ayoub N, Noma K, Isaac S, Kahan T, Grewal SI, Cohen A. A novel jmjC domain protein modulates heterochromatization in fission yeast. Mol Cell Biol. 2003; 23: 4356–4370. pmid:12773576 doi: 10.1128/mcb.23.12.4356-4370.2003

31. Braun S, Garcia JF, Rowley M, Rougemaille M, Shankar S, Madhani HD. The Cul4-Ddb1$^{Cdt2}$ ubiquitin ligase inhibits invasion of a boundary-associated antisilencing factor into heterochromatin. Cell 2011; 144: 41–54. doi: 10.1016/j.cell.2010.11.051. pmid:21215368

32. Agarwal S, van Cappellen WA, Guénolé A, Eppink B, Linsen SE, Meijering E, et al. ATP-dependent and independent functions of Rad54 in genome maintenance. J Cell Biol. 2011; 192: 735–750. doi: 10.1083/jcb.201011025. pmid:21357745

# Chapter 8

# PURIFICATION AND CHARACTERIZATION OF A DNA-BINDING RECOMBINANT PREP1:PBX1 COMPLEX

Lisa Mathiasen[1], Chiara Bruckmann[1], Sebastiano Pasqualato[2], Francesco Blasi[1]

[1] FIRC (Foundation for Italian Cancer Research) Institute of Molecular Oncology (IFOM), via Adamello 16, 20139, Milan, Italy

[2] Crystallography Unit, Department of Experimental Oncology, European Institute of Oncology, Via Adamello 16, Milan, 20139, Italy

## ABSTRACT

Human PREP1 and PBX1 are homeodomain transcriptional factors, whose biochemical and structural characterization has not yet been fully described. Expression of full-length recombinant PREP1 (47.6 kDa) and PBX1 (46.6 kDa) in *E. coli* is difficult because of poor yield, high instability and insufficient purity, in particular for structural studies. We cloned the cDNA of both proteins into a dicistronic vector containing an N-terminal glutathione S-transferase (GST) tag and co-expressed and co-purified a stable PBX1:PREP1 complex. For structural studies, we produced two C-terminally truncated complexes that retain their ability to bind DNA and are more stable than the full-length proteins through various purification steps. Here we report the production of large amounts of soluble and pure recombinant human PBX1:PREP1 complex in an active form capable of binding DNA.

## INTRODUCTION

Homeodomain TALE (three amino acids loop extension) proteins constitute a large class of eukaryotic DNA-binding proteins that regulate transcription of a broad range of developmentally important genes [1]. These proteins share a 60 amino acid DNA-binding domain which has been conserved in sequence, structure and mechanism of DNA-binding. While monomeric homeodomain proteins exhibit a limited ability to discriminate between different DNA sequences, their specificity is significantly enhanced through the cooperative binding with other DNA binding partners. PBX1 (pre-B-cell leukemia homeobox 1) [2,3], and PREP1 (PBX-regulating protein 1) also known as

PKNOX1 [4] both belong to the TALE family of homeodomain proteins and form a strong and stable DNA-independent complex [5]. PBX1 contains a nuclear localization signal and carries PREP1 into the nucleus while in turn PREP1 prevents PBX1 nuclear export [6]. PREP1 and PBX1 form trimeric complexes with HoxB1 on target enhancers which play an important role in development [7,8].

PBX1 has a dynamic subcellular localisation. It contains two nuclear localisation signals very close to the homeodomain [6,9] and two nuclear export signals (NES) within the PBC-A domain. Deletion of these Leu/Ile-rich signals impairs nuclear export, although the two NESs [10] were shown to function independently of each other, as deletion of either one did not impair nuclear export. It was suggested that binding of PREP1 masks the NESs and thereby favours retention into the nucleus [6]. The structural knowledge of these transcription factors is limited to NMR structures of PBX1 homeodomain free in solution and bound to DNA [11–13], the crystallographic structure of HoxB1-PBX1 homeodomains and flanking residues bound to DNA [14], and to the NMR structure of free PREP1 homeodomain (PDB: 1X2N). Very little is known of the interaction between PREP1 and PBX1, except that it is lost when the HR1 and HR2 regions are deleted [5]. The three-dimensional structure of this region is not known, nor are the details of the interaction. This interaction is also important because it does not only occur in PREP1, but also in its homolog MEIS1 that likewise is able to form dimers with PBX1 [15]. PREP1 and MEIS1 share identical HR1 and HR2 regions, which in both cases appear to be required to interact with PBX1. Since the number of proteins involved in these interactions is high (four PBX, two PREP and three MEIS, counting only the full length gene products and none of the known alternatively spliced forms), this surface of interaction is worth exploring.

In many cases, structural exploration is made difficult by inherent structural properties of the proteins, like instability. In this paper we report studies aimed at purifying and characterizing a recombinant DNA-binding PREP1:PBX1 complex, and two stable and DNA-binding carboxy-terminally truncated PBX1:PREP1 complexes.

## RESULTS

### Computational Analysis Predicts that PBX1 Amino- and Carboxy-termini are disordered, while PREP1 Displays low Complexity only in its Amino-terminus

Secondary structure predictions were performed by using the JPred3 server [16], a web server that in a protein sequence defines each amino acid residue into either α-helix, β-sheet or random coil secondary structures. Identification of low-complexity regions was done using a computer algorithm implemented by the program SEG [17]. This program reports regions of low complexity if there is a continuous stretch of a sequence with an entropy score below a defined threshold. Results from JPred and SEG for PREP1 and PBX1 are summarised. PREP1 is predicted to be composed of α-helices and random coils, without β-strands. The conserved regions, HR1 and HR2, are predicted to be predominantly helical in their structure. The homeodomain is predicted to be composed of three α-helices, of which the third is relatively long, compared to other homeodomains. The non-conserved regions of PREP1 are dominated by random coils and stretches of amino acids of low complexity are found in these regions.

The predicted structural organisation of PBX1 is similar: the PBC-A and PBC-B conserved regions are composed of helices and non-conserved regions are dominated by random coils. The region between PBC-A and PBC-B contains an alanine-rich stretch of low complexity. This region has been suggested to function as a flexible linker in complex formation. The homeodomain of PBX1 is predicted to be composed of three α-helices, however from the available structures of PBX1 we know that the third helix is split in two, forming a turn of a $3_{10}$ helix and a short fourth helix [14].

The sequences of PREP1 and PBX1 were analyzed with GlobPlot2.1 [18] for prediction of the proteins propensity for order/disorder. An increase of the disorder propensity sum indicates disorder, whereas a decrease is indicative of ordered/globular structure of the protein. The N-termini of both PREP1 and PBX1 were predicted to be disordered, in the region including the first 50 residues. The C-terminal region to the homeodomain of PBX1 appeared to be disordered from residue 317 just outside of the homeodomain to the C-terminus of the protein.

## Limited Proteolysis on Singly Expressed Proteins

We used limited proteolysis, N-terminal amino acid sequencing and mass spectrometry to determine the more stable PREP1 and PBX1 constructs. When a proteolytic cleavage is observed, it usually occurs in non-conserved or disordered regions, the removal of which may improve the stability of the protein. Full-length PREP1 ($PREP1_{1-436}$) and PBX1 ($PBX1_{1-430}$) were subjected to limited proteolysis with trypsin. PREP1 (Fig 1A) was degraded to a distinct ~40 kDa band (band 1) visible on a Coomassie stained SDS PAGE gel. This fragment was stable and persisted after 3 hours of incubation with a 1:500 dilution of trypsin. N-terminal sequencing of this 40 kDa fragment identified the sequence Gly-Pro-Leu-Gly-Ser-Met-Met, corresponding to the PreScission cleavage site in the pGEX6p-2rbs vector and the first two residues of PREP1, indicating that the fragment contained the N-terminus of the protein. Judging from the migration of the fragment, the pattern of trypsin cleavage sites in the PREP1 sequence, and the mass spectromery results (Fig 1D), we estimated that the fragment corresponded to PREP1$_{1-344}$. Therefore, we generated a construct corresponding to this sequence. Another stable fragment of ~28 kDa was visible on SDS PAGE gel (Fig 1A) and was identified in MS as PREP1 fragment 1–230 (band 2). Since this portion of PREP1 does not contain the homeodomain, a DNA construct was not produced.

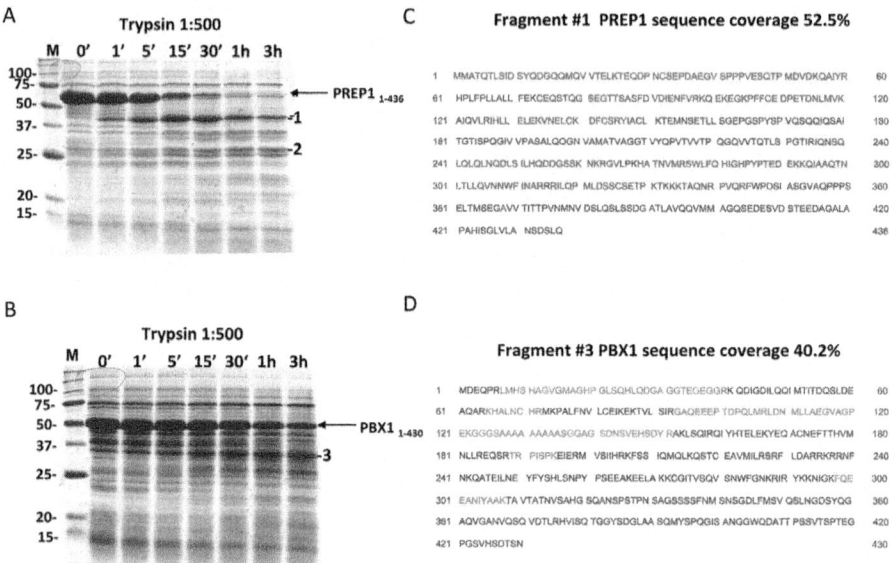

**Figure 1:** Limited proteolysis analysis of recombinant PREP1 and PBX1.

Full length PREP1 and PBX1 were subjected to limited proteolysis with trypsin. The reactions (total volume 100 μl) were performed at room temperature, 10 μl volumes were taken out at the indicated time points, supplemented with sample buffer and boiled prior to loading onto SDS PAGE. The gels were Coomassie stained. A. Limited proteolysis of PREP1. Lane M, Bio-Rad size standard; two bands of ~40 kDa (1) and ~28 kDa (2) were chosen for subsequent N-terminal sequencing. B. Limited proteolysis of PBX1 with trypsin. Lane M, Bio-Rad size standard; band 3 is the proteolysis fragment chosen for mass spectrometry analysis. C and D. Identification of PREP1 and PBX1 fragments by MALDI-TOF mass spectrometry analysis. Peptides of PREP1 and PBX1 were identified by MALDI-TOF analysis after digestion of fragments 1–3 with trypsin. Fragment 1 contained PREP1 and the matching peptides (red) covered 52.5% starting from the N-terminus and ending at residue 344. Fragment 2 contains the N-terminal part of PREP1 excluding the homeodomain. Fragment 3 contains PBX1, and the peptides (blue) covered 40.2% of the sequence, from residue 7 to 308.

In the case of PBX1, the largest trypsin-resistant fragment corresponded (Fig 1C) to the region 7–308 (band 3). Since PBX1 contains an arginine at position 6, it is likely that the fragment in fact spans the region 1–308, containing the homeodomain.

Therefore, based on computationally identified disordered region starting at residue 317 and proteolysis data that identified a stable 8–308 fragment, we generated both $PBX1_{1-308}$ and $PBX1_{1-317}$ DNA constructs.

## Purification of the Full-length and C-terminally Truncated PREP1 and PBX1

Expression and purification of PREP1 and PBX1 during the various steps of purification was evaluated by SDS PAGE and Coomassie staining as shown in Fig 2. Samples from the different purification steps were run in the various lanes as indicated in the figure legend. The theoretical molecular mass of full-length PREP1 is 47.6 kDa, but it runs in SDS PAGE (Fig 2A) with an apparent $M_r \sim 60$ kDa [19]. Full length PREP1 was expressed at relatively low levels <1mg/liter of bacterial culture, whereas the C-terminal deletion mutant $PREP1_{1-344}$ (Fig 2C) was expressed at higher levels as it seemed to be less prone to degradation. Also full length PBX1 showed relatively low expression as compared to its C-terminal deletion mutant $PBX1_{1-317}$ (Fig 2B and 2D). Expression yields are summarized in Table 1.

**Table 1:** Overview of PREP1 and PBX1 constructs

| # | Construct | Molecular weight (kDa) | Yield (mg/L culture) | Comments |
|---|---|---|---|---|
| 1 | GST-PREP1(1–436) full-length | 47.6 | <1 mg | C-terminal degradation |
| 3 | GST-PREP1(1–344) | 38.2 | ~1mg | - |
| 13 | GST-PBX1(1–430) full-length | 46.6 | ~0.5 mg | C-terminal degradation |
| 15 | GST-PBX1(1–317) | 35.3 | ~1 mg | - |
| 16 | GST-PBX1(1–308) | 34.5 | ~1 mg | - |
| 25 | GST-PREP1(1–436):PBX1(1–430) | 47.6 + 46.6 | - | Proteins seem highly unstable |
| 35 | GST-PBX1(1–430):PREP1(1–436) | 46.6 + 47.6 | ~1 mg | - |
| 38 | GST-PBX1(1–317):PREP1(1–344) | 35.3 + 38.2 | >2 mg | High expression, no degradation |
| 39 | GST-PBX1(1–308):PREP1(1–344) | 34.5+ 38.2 | >2 mg | High expression, no degradation |

doi:10.1371/journal.pone.0125789.t001

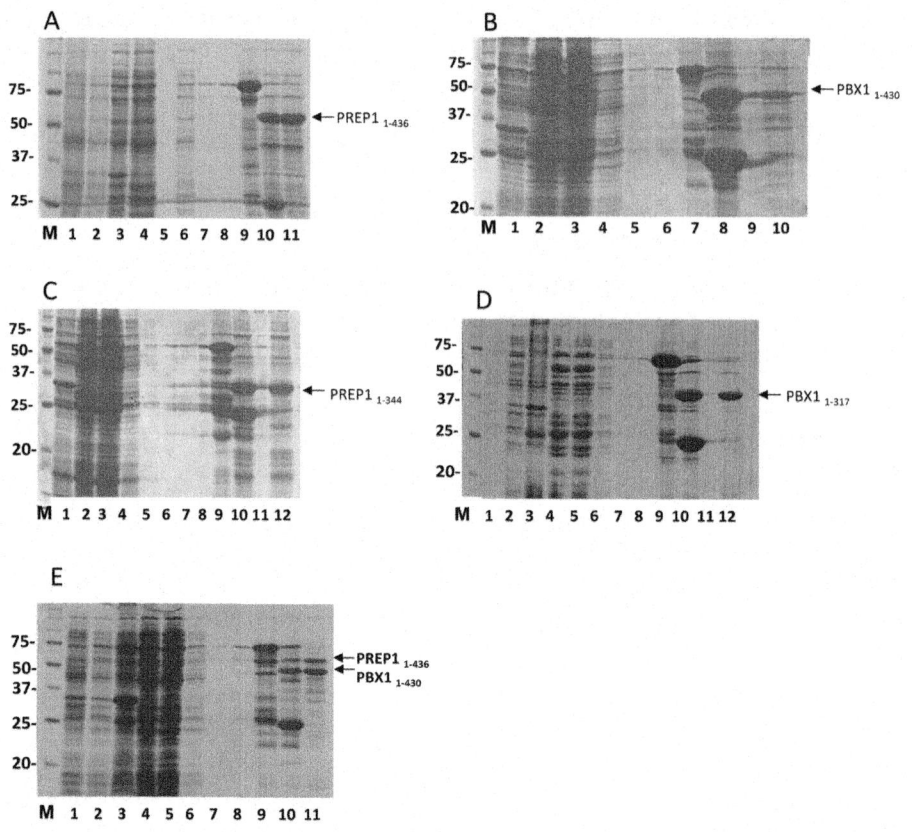

**Figure 2:** Expression and purification of recombinant PREP1 and PBX1.

A. Expression and purification of full-length PREP1. SDS PAGE analysis of proteins at different purification stages. Lane M, Bio-Rad size standard; Lane 1, total lysate before IPTG induction; Lane 2, total lysate after IPTG induction; Lane 3, pellet; Lane 4, cleared lysate; Lane 5,empty; Lane 6, first wash after incubation with glutathione beads (Tris-buffer with 0.5 M NaCl); Lane 7, second wash (Tris-buffer with 0.3 M NaCl); Lane 8, third wash (Tris-buffer with 0.3 M NaCl); Lane 9, glutathione beads before elution; Lane 10; glutathione beads after elution; Lane 11, supernatant from glutathione beads, containing the eluted protein. B. Expression and purification of full-length PBX1.Lane M, Bio-Rad size standard; Lane 1, total lysate before IPTG induction; Lane 2, total lysate after IPTG induction; Lane 3, cleared lysate; Lane 4, flow-through after incubation with glutathione beads; Lane 5, first wash (Tris-buffer with 0.5 M NaCl); Lane 6, second wash (Tris-buffer with 0.3 M NaCl); Lane 7, glutathione beads before elution; Lane 8; glutathione beads after elution; Lane 9, empty; Lane 10, supernatant from glutathione beads, containing the eluted protein. C. Expression and purification of $PREP1_{1-344}$. Lane M, Bio-Rad size standard; Lane 1, total lysate before IPTG induction; Lane 2, total lysate after IPTG induction; Lane 3, cleared lysate; Lane 4, flow-through after incubation with glutathione beads; Lane 5, first wash (Tris-buffer with 0.5 M NaCl); Lane 6, second wash (Tris-buffer with 0.3 M NaCl); Lane 7, third wash (Tris-buffer with 0.3 M NaCl); Lane 8, empty; Lane 9; glutathione beads before elution; Lane 10, glutathione beads after elution; Lane 11, empty; Lane 12, supernatant from glutathione beads, containing the eluted protein. D. Expression and purification of $PBX1_{1-317}$. Lane M, Bio-Rad size standard; Lane 1, total lysate before IPTG induction; Lane 2, total lysate after IPTG induction; Lane 3, pellet; Lane 4, cleared lysate; Lane 5, flow-through after incubation with glutathione beads; Lane 6, first wash (Tris-buffer with 0.5 M NaCl); Lane 7, second wash (Tris-buffer with 0.3 M NaCl); lane 8, third wash (Tris-buffer with 0.3 M NaCl); Lane 9, glutathione beads before elution; Lane 10, glutathione beads after elution; Lane 11, empty; Lane 12, supernatant from glutathione beads, containing the eluted protein. E. Co-expression and purification of $GST-PBX1_{1-430}PREP1_{1-436}$. Lane M, Bio-Rad size standard; Lane 1, total lysate before IPTG induction; Lane 2, total lysate after IPTG induction; Lane 3, pellet; Lane 4, cleared lysate; Lane 5, flow-through after incubation with glutathione beads; Lane 6, first wash (Tris-buffer with 0.5 M NaCl); Lane 7, second wash (Tris-buffer with 0.3 M NaCl); Lane 8, third wash (Tris-buffer with 0.3 M NaCl); Lane 9, glutathione beads before elution; Lane 10, glutathione beads after elution; Lane 11, supernatant containing the eluted proteins.

## High-level Expression of PREP1:PBX1 Complex from Dicistronic Vector

We then tested the effect of co-expression of PREP1 and PBX1 and its deletion mutants as GST-fusion in the dicistronic expression vector pGEX6p-2rbs. The order in which PREP1 and PBX1 were cloned in the expression vector turned out to be crucial for expression. In the GST-PREP1:PBX1 construct, in which PREP1 is fused to GST, expression was low, whereas PBX1 and PREP1 expressed at significantly higher levels in the construct GST-PBX1:PREP1, where PBX1 is fused to GST (Fig 2E). Expression yields of the various constructs are shown in Table 1.

## Three Chromatographic Steps for the Purification of the PBX1:PREP1 Complex

To obtain protein samples of high purity, we employed three chromatographic steps and optimised various parameters in each step. Fractionation was monitored by SDS PAGE followed by Coomassie staining. In order to eliminate heat shock protein 70 and DNA contaminants we added 5 mM $MgCl_2$–5 mM ATP in the buffers of the first wash of the glutathione beads and 1 M NaCl in the lysis buffer. The washing buffer used in the last two washes of the glutathione beads had a NaCl concentration of 0.3 M. Therefore, the eluted protein samples were diluted to 0.1 M NaCl for the subsequent ion exchange chromatography step.

In the case of the full-length PBX1:PREP1 complex, the anion exchange Resource Q column was employed. The majority of the PBX1:PREP1 eluted at 165–200 mM NaCl in a 0.1 M to 1 M NaCl gradient (Fig 3A). A smaller peak which eluted around 210–225 mM NaCl contained a 70 kDa protein contaminant (pointed out in Fig 3B) in SDS PAGE of the fractions.

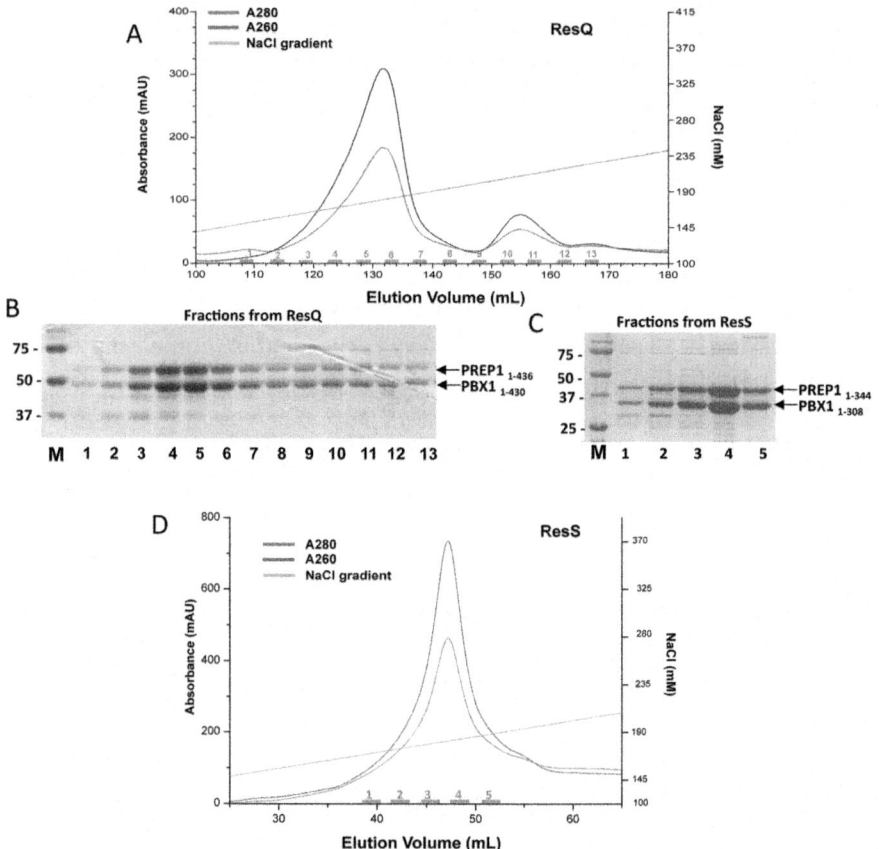

**Figure 3:** Ion exchange purifications of full-length PBX1:PREP1 and C-terminally truncated $PBX1_{1-308}$:$PREP1_{1-344}$ complexes.

A. Co-expressed and purified full-length PBX1:PREP1 complex was loaded onto a Res Q anion exchange column and eluted with a 0.1–1 M NaCl gradient. B. SDS PAGE of fractions from anion exchange of full-length PBX1:PREP1. Lane M, Bio-Rad size standard; lanes 1–13, fractions indicated in cyan in the above chromatogram; fraction volume was 1.5 ml, and on SDS PAGE were loaded 10 µl of each fraction C. SDS PAGE of fractions from anion exchange column of $PBX1_{1-308}$:$PREP1_{1-344}$. Lane M, Bio-Rad size standard; lanes 1–5, fractions indicated in cyan in the chromatogram below; fraction volume was 1.5 ml, and on SDS PAGE were loaded 10 µl of each fraction D. Co-expressed and purified $PBX1_{1-308}$:$PREP1_{1-344}$ complex was loaded onto a Res S cation exchange column and eluted with a 0.1–1 M NaCl gradient.

For the co-purification of the C-terminal deletion mutants $PBX1_{1-317}$:$PREP1_{1-344}$ and $PBX1_{1-308}$:$PREP1_{1-344}$, the cation exchange, Resource S, column was used. $PBX1_{1-308}$:$PREP1_{1-344}$ eluted at 165–200 mM NaCl in a 0.1 M to 1 M NaCl gradient as shown in Fig 3C and 3D, and $PBX1_{1-317}$:$PREP1_{1-344}$ at 220–235 mM NaCl, which was evident in the following size exclusion chromatography step and could be avoided only by excluding the first fractions of the main peak from the pool. Fractions from the main peak were pooled and concentrated using Vivaspin concentrators.

As a final purification step, the concentrated protein fractions from the ion exchange were loaded onto a Superose 6 10/300 gel filtration column. The full-length $PBX1_{1-430}$:$PREP1_{1-436}$ complex, which has an estimated $M_r$ of 94.2 kDa, eluted between 158 and 670 kDa markers, indicating that the complex could be of a higher order stoichiometry (Fig 4A). Fractions from the main peak were pooled and concentrated using Vivaspin concentrators and migrated in SDS PAGE with the expected rate (Fig 4B).

A. Size exclusion chromatography on a Superose 6 10/300 column of the full-length PBX1:PREP1 complex purified by anion exchange chromatography. Markers were thyroglobulin ($M_r$ 670,000), bovine gamma globulin ($M_r$ 158,000), chicken ovalbumin ($M_r$ 44,000), equine myoglobin ($M_r$ 17,000), and vitamin $B_{12}$ ($M_r$ 1,350). B. SDS PAGE of fractions from full-length PBX1:PREP1 gel filtration. Lane M, Bio-Rad size standard; lanes 1–7, fractions indicated in cyan in the above chromatogram; fractions volume was 0.5 ml, and on SDS PAGE were loaded 10 µl of each fraction C. SDS PAGE of fractions from $PBX1_{1-308}$:$PREP1_{1-344}$ gel filtration. Lane M, Bio-Rad size standard; lanes 1–7, fractions indicated in cyan in the chromatogram below; fractions volume was 0.5 ml, and on SDS PAGE were loaded 10 µl of each fraction D. Size exclusion chromatography on a Superose 6 10/300 column of the truncated $PBX1_{1-308}$:$PREP1_{1-344}$ complex purified by cation exchange chromatography. Markers were as in panel A.

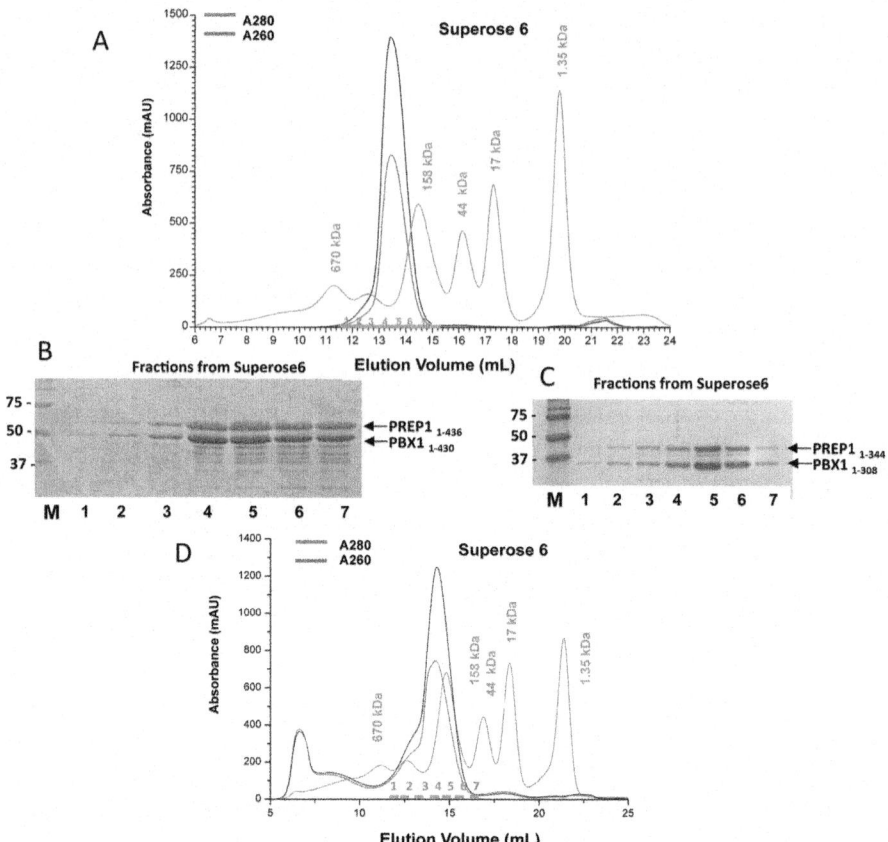

**Figure 4:** Gel filtrations of full-length PBX1:PREP1 and PBX1 $_{1-308}$:PREP1 $_{1-344}$ complexes.

The double C-terminal deletion mutants PBX1 $_{1-317}$:PREP1 $_{1-344}$, and PBX1 $_{1-308}$:PREP1 $_{1-344}$ which have an estimated $M_r$ of respectively 73.5 kDa and 72.7 kDa, eluted in Superose 6 as a single peaks with an apparent mass of ~200 kDa, again indicating a complex of higher order stoichiometry.

Full length PBX1 $_{1-430}$:PREP1 $_{1-436}$ complex, the C-terminal deletion PBX1 $_{1-317}$:PREP1 $_{1-344}$ and PBX1 $_{1-308}$:PREP1 $_{1-344}$ complexes eluted from Superose 6, were concentrated to 5–10 mg/ml and the SDS PAGE of the samples is shown in Fig 5A,5B and 5C. Fig 5D and 5E shows immunoblottings in which a sample of PBX1 $_{1-430}$:PREP1 $_{1-436}$ was analysed after purification by SDS PAGE by immunoblotting with anti-PBX1 antibodies directed towards the N-terminus (panel D, left) or the C-terminus (panel D, right). Apart from the band containing the full length PBX1, several bands containing the N-terminal part were present, indicating that degradation/truncation occurs in the C-terminal region of PBX1. PREP1 stability in the PBX1 $_{1-430}$:PREP1 $_{1-436}$ construct was also analysed by immunoblotting (Fig 5E), using a monoclonal antibody raised against the N-terminus (amino acids 1–155) shown in the panel on the left, and a polyclonal antibody raised against the whole protein (residues 15–436), in the right panel. Degradation of PREP1 is less severe than in PBX1, but we can also observe C-terminal degradation, that leads to a ~40 kDa band.

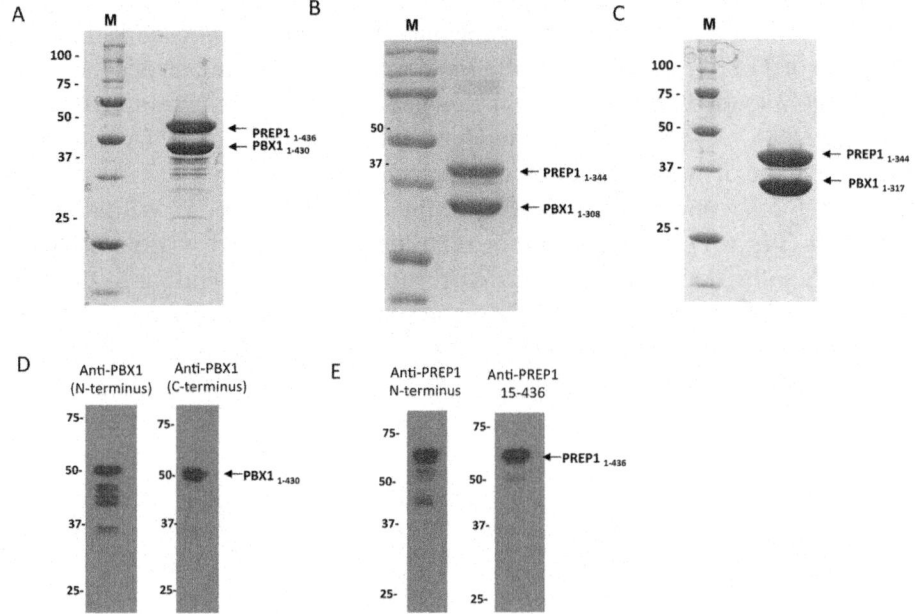

**Figure 5:** Electrophoretic migration of protein samples after three chromatographic steps.

A, B, and C. 2 µl of concentrated samples (at 10 mg/ml) of $PBX1_{1-430}$:$PREP1_{1-436}$, $PBX1_{1-308}$:$PREP1_{1-344}$ and $PBX1_{1-317}$:$PREP1_{1-344}$ were loaded onto SDS PAGEs. The full-length proteins are shown in panel A and some degradation products are present. The C-terminal deletion mutants of high purity are shown on the panels B and C. D and E. Immunoblots of purified $PBX1_{1-430}$:$PREP1_{1-436}$ complex, where the N-terminal degradation of both PBX1 and PREP1 is evident.

## Recombinant PBX1:PREP1 Complexes are Capable of Binding DNA

PREP1 was originally identified because it bound to an oligonucleotide corresponding to 31 bp of the Plau gene enhancer (O1, see Table 2); methylation interference studies have indicated that the TGACAG sequence of the human *Urokinase* enhancer was the core binding motif of the PBX1:PREP1 complex

[19]. PCR mediated binding site selection *in vitro* and ChIP-seq *in vivo* have identified the motif TGATTGACAG as the optimal binding site for PREP1-PBX1 dimers [20,21] which indeed includes part of the core binding motif TGACAG. We tested the binding of the full length $PBX1_{1-430}$:$PREP1_{1-436}$ complex to two different lengths of the O1 oligonucleotide (Table 2), an 11 bp and a 22 bp fragments, in an electrophoretic mobility shift assay (EMSA). The PBX1:PREP1 complex was incubated at 1:1 molar ratio with the 22 bp (Fig 6A) or 11 bp (Fig 6B) DNA oligonucleotides and in the presence of excess poly(dIdC), loaded onto native gel, alongside a lane with the protein complex without DNA. A band shift was evident in the presence of DNA and only a small amount of unbound DNA was detected in the native gel stained with GelRed (Fig 6A and 6B, left panels), at the bottom of lanes 2. With the 22 bp fragment (Fig 6A), a second minor retarded band formed, visible both with the DNA and Coomassie stainings corresponding to the binding to the DNA fragment of a second PBX1:PREP1 complex, possibly because of a second lower affinity binding site in the 22 bp DNA oligo or to the minor formation of a higher order protein structure.

**Table 2:** O1 oligos

| Oligo | Length | Sequence |
|-------|--------|----------|
| O1 | 21 bp | 5'- TCCTGAGGTGACAGAAGGAAG -3' |
| O1 | 11 bp | 5'- TAGTGACAGAA -3' |

doi:10.1371/journal.pone.0125789.t002

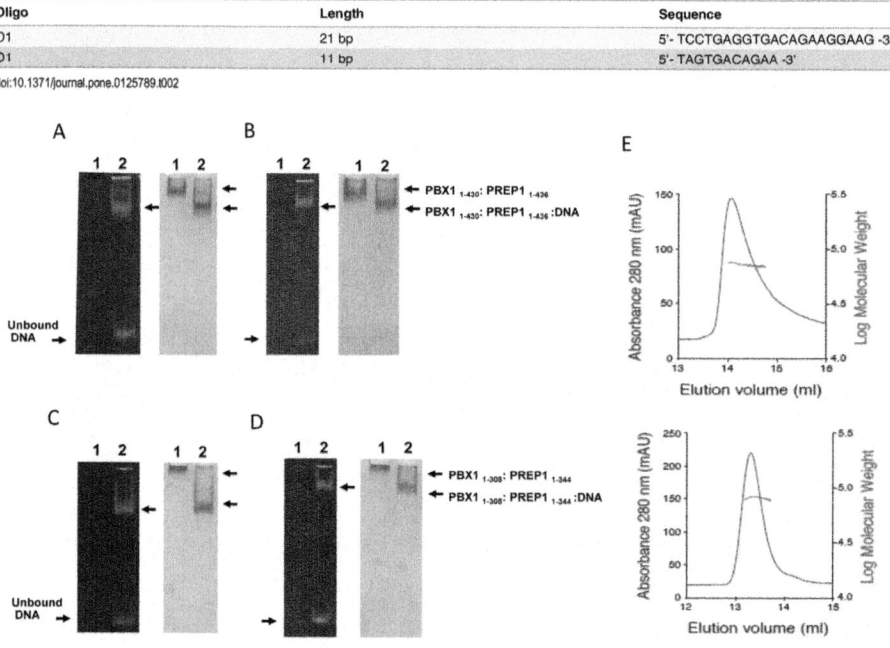

**Figure 6:** DNA binding of purified recombinant PBX1:PREP1 complexes.

A and B. EMSA of PBX1:PREP1 complex binding to O1 oligonucleotides. Co-purified full-length PBX1:PREP1 complex (2.5 µM) was incubated in the presence of 1 µg of poly(dIdC) at 1:1 molar ratio with a 22 bp DNA fragment (Panel A) or an 11 bp DNA fragment (Panel B) for 30 minutes at 4°C before loading on gel (Lanes 2). The sequences of DNA are reported in Table 2. As a reference the complex without DNA was also loaded on gel (Lanes 1). Note that with the 22 bp oligo (Panel B) at the highest ratios an additional upper band is visible, possibly corresponding to the binding of a second PREP1:PREP1 complex to the DNA. C and D. EMSA of PBX1$_{1-308}$:PREP1$_{1-344}$:DNA complex. Co-purified PBX1$_{1-308}$:PREP1$_{1-344}$ complex (2.5 µM) was incubated in the presence of 1 µg of poly(dIdC) at 1:1 molar ratio with the 22 bp (Panel C) or the 11 bp DNA fragment (Panel D) for 30 minutes at 4°C before loading on gels (Lanes 2). In Lanes 1 were loaded the protein complexes without DNA. E. Static light scattering analysis of the PBX1$_{1-308}$:PREP1$_{1-344}$ complex. The chromatograms show the UV absorbance in blue (scale on the left) and the calculated molecular mass in red (scale on the right). PBX1$_{1-308}$:PREP1$_{1-344}$ complex was analyzed in the absence (top) and in the presence (bottom) of the 11 bp DNA oligonucleotide.

It is noticeable that the PBX1:PREP1 complex migrates slower in the absence of DNA than when bound to DNA, as visible in the Coomassie staining (lanes 1 and 2 of Fig 6A and 6B). Furthermore in the presence of 22 bp DNA the migration is greater that with 11 bp DNA. This is easily explainable because in native gels proteins separate on the basis of their charge to mass ratio; when bound to the same protein a 22 bp DNA fragment adds more negative charges to the complex than an 11 bp one.

We also tested on EMSA the ability of C-terminal truncated PBX1$_{1-308}$:PREP1$_{1-344}$ to bind the 11 bp and 22 bp O1 oligos (Table 2) and the results (Fig 6C and 6D) are similar to those obtained for the full-length PBX1:PREP1 complex. Interestingly, the truncated PBX1:PREP1 complex in the absence of DNA does not migrate at all in the gel, due to the high positive charge of the protein complex under the assay conditions. Moreover, the second minor band observed with the full-length complex is absent. This result indicates that deletion of the C-termini does not affect DNA binding.

Overall, the results show that the recombinant complexes retain the DNA binding activity and specificity of the native proteins.

## PBX1$_{1-308}$:PREP1$_{1-344}$ Size Analysis by Static Light Scattering

The molecular weight of the purified heterodimeric PBX1$_{1-308}$: PREP1$_{1-344}$ complex was determined by static light scattering using the three-detector method [22]. The chromatograms in Fig 6E show the detector UV readings in blue, while the red line indicates the calculated molecular mass of the two complexes. The panel on top corresponds to the $PBX1_{1-308}:PREP1_{1-344}$ apo-complex, while at the bottom is the $PBX1_{1-308}:PREP1_{1-344}:DNA$ complex. Both complexes were monodispersed and displayed a heterodimeric behaviour under these conditions. $PBX1_{1-308}:PREP1_{1-344}$ produced an asymmetric single peak eluting at 14.0 ml, and the extrapolated molecular weight (72 kDa) was in perfect agreement with the theoretical molecular weight (72631.2 Da). $PBX1_{1-308}:PREP1_{1-344}:DNA$ peak eluted symmetrically around 13.3 ml, and indicated an 80 kDa species, consistent with the expected molecular weight of 79302.7 Da.

## Discussion

High-yields of soluble and active proteins are required for structural and biochemical characterizations. Homeodomain transcription factors are not well structurally characterized apart from their DNA binding motif. Expression in *E. coli* of sufficient amounts of recombinant full-length homeodomains has surely been a bottleneck, because of their poor yield and insufficient purity.

We singly expressed full-length PREP1 and PBX1, using an N-terminal GST tag to enhance their solubility. The expression was quite low (<1 mg/liter of culture) and we observed C-terminal degradation, resulting in a poor purification yield and low quality. The insertion of both PREP1 and PBX1 in the same dicistronic vector for co-expression and co-purification significantly improved their expression yields. However, PBX1 seems to be more prone to degradation, since we effectively improved the purification yield of the whole complex only by directly fusing PBX1 to the GST tag. Importantly the PBX1:PREP1 complex is resistant to dissociation, because purification steps at high-salt concentration did not affect the stability of the complex. This suggests a hydrophobic nature of the PBX1:PREP1 interaction.

By co-expressing PBX1:PREP1, we increased the purification yield to >1 mg/litre of culture, but the full-length complex, although able to bind DNA as demonstrated by EMSA, was unstable and both PBX1 and PREP1 displayed extensive protein degradation in the C-terminus. Even though we obtained a sufficient amount of protein for biochemical characterization, the purified protein was not homogeneous enough for future structural studies.

Computational analysis predicted that the N-terminal regions of both PREP1 and PBX1 are disordered, while partial proteolysis analysis and western blots pointed out that both recombinant proteins are more prone to C-terminal degradation. PBX1:PREP1 dimerization surface is located in their N-termini, and indeed when the complex is co-expressed N-terminal sequencing of partially trypsinized samples indicated intact N-termini. Instead, PREP1 is sensitive to trypsin proteolysis after residue 344, while PBX1 after residue 308, just downstream of the homeodomains. We cloned into the dicistronic vector the C-terminal truncated forms, including also a $PBX1_{1-317}$ truncated version, according to computational analysis results. Co-expression and co-purification of the C-terminal truncated PBX1:PREP1 increased the yield to >2 mg/liter of culture and we observed no degradation during purification. The deletions end very close to the homeodomains, but the C-terminally truncated $PBX1_{1-308}$:$PREP1_{1-344}$ complex binds DNA as well as the full-length. Interestingly, on gel filtration all the PBX1:PREP1 complexes run with an apparent molecular weight of 200 kDa, twice more than expected. Static light scattering measurements demonstrates that the experimental molecular weight of purified $PBX1_{1-308}$:$PREP1_{1-344}$ complex corresponds exactly to the theoretical one, even when $PBX1_{1-308}$:$PREP1_{1-344}$ is bound to DNA. This discrepancy might be explained because the gel filtration column fractionates proteins on the basis of their hydrodynamic radius (Stokes radius) and not on their molecular weight. The PBX1:PREP1 complex might therefore fold into an elongated complex in solution, or might not have a rigid structure.

Our co-expression and co-purification strategy of PBX1:PREP1 complex, besides improving the expression yields, provides an efficient method for PBX1:PREP1 complex preparation. Proteins produced in this manner are homogeneous and have a purity of >99% on SDS PAGE, ideal for structural studies.

# MATERIALS AND METHODS

## Protein Prediction Software

Secondary structure analyses were performed with the web based programs JPred2 and 3 [16], SEG [17] and GlobPlot2.1 [18].

## Cloning of Recombinant Proteins

Protein constructs were designed by computational sequence analyses aided by limited proteolysis results. The DNA sequence encoding PREP1 contains a BamHI restriction site which was eliminated by QuikChange (Agilent Technologies, Santa Clara, CA) mutagenesis (G→A in position 705) using the primers

PREP_fwd (5'-CCTGGGACAATTAGaATCCAGAACTCCCAGC-3') and

PREP_bck (5'- GCTGGGAGTTCTGGATtCTAATTGTCCCAGG-3')

According to the manufacturer's protocol. DNA encoding full length

PREP1 or PBX1, was amplified by PCR with primers containing BamHI and XhoI restrictions sites. Primers used for cloning are shown in Table 3. The amplified PCR products were inserted in pGEX6p-2rbs via BamHI and SalI restriction sites [23]. The mutated DNA was transformed into One-Shot chemically competent TOP10 *E. coli* cells (Invitrogen). The protein inserts and the flanking cloning sites were validated by sequencing of the DNA clones.

**Table 3:** Primers used for cloning of PREP1 and PBX1 constructs into PGEX-6p-2rbs

| Oligo | Sequence 5'-3' | Restriction site |
|---|---|---|
| PREP1> | cgcggatccATGATGGCTACACAGACATTAAGTATAG | BamHI |
| PREP1<344 | ccgctcgagttaCCTCTGAACTGGCCGGTTC | XhoI |
| PREP1<436 | ccgctcgagCTACTGCAGGGAGTCACTGTTC | XhoI |
| PBX1> | cgcggatccATGGACGAGCAGCCCAGG | BamHI |
| PBX1<308 | ccgctcgagttaTTTGGCAGCATAAATATTGGCTTC | XhoI |
| PBX1<317 | ccgctcgagttaTGACACATTGGTAGCAGTGAC | XhoI |
| PBX1<430 | ccgctcgagttaTCAGTTGGAGGTATCAGAGTGAAC | XhoI |

doi:10.1371/journal.pone.0125789.t003

## Expression Vector

PBX1:PREP1 complex was subcloned in a modified pGEX-6P vector (pGEX-6P-2RBS) to support dicistronic expression. This vector allows cloning of both genes under the control of a single promoter, each gene having its own ribosome-binding site. The same vector was also used for single protein expression.

## Expression and Purification of Proteins

Expression was performed in the *E. coli* strain BL21(DE3)pLysS (Promega, Madison, WI) designed to enhance the expression of eukaryotic proteins that contain codons rarely used in E. coli (i.e. AUA, AGG, AGA, CGG, CUA, CCC and GGA).

Protein expression was induced with 0.1 mM isopropyl-β-D-thiogalactopyranoside (IPTG). Expression was continued for 16–20 h at 16°C. Cells were harvested by centrifugation at 4,000 rpm for 15 minutes in a Beckman JLA rotor and resuspended in lysis buffer (20 mM Tris pH 7.4, 1 M NaCl, 10% glycerol, 0.5 mM EDTA and 1 mM DTT) supplemented with Protease Inhibitor Cocktail Set III Calbiochem (Billerica, MA) and 1 mg/ml lysozyme per litre of *E. coli* culture. Sonication was done with a Bandelin Sonopuls (Berlin, Germany) sonicator for 3 × 45 seconds with 5 pulses at 30–40% of max power. After sonication, bacterial lysates were cleared by centrifugation at 40,000 × g for 1–2 hours using a Beckman JA-20 rotor.

## DNA Oligonucleotides

DNA oligonucleotides used in this study were purchased from Sigma-Aldrich Biotechnology (Milano, Italy), oligonucleotides labelled with fluorophores were purified by HPLC whereas all other oligonucleotides were purified by desalting. Oligonuclotides for binding studies were annealed by dissolving the lyophilised DNA (0.1–1 mM) in annealing buffer (10 mM Tris pH 7.6, 50 mM NaCl, 0.5 mM EDTA, 5 mM $MgCl_2$) and incubating at 95°C in a heat block. After 5–10 minutes the samples were removed from the heating block and allowed to cool slowly at room temperature.

## N-terminal Sequencing and Mass Spectrometry

N-terminal sequencing was performed by automated Edman degradation. Bands for mass spectrometry analysis were cut from the Coomassie stained SDS PAGE gels. Protein samples were digested with trypsin and protein fingerprinting performed by mass spectrometry analysis (MALDI-TOF).

## Antibodies

Anti-PREP1 monoclonal antibody, CH12.2, which was raised against residues 1–155 of human PREP1 was produced by the Antibody and Protein Unit-Cogentech (Milan, Italy). Anti-PREP1 polyclonal raised against the whole protein (residues 15–436) was from Santa Cruz (Dallas, TX) (sc-6245). Anti-PBX1 polyclonal antibody raised against the N-terminal region of human PBX1a was from Cell Signalling (Euroclone, Milan, Italy) (#4342). Anti-PBX1 polyclonal raised against the C-terminus of PBX1/2/3 was from Santa Cruz (sc-888).

## Affinity Chromatography

GST-fused proteins were purified using glutathione-sepharose 4B beads (GE Healthcare) according to manufacturer's instructions. GST was cleaved off with 10 µg/ml of preScission protease (GE Healthcare, Milano, Italy) for 16 hours at 4°C.

## Ion Exchange Chromatography

GST-free proteins were diluted into ion exchange buffer (20 mM Tris pH 7.4, 0.1 M NaCl, 10% glycerol, 0.5 mM EDTA, 0.5 mM EGTA and 1 mM DTT) to a final NaCl concentration of 0.1 M and run on a Resource Q (GE Healthcare) anion exchange column. Deletion mutants were run on a Resource S (GE Healthcare) cation exchange column. In both cases the proteins were eluted using a 0.1–1.0 M NaCl gradient. The choice of the column was based on the estimated pI values of the proteins.

## Size Exclusion Chromatography

Proteins were purified by size exclusion chromatography on a Superose 6 10/300 column (GE Healthcare) equilibrated in 20 mM Tris pH 7.4, 0.3 M NaCl, 5% glycerol, 0.5 mM EDTA and 1 mM DTT at a flow rate of 0.3 ml/min. Protein markers used for size exclusion chromatography were the gel filtration standards from Bio-Rad (Hercules, CA). Protein markers used for SDS PAGE were Precision Plus Protein Dual Standards from Bio-Rad.

## Electrophoretic Mobility Shift Assays

Non denaturing gels were prepared as one-step 5–15% gradient in a final volume of 15 ml with 5 or 15% Acrylamide:bisacrylamide solution, 0.8% glycerol, 0,5x TBE (45 mM Tris base, 45 mM boric acid, 1 mM EDTA, pH 8.3), 0.1% ammonium persulfate and 6 μl TEMED (Euroclone). Binding reactions were assembled at room temperature in a total volume of 15 μl (in 20 mM Tris pH 7.4, 150 mM NaCl, 5% glycerol and 1 mM DTT) and incubated 15 minutes before loading. The gel was pre-electrophoresed, at 4°C, for 20 minutes at 90 V. After loading, electrophoresis continued for 2h at 4°C before the gel was stained for 30 minutes with GelRed (Hayward, CA) in 0.5 x TBE for DNA detection. Then the gel was stained in Coomassie Blue for proteins detection. Poly(dIdC) was purchased from Roche Diagnostics S.p.A. (Milano, Italy).

## Limited Proteolysis

Trypsin used for limited proteolysis was from Roche Diagnostics. Limited proteolysis was perfomed in 20 mM Tris pH 7.4, 0.3 M NaCl, 10% glycerol, 0.5 mM EDTA and 1 mM DTT at 25°C. The reactions (100 μl) contained 0.150 mg/ml PREP1, PBX1, PREP1:PBX1 complex or PREP1/PBX1/DNA complex. Trypsin, was added to a final concentration of 4 μg/ml or 80 ng/ml and 10 μl volumes were removed at the indicated time points. Reactions were quenched with 10 μl of 5 × SDS PAGE loading buffer, heated at 95°C for 5 minutes, electrophoresed on a 12.5% SDS PAGE gel and stained with Coomassie Blue.

## Static Light Scattering

Static Light Scattering analysis was performed on a Viscotek GPCmax (Malvern, UK) instrument. In our setup, the detectors were connected with two TSKgel G3000PWxl size-exclusion chromatography columns (Tosoh bioscience, King of Prussia, PA) in series. The system was equilibrated in 20 mM Tris buffer pH 7.2, 200 mM NaCl, 5% glycerol, and 1 mM TCEP and calibrated with BSA. $PBX1_{1-308}:PREP1_{1-344}$ and $PBX1_{1-308}:PREP1_{1-344}:DNA$ complexes were both loaded at 1.5mg/ml and eluted isocratically.

## ACKNOWLEDGMENTS

We are grateful to Valentina Cecatiello (IFOM, Milan) for help with static light scattering, to Anna De Antoni (IFOM, Milan) for kindly providing pGEX-6P-2RBS vector, to Jeff Keen, University of Leeds, UK, for performing N-terminal sequencing by Edman degradation, to Angela Bachi (IFOM, Milan) for mass spectrometry analysis, and to Andrea Musacchio for useful discussions.

## AUTHOR CONTRIBUTIONS

Conceived and designed the experiments: LM CB FB. Performed the experiments: LM CB. Analyzed the data: LM CB SP FB. Wrote the paper: CB LM FB.

## REFERENCES

1. Longobardi E, Penkov D, Mateos D, De Florian G, Torres M, Blasi F (2014) Biochemistry of the tale transcription factors PREP, MEIS, and PBX in vertebrates. Developmental dynamics: an official publication of the American Association of Anatomists 243.

2. Kamps MP, Look AT, Baltimore D (1991) The human t(1;19) translocation in pre-B ALL produces multiple nuclear E2A-Pbx1 fusion proteins with differing transforming potentials. Genes & development 5: 358–368.

3. LeBrun DP, Cleary ML (1994) Fusion with E2A alters the transcriptional properties of the homeodomain protein PBX1 in t(1;19) leukemias. Oncogene 9: 1641–1647. pmid:8183558

4. Berthelsen J, Zappavigna V, Mavilio F, Blasi F (1998) Prep1, a novel functional partner of Pbx proteins. The EMBO journal 17: 1423–1433. pmid:9482739

5. Berthelsen J, Zappavigna V, Ferretti E, Mavilio F, Blasi F (1998) The novel homeoprotein Prep1 modulates Pbx-Hox protein cooperativity. The EMBO journal 17: 1434–1445. pmid:9482740

6. Berthelsen J, Kilstrup-Nielsen C, Blasi F, Mavilio F, Zappavigna V (1999) The subcellular localization of PBX1 and EXD proteins depends on nuclear import and export signals and is modulated by association with PREP1 and HTH. Genes & development 13: 946–953.

7.    Ferretti E, Marshall H, Popperl H, Maconochie M, Krumlauf R, Blasi F (2000) Segmental expression of Hoxb2 in r4 requires two separate sites that integrate cooperative interactions between Prep1, Pbx and Hox proteins. Development (Cambridge, England) 127: 155–166. pmid:10654609

8.    Tumpel S, Cambronero F, Ferretti E, Blasi F, Wiedemann LM, Krumlauf R (2007) Expression of Hoxa2 in rhombomere 4 is regulated by a conserved cross-regulatory mechanism dependent upon Hoxb1. Developmental biology 302: 646–660. pmid:17113575

9.    Saleh M, Huang H, Green NC, Featherstone MS (2000) A conformational change in PBX1A is necessary for its nuclear localization. Experimental cell research 260: 105–115%* Copyright 2000 Academic Press. pmid:11010815

10.   Kilstrup-Nielsen C, Alessio M, Zappavigna V (2003) PBX1 nuclear export is regulated independently of PBX-MEINOX interaction by PKA phosphorylation of the PBC-B domain. The EMBO journal 22.

11.   Farber PJ, Mittermaier A (2011) Concerted dynamics link allosteric sites in the PBX homeodomain. Journal of molecular biology 405: 819–830%* Copyright A(c) 2010 Elsevier Ltd. All rights reserved. doi: 10.1016/j. jmb.2010.11.016. pmid:21087615

12.   Jabet C, Gitti R, Summers MF, Wolberger C (1999) NMR studies of the pbx1 TALE homeodomain protein free in solution and bound to DNA: proposal for a mechanism of HoxB1-Pbx1-DNA complex assembly. Journal of molecular biology 291: 521–530%* Copyright 1999 Academic Press. pmid:10448033

13.   Sprules T, Green N, Featherstone M, Gehring K (2000) Conformational changes in the PBX homeodomain and C-terminal extension upon binding DNA and HOX-derived YPWM peptides. Biochemistry 39: 9943–9950. pmid:10933814

14.   Piper DE, Batchelor AH, Chang CP, Cleary ML, Wolberger C (1999) Structure of a HoxB1-Pbx1 heterodimer bound to DNA: role of the hexapeptide and a fourth homeodomain helix in complex formation. Cell 96: 587–597. pmid:10052460 doi: 10.1016/s0092-8674(00)80662-5

15.   Chang CP, Jacobs Y, Nakamura T, Jenkins NA, Copeland NG, Cleary ML (1997) Meis proteins are major in vivo DNA binding partners for wild-type but not chimeric Pbx proteins. Molecular and cellular biology 17: 5679–5687. pmid:9315626

16. Cole C, Barber JD, Barton GJ (2008) The Jpred 3 secondary structure prediction server. Nucleic acids research 36. doi: 10.1093/nar/gkn238

17. Wootton JC, Federhen S (1996) Analysis of compositionally biased regions in sequence databases. Methods in enzymology 266: 554–571. pmid:8743706 doi: 10.1016/s0076-6879(96)66035-2

18. Linding R, Russell RB, Neduva V, Gibson TJ (2003) GlobPlot: Exploring protein sequences for globularity and disorder. Nucleic acids research 31: 3701–3708. pmid:12824398 doi: 10.1093/nar/gkg519

19. Berthelsen J, Vandekerkhove J, Blasi F (1996) Purification and characterization of UEF3, a novel factor involved in the regulation of the urokinase and other AP-1 controlled promoters. The Journal of biological chemistry 271: 3822–3830. pmid:8632000 doi: 10.1074/jbc.271.7.3822

20. Knoepfler PS, Kamps MP (1997) The highest affinity DNA element bound by Pbx complexes in t(1;19) leukemic cells fails to mediate cooperative DNA-binding or cooperative transactivation by E2a-Pbx1 and class I Hox proteins—evidence for selective targetting of E2a-Pbx1 to a subset of Pbx-recognition elements. Oncogene 14: 2521–2531. pmid:9191052 doi: 10.1038/sj.onc.1201097

21. Penkov D, Mateos San Martin D, Fernandez-Diaz LC, Rossello CA, Torroja C, Sanchez-Cabo F, et al. (2013) Analysis of the DNA-binding profile and function of TALE homeoproteins reveals their specialization and specific interactions with Hox genes/proteins. Cell reports 3: 1321–1333%* Copyright (c) 2013 The Authors. Published by Elsevier Inc. All rights reserved. doi: 10.1016/j.celrep.2013.03.029. pmid:23602564

22. Wen J, Arakawa T, Philo JS (1996) Size-exclusion chromatography with on-line light-scattering, absorbance, and refractive index detectors for studying proteins and their interactions. Analytical biochemistry 240: 155–166. pmid:8811899 doi: 10.1006/abio.1996.0345

23. Sironi L, Melixetian M, Faretta M, Prosperini E, Helin K, Musacchio A (2001) Mad2 binding to Mad1 and Cdc20, rather than oligomerization, is required for the spindle checkpoint. The EMBO journal 20: 6371–6382. pmid:11707408 doi: 10.1093/emboj/20.22.6371

# Chapter 8

## ATR SUPPRESSES ENDOGENOUS DNA DAMAGE AND ALLOWS COMPLETION OF HOMOLOGOUS RECOMBINATION REPAIR

Adam D. Brown[1,2], Brian W. Sager[1,2], Aparna Gorthi[1,2], Sonal S. Tonapi[1,2], Eric J. Brown[3], Alexander J. R. Bishop[1,2,4]

[1] Department of Cellular and Structural Biology, University of Texas Health Science Center at San Antonio, San Antonio, Texas, United States of America

[2] Greehey Children's Cancer Research Institute, University of Texas Health Science Center at San Antonio, San Antonio, Texas, United States of America

[3] Abramson Family Cancer Research Institute, Department of Cancer Biology, University of Pennsylvania School of Medicine, Philadelphia, Pennsylvania, United States of America

[4] Cancer Therapy and Research Center, University of Texas Health Science Center, San Antonio, Texas, United States of America

## ABSTRACT

DNA replication fork stalling or collapse that arises from endogenous damage poses a serious threat to genome stability, but cells invoke an intricate signaling cascade referred to as the DNA damage response (DDR) to prevent such damage. The gene product ataxia telangiectasia and Rad3-related (ATR) responds primarily to replication stress by regulating cell cycle checkpoint control, yet it's role in DNA repair, particularly homologous recombination (HR), remains unclear. This is of particular interest since HR is one way in which replication restart can occur in the presence of a stalled or collapsed fork. Hypomorphic mutations in human *ATR* cause the rare autosomal-recessive disease Seckel syndrome, and complete loss of *Atr* in mice leads to embryonic lethality. We recently adapted the *in vivo* murine pink-eyed unstable ($p^{un}$) assay for measuring HR frequency to be able to investigate the role of essential genes on HR using a conditional Cre/loxP system. Our system allows for the unique opportunity to test the effect of ATR loss on HR in somatic cells under physiological conditions. Using this system, we provide evidence that retinal pigment epithelium (RPE) cells lacking ATR have decreased density with abnormal morphology, a decreased frequency of HR and an increased level of chromosomal damage.

# INTRODUCTION

DNA damage is an unavoidable consequence of life resulting from both endogenous and exogenous sources. Dividing cells are particularly susceptible to DNA damage as many lesions can cause replication forks to stall and/or collapse. Without an appropriate response, such interruptions to DNA replication can lead to genome instability. To ensure that chromosomes are accurately and faithfully duplicated, cells have evolved an elaborate set of DDR mechanisms in which DNA replication slows allowing for the recruitment of DNA repair factors while also preventing potentially deleterious progression through cell cycle [1]. One such response to replication stress involves the protein kinase ATR. The ATR protein kinase is a member of the phosphoinositide 3 kinase (PIKK) family, and is the orthologue of *Saccharomyces cerevisiae* Mec1. Activation of ATR by replication-blocking DNA damage elicits a pleiotropic signal transduction pathway that includes numerous transducer and effector proteins [2].

Similar to many other DDR proteins, *ATR* is an essential gene and its absence leads to early embryonic lethality in mice prior to embryonic day 7.5 (E7.5) [3]. Furthermore, loss of ATR function via the disruption of the kinase domain also results in early embryonic lethality before E8.5 [4]. It is interesting to note that *bona fide* ATR heterozygous mice exhibited an increase incidence of tumors [3] and that ATR heterozygosity results in a decreased S-phase arrest [5]. Conclusions drawn from these earlier studies were that the lethality is likely due to a defect in checkpoint control in the presence of replication stress during a time of rapid cellular proliferation. Supporting this conclusion, subsequent studies have demonstrated that replication inhibition and DNA damage induce the formation of ATR foci [6], and ATR prevents the formation of DNA double strand breaks (DSBs) in response to stalled replication forks [7]–[9].

One of the mechanisms used by cells to resolve replication stress or a stalled replication fork, such that they can continue dividing, is HR repair [10]. This type of repair is considered a high fidelity process since it utilizes homologous sequences as an accurate template. This template is typically provided by the sister chromatid during S- or G2-phase before mitosis. Due to the correlation between replication stress-induced ATR activation and initiation of HR, a number of studies have investigated the role of ATR in regulating HR. Indirect evidence suggesting that ATR promotes HR include the findings that ATR phosphorylates a number of substrates known to directly affect HR (*e.g.* CHK1, BRCA1 and BLM) [6], [11]–[14]. Additionally, the PIKK inhibitor caffeine, which inhibits ATR kinase activity as well as other PIKKs such as ATM and DNA-PK, decreased site directed DSB-induced

HR [15]. Further, reduced levels of ATR rendered cells sensitive to PARP1 inhibition [16], a phenomenon often associated with a HR defect. More direct studies addressing the role of ATR in HR have also been conducted. Following expression of a kinase dead ATR mutant protein, Wang *et al.*, found that HR frequency decreased following a restriction enzyme-mediated site-directed DSB, presumably in a dominant-negative fashion[17]. This would suggest that ATR promotes HR following DNA damage. In contrast, Chanoux *et al.* found that the conditional deletion of ATR in mouse embryonic fibroblasts resulted in an approximate two-fold increase in spontaneous-induced RAD51 foci (a surrogate marker of an early and essential step in HR), and that this result was further increased in the presence of the replication stress inducing agent aphidicolin [18]. However, it should be noted that though RAD51 protein is necessary in an early step of HR, its accumulation in nuclear foci does not necessarily indicate completion of the HR repair event.

In light of these discrepancies, we set out to investigate whether ATR deficiency affects the frequency of HR that is observed through normal development. For this study, we utilized the $p^{un}$ *in vivo* mouse system for measuring HR. The basis of the $p^{un}$ system is a 70 kilobase tandem repeat of genomic material within the pink-eyed dilution ($p$; also known as *Oca2*) gene, rendering it functionless [19]–[21]. The $p$ gene is involved in pigmentation, so the $p^{un}$ mouse is identified phenotypically by the appearance of a light grey coat and pink eyes (due to a colorless RPE cell layer). Within the eye, the RPE is normally a pigmented cell type, and restoration of this pigmentation in otherwise transparent cells is the basis of our assay. To produce a functional $p$ gene in RPE cells carrying homozygous $p^{un}$ mutation, a deletion-mediated recombination event must occur between the duplicated region, deleting one copy of the 70 kb region and establishing the correct intron/exon format of the wild-type gene. Based upon studies using an analogous system in yeast [22], as well as genetic and exposure studies with the $p^{un}$ assay [23]–[25] conducted by others and our laboratory, it seems that these recombination events could occur via single strand annealing (SSA), unequal crossing over or gene conversion either between sister chromatids or homologues or via a template switch event during DNA replication. To assess the role of essential genes like *Atr* on HR, we recently modified the $p^{un}$ assay, to include the tissue-specific*Cre/loxP* system [24] and now extend this system to investigate the *in vivo* conditional loss of ATR on HR. Our findings suggest a role for ATR in promoting HR and that its absence results in chromosomal instability and cellular abnormalities.

# MATERIALS AND METHODS

## Mouse Lines

C57BL/6J pink-eyed unstable ($p^{un/un}$) mice were obtained from Jackson Laboratory. The $p^{un}$ mutation is a recessive mutation, so genotyping homozygosity of this allele is through the appearance of a dilute (*i.e.* grey) fur coat and pink eyes. $Atr^{+/neo}$ [3] and $Atr^{flox/flox}$ [9] were obtained from E.J. Brown on a C57BL6 background and backcrossed two times to $p^{un/un}$ mice. Cre expressing ($Trp1$-$Cre^{tg/tg}$ $p^{un/un}$) mice and Cre activity reporter based on expression of nuclear localized beta-galactosidase ($\beta$-gal) ($RC::PFwe^{ki/ki}$ $p^{un/un}$) [26] were used to establish two cohorts of mice in a manner similar to the study by Brown *et al.* [24]; 1) $Atr$ constitutive ($Atr^{+/-}$ $Trp1$-$Cre^{tg/tg}$ $p^{un/un}$) and 2) $Atr$ floxed ($Atr^{flox/flox}$ $RC::PFwe^{ki/ki}$ $p^{un/un}$). Animals from each cohort were crossed to generate $Atr$ conditional heterozygous ($Atr^{cond/+}$ $Trp1$-$Cre^{tg/o}$ $RC::PFwe^{ki/+}$ $p^{un/un}$) and $Atr$ conditional null ($Atr^{cond/-}$ $Trp1$-$Cre^{tg/o}$ $RC::PFwe^{ki/+}$ $p^{un/un}$). The study was approved by the University of Texas Health Science Center at San Antonio Institutional Animal Care and Use Committee (IACUC) policy as outlined in our protocol number 07005-34-02-A,B1,C. The facility is operated in compliance with the Public Law 89–544 (Animal Welfare Act) and its amendments, Public Health Services Policy on Humane Care and Use of Laboratory Animals (PHS Policy) using the Guide for the Care and Use of Laboratory Animals (Guide) as the basis of operation. Periodic inspections are conducted by the United States Department of Agriculture (USDA). The University is accredited by the Association for Assessment and Accreditation of Laboratory Animal Care, International (AAALAC). Mice were euthanized using procedures recommended by the Panel of Euthanasia of the American Veterinary Medical Association, namely $CO2$ asphyxiation with $CO2$ delivered from a compressed gas cylinder by inhalation to effect in a chamber that has not been recharged.

## Retinal Pigment Epithelium Dissection and Whole Mount Staining

Eyes were removed, rinsed in phosphate buffered saline (PBS), fixed in 4% paraformaldehyde (PFA), rinsed again in PBS and stained for $\beta$-galactosidase activity as previously described [24]. Stained eyes were dissected and RPE whole mounts were mounted according to Claybon and Bishop [27]. For those RPE whole mounts that were stained with phalloidin, Alexa Fluor 546 Phalloidin (Molecular Probes, Life Technologies) was used according to manufacture instructions. In brief, PFA fixed RPE whole mounts were rinsed in PBS and then blocked in 1% bovine serum albumin for twenty minutes at room temperature. Phalloidin stock reagent was diluted 1:40 in blocking solution and incubated in the dark for an additional twenty minutes. Samples were washed three times in PBS, mounted onto glass microscope slides and imaged.

## Visualization and Analysis of Whole Mounts

All RPE whole mounts were visualized using a Zeiss Lumar V.12 stereomicroscope and Zeiss Axiovision 4.6 software. Measurements of RPE length were performed using Adobe Photoshop, and the reported RPE petal length was defined as the distance from the optic nerve to the distal edge of the RPE. Relative RPE petal length was calculated by dividing the petal length for each sample per genotype by the average petal length of the $p^{un/un}$ samples. To determine cell density, a 200 $\mu m^2$ area was drawn using Adobe Photoshop at the position that is approximately 0.6 of the petal length from the optic nerve head to the edge of the RPE and the total number of cells located completely inside this area were counted. If any portion of a cell was found to be outside of this area, then it was excluded from counting. The same 200 $\mu m^2$ area used for cell density was also used for quantifying micronuclei (see results for description) with the exception that any cell (*i.e.* touching or entirely inside the area) was used. Therefore, we have used this region as a representation of the entire RPE and reported the data as the percentage of cells with micronuclei. The observation of cellular morphology and quantifications of cell density and micronuclei were done using the phalloidin-stained images in order to accurately identify cell boundaries. Using these fluorescent-based images meant that β-galactosidase positive staining nuclear material (*i.e.* blue nuclei and blue micronuclei) appears black and not blue.

RPE HR reversion events (*i.e.* eye spots) were scored by the phenotypic appearance of a cell with their cytoplasm packed with brown melanosomes. Unless otherwise noted, the number of pigmented cells or groups of pigmented cells are counted for the entire RPE whole mount by using a stereomicroscope with criteria for what constitutes a single eye spot (single HR event) set forth by Bishop *et al.* [28]. To define Cre activity in our system we used a nuclear localized β-galactosidase Cre reporter system [24], [26]. The nuclear localization of this enzymatic reporter is paramount to our system because it allows for the detection of both Cre activity (nuclear blue stain) and HR (cytoplasmic brown pigmentation) in the same cell. For each eye spot counted, the presence of a β-galactosidase positive stained nucleus was also recorded. Therefore, the frequency of eye spots per RPE with or without β-galactosidase staining was used to calculate the frequency of HR (*i.e.* number of eye spots per RPE). Additionally, the overall percentage of β-galactosidase staining for each RPE whole mount was visually assessed and assigned a percentage.

## I-SceI DR-GFP HR Assay

The U2OS cell line containing a stably integrated copy of the direct repeat-green fluorescent protein (DR-GFP) construct was kindly provided by Dr. Maria Jasin, and the frequency of HR was measured according to previous publications [29], [30]. Briefly, Liopfectamine RNAiMAX (Invitrogen) was used to transfect 75 picomoles of either scramble control siRNA (Santa Cruz sc-37007) or ATR siRNA (Santa Cruz sc-29763) into $1 \times 10^5$ DR-GFP U2OS cells using the reverse transfection method according to the manufacturer's protocol. Cells were then washed 24 hours later and transfected with I-SceI expression vector, empty vector or a GFP expression vector using Lipofectamine 2000 (Invitrogen) as an internal control for transfection efficiency. After 72 hours, cells were trypsinized and HR frequency was quantified as percentage GFP$^+$ cells via flow cytometry. The experiment was run in triplicate for each condition. Knockdown efficiency was measured by western blot using standard methods (1:1000 dilution of ATR antibody, Abcam ab2905). All statistical analyses were performed using GraphPad Prism (GraphPad Software, Inc.).

# RESULTS

## Size Reduction in ATR Conditional Null Eyes

It has been demonstrated that ATR is essential for embryonic development [3], [4], as well as being necessary for maintenance and homeostasis of tissues in adult mice using a ubiquitously expressing inducible Cre system [31]. We recently developed a conditional system to assess the role of essential genes on HR frequency by excising the gene of interest only in the RPE [24]. Expression of the Cre transgene is driven by the tyrosinase related protein 1 (Trp1) promoter, whose expression is restricted to the RPE during mouse embryonic development [32]. Therefore, we wanted to assess the effect of ATR loss on RPE development and HR frequency. We initially observed that the well-stained *Atr* conditional null eyes were markedly reduced in size (Fig. 1a) and further observation of RPE whole mounts confirmed this finding (Fig. 1b). We quantified this reduction in size by determining the change of petal length relative to $p^{un/un}$ samples (Fig. 1c). In doing so, we observed an approximate 45% reduction in RPE petal length of *Atr* conditional nulls compared to $p^{un/un}$ samples ($P<0.0001$, ANOVA with Tukey's multiple comparison test). The *Trp1-Cre*$^{tg/o}$ and *Atr* conditional heterozygous eyes were also smaller than $p^{un/un}$ samples ($P<0.0001$, ANOVA with Tukey's multiple comparison test), yet both were larger than the *Atr*conditional nulls. Therefore, the reduction in RPE petal length that can be attributed solely to loss of ATR is approximately 30%. In addition to the small ATR null eyes, we also observed a number of eyes of this same genotype that are apparently normal in size (Fig. 1b). Interestingly though, the proportion of β-galactosidase positive cells (*i.e.* denoting Cre activity) is substantially decreased in these RPE (Fig. 1b). Petal length of these RPE is similar to the *Trp1-Cre^{tg/o}* and *Atr* conditional heterozygous eyes and larger than the small *Atr* conditional null samples (*P*<0.0001, ANOVA with Tukey's multiple comparison test) (Fig. 1c).

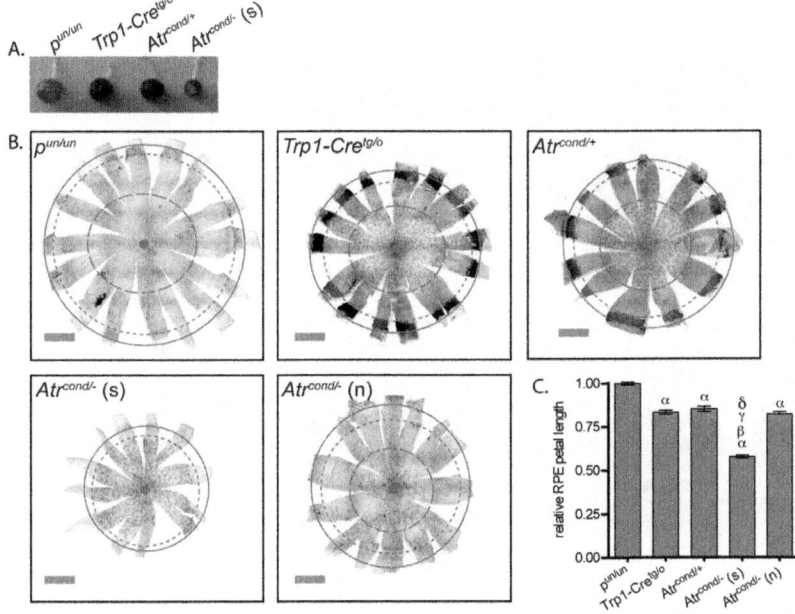

**Figure 1:** Conditional deletion of *Atr* leads to a reduction in the size of mouse eyes. Extracted eyes (**A**) and dissected RPE whole mounts (**B**) from 30 day old mice following nuclear localized β-gal activity staining. Conditional deletion of *Atr* often resulted in the significant reduction of eye size (A and B). The conditional loss of ATR significantly reduced RPE petal length as defined by the distance from the optic nerve to the distal edge of the RPE (C). i *p^{un/un}* (n = 7); ii *Trp1-Cre^{tg/o}* (n = 6); iii *Atr^{cond/+}* (n = 6); iv (s) *Atr^{cond/-}* small (n = 10); iv (n) *Atr^{cond/-}* normal (n = 6) (A, B and C). Solid red lines indicate the distal edge of the RPE, small dashed red lines indicate the proximal edge of the RPE, large red dashed lines denote the region that is 0.6 of the petal length (petal length equals the distance between the optic nerve and the proximal edge of the RPE) and solid red circles indicate the optic nerve. Scale bar: 1 mM (B). For (C) α: comparison to *p^{un/un}* (*P<0.0001*); β: comparison to *Trp1-Cre^{tg/o}* (*P<0.0001*); γ: comparison to *Atr^{cond/+}* (*P<0.0001*) and δ: comparison to *Atr^{cond/−}* normal (*P<0.0001*); error bars indicate S.E.M.

## Loss of ATR Results in Abnormal RPE Cellular Morphology and Increase in Chromosomal Instability

The cells of $p^{un/un}$ RPE exhibited a uniform cobble stone morphology, yet we observed varying degrees of differences from this norm in cells of *Trp1-Cre$^{tg/o}$*, *Atr* conditional heterozygous and*Atr* conditional null RPEs, and these differences were most apparent in the *Atr* conditional null in which the cells were markedly increased in size with abnormal morphology (Fig. 2a). We therefore quantified cell density for each genotype and found that the loss of ATR was associated with a decrease in cell density and increase in cell size (Fig. 2b).

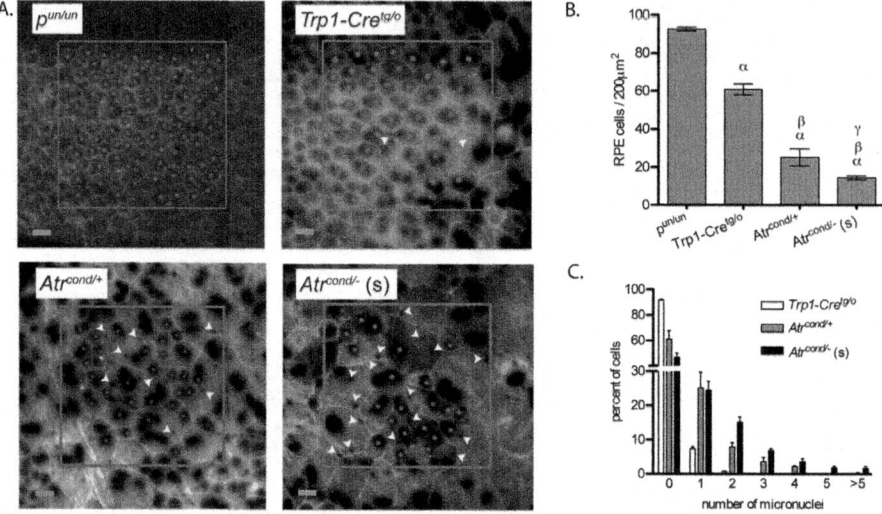

**Figure 2:** Morphological abnormalities of the RPE monolayer and increased chromosomal damage following the conditional deletion of *Atr*. Loss of either a single copy of *Atr* or both copies of *Atr* lead to morphological abnormalities of the RPE (A), a significant decrease in cell density (B) and a significant increase in micronuclei formation (C). i $p^{un/un}$ (n = 5); ii *Trp1-Cre$^{tg/o}$* (n = 7); iii *Atr$^{cond/+}$* (n = 9); and iv (s) *Atr$^{cond/-}$* small (n = 23); (A, B and C). RPE whole mounts were stained for nuclear localized β-gal activity (black spots) and phalloidin (yellow) to identify nuclear material and cell boundaries, respectively. Red boxes indicate the 200 μm² region used for cell counting, solid red circles and white arrowheads mark individual cells and micronuclei, respectively (A). Error bars indicate S.E.M. (B and C). Scale bar: 25 μm (A). For (B) α: comparison to $p^{un/un}$ ($P<0.001$); β: comparison to*Trp1-Cre$^{tg/o}$* ($P<0.001$) and γ: comparison to *Atr$^{cond/+}$* ($P<0.01$).

In order to compare across the genotypes, we used a 200 μm² area at a position that is approximately 0.6 of the petal length from the optic nerve head to the edge of the RPE (Fig. 1b). This region was chosen because it is approximately the point in which the cells of the *Atr* conditional heterozygous RPEs regained a more wild-type (*e.g.* cobblestone-like) morphology. It is interesting to note that previous work using the $p^{un}$ system found that this region corresponded to the onset of the last third of embryonic RPE development [33], and this stage of embryonic development is when *Trp-1 Cre* expression within the RPE was previously noted to begin decreasing [32]. Furthermore, the distance that is 0.6 of the length of the *Atr* conditional heterozygous or *Trp1-Cre*$^{tg/o}$ RPE is approximately the same as the total petal length of the small *Atr* conditional null RPE (optic nerve head to the proximal edge) and where the cells of the more normal-sized *Atr* conditional null RPE no longer stained positive for Cre activity (*i.e.* lack of blue β-galactosidase stain) (Fig. 1b), suggesting that Cre is most likely not active from this point onward. The greatest loss in cell density was in RPE of the small *Atr* conditional null samples when compared to either $p^{un/un}$ ($P<0.001$, ANOVA with Tukey's multiple comparison test), *Trp1-Cre*$^{tg/o}$ ($P<0.001$, ANOVA with Tukey's multiple comparison test) or *Atr* conditional heterozygous ($P<0.01$, ANOVA with Tukey's multiple comparison test) (Fig. 2b). Cell density appears to be ATR dose dependent as *Atr* conditional heterozygous RPE were significantly decreased from *Trp1-Cre*$^{tg/o}$ ($P<0.001$, ANOVA with Tukey's multiple comparison test). Normal sized *Atr* conditional null RPE were excluded from this analysis due to the lack of Cre activity in this region, as this indicates that these cells are actually *Atr* heterozygous (*i.e.* $Atr^{flox/-}$).

The reporter employed in our system is a Cre-mediated nuclear localized β-galactosidase [26]. Unlike our previous study and unpublished data in which blue stain was restricted to the nuclei of these now fully differentiated post-mitotic RPE cells, we observed small cytoplasmic bodies of blue stain in addition to the more usual large blue nuclei (Fig. 2a, white arrow heads) in the ATR conditionally deleted RPE cells. A previous report using an ATR hypomorphic Seckel syndrome cell line described an approximate two-fold increase of spontaneous micronuclei compared to wild-type and this difference was further augmented following treatment with the replication stress-inducing drug hydroxyurea [34]. Based on the location and size of the small blue staining bodies, as well as the study by Alderton *et al.*, we believe that it is reasonable to assume that these small blue stained bodies are micronuclei. We next quantified these blue bodies in the RPE of our conditional samples and found an inverse correlation between the amount of ATR and levels of micronuclei ($P<0.0001$, Chi-square test) (Fig. 2c). Interestingly, much like the observed change in cell density, *Atr* conditional heterozygosity resulted in a micronuclei haploinsufficiency phenotype and may reflect the decreased S-phase checkpoint previously observed with *ATR* heterozygous cells [5].

## ATR Promotes Homologous Recombination in Mouse RPE Cells

We next assessed the frequency of $p^{un}$ reversion events (Fig. 3a) in our samples by quantifying the presence of pigmented eye spots per RPE. We previously quantified HR frequency using three different criteria: 1) all eye spots regardless of nuclei color; 2) eye spots with blue nuclei (*i.e.* β-galactosidase positive stained nuclei to denote Cre activity occurred) and 3) eye spots with blue nuclei from RPE with an overall blue stain of ≥80% [24]. Using the most stringent criteria, we found that the frequency of HR is significantly decreased in the *Atr* conditional null RPEs compared to *Trp1-Cre*$^{tg/o}$ and *Atr* conditional heterozygous (*P*<0.001, Kruskal-Wallis with Dunn's multiple comparison test) (Fig. 3bi and 3c). Whereas *Atr* haploinsufficiency impacted cell density and occurrence of micronuclei, the frequency of HR in *Atr* conditional heterozygotes was not different from *Trp1-Cre*$^{tg/o}$ (Fig. 3b). To account for the different sized conditional null samples we separated out small and normal sized RPE and reanalyzed them. In order to do this, we had to reduce the stringency criteria to include all eye spots regardless of nuclei color. Though this reduced stringency likely results in the inclusion of events that occurred without deletion of Atr (thus inflating the HR frequency), it was necessary to allow inclusion of additional samples because the majority of the normal sized null samples had limited blue staining. The frequency of HR for both the small (*P*<0.001, Kruskal-Wallis with Dunn's multiple comparison test) and normal (*P*<0.05, Kruskal-Wallis with Dunn's multiple comparison test) sized *Atr* conditional null samples were significantly decreased compared to *Trp1-Cre*$^{tg/o}$ and *Atr* conditional heterozygous (*P*<0.0001, Kruskal-Wallis with Dunn's multiple comparison test) (Fig. 3bii), but not different from each other. In order for this analysis to be valid, we assume that all HR eye spots lacking blue nuclei were actually conditionally null rather than remaining heterozygous. Our previous work though, found that the amount of blue stain strongly correlated with excision of our gene of interest [24]. Therefore, we next asked if in the approximate outer one third of the normal sized null eyes whether the HR frequency is similar to that of *Trp1-Cre*$^{tg/o}$ and *Atr* conditional heterozygous samples (Fig. 1b). This region was found to be devoid of blue stain suggesting that the floxed allele of *Atr* in these cells was not excised (*i.e.* the cells in this region are heterozygous for *Atr*). As expected from our previous analysis, the HR frequency in this region was similar to that of the *Trp1-Cre*$^{tg/o}$ and *Atr* conditional heterozygous groups (Fig. 3biii).

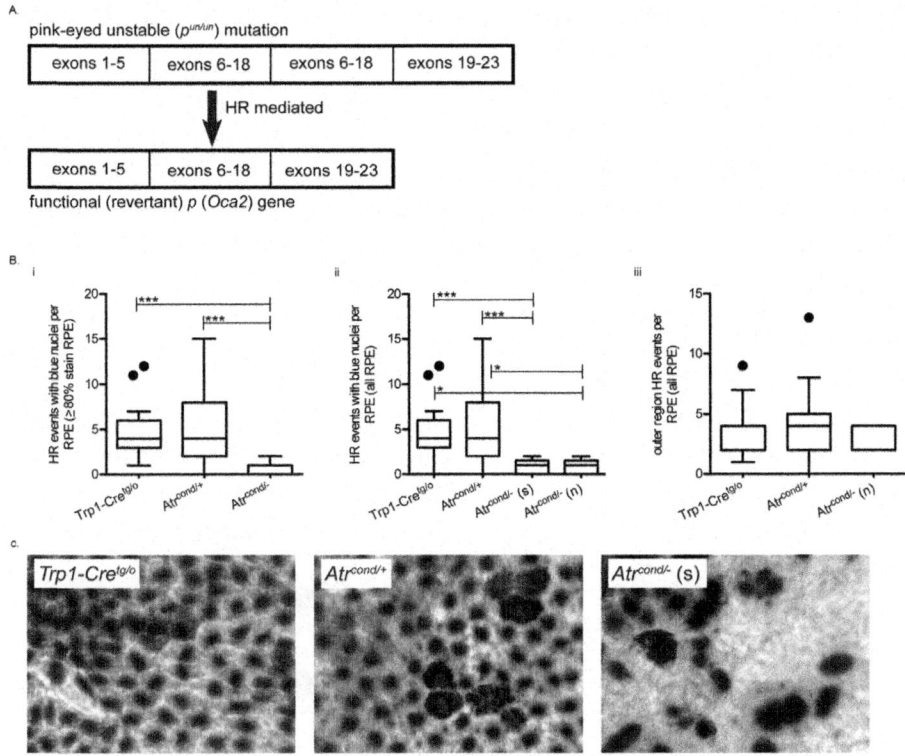

**Figure 3:** Spontaneous homologous recombination repair is decreased in the absence of ATR *in vivo*.

Schematic representation of the $p^{un}$ mutation (tandem duplication of exons 6–18) in which a HR-mediated event facilitates the deletion of one copy of the repeat resulting in the reversion of the mutant allele to a functional $p$ gene. Reversion (HR) events are scored phenotypically by counting the numbers of single or groups of cells in the RPE with brown pigmentation in their cytoplasm. (A). Loss of one copy of *Atr* does not affect HR frequency (*i.e.* the number of $p^{un}$ reversion events per RPE) (Bi and ii), whereas complete loss of ATR resulted in the significant decrease of HR frequency (Bi and ii). Within the outer clear region of $Atr^{cond/-}$ normal eyes (*i.e.* no β-gal activity suggesting the presence of a single copy of *Atr*), the HR frequency was not different from $Trp1$-$Cre^{tg/o}$ and $Atr^{cond/+}$ in similar regions (Biii). Representative eye spot images from different genotypes in B. Pigmentation can be observed in the cytoplasm following a $p^{un}$ reversion event in the RPE. Blue nuclei indicate nuclear-localized Cre activity (**C**). For (**B** and **C**) (s) $Atr^{cond/-}$ small and (n) $Atr^{cond/-}$ normal eyes; *$P<0.05$ and ***$P<0.001$.

## ATR Promotes Homologous Recombination in a Human Tissue Culture Assay

Finally, to corroborate our *in vivo* data, we used the well-established DR-GFP *in vitro* HR reporter assay [29], [30] to examine HR in cells lacking ATR. ATR expression was decreased using siRNA targeted to human *ATR* in the U2OS cell line that stably expresses the DR-GFP construct [35]. The frequency of I-*SceI*-induced HR was two-fold lower in cells with decreased ATR expression compared to scramble control siRNA cells ($P$ = 0.0009, unpaired t-test) (Figs. 4a and 4b), matching what was previously reported with either caffeine exposure [15] or expression of a presumably dominant negative ATR kinase dead mutant [17]. This result also suggests that ATR may have a more general role in controlling homologous recombination in response to a double strand break, having already been shown to be involved in some of the latter parts of break processing [36].

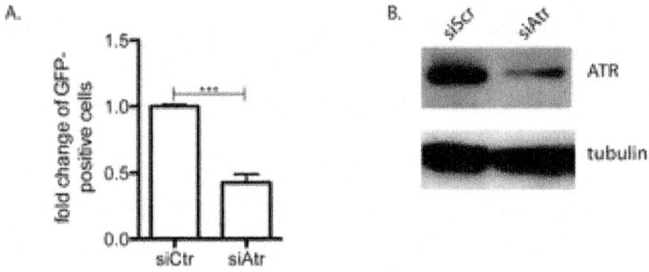

**Figure 4:** ATR promotes homologous recombination *in vitro*.

I-*SceI*-induced HR-mediated repair was significantly decreased in the DR-GFP U2OS cell line with reduced ATR expression (A). For (A) ****P<0.001* and n = 3. (B) A representative image of ATR expression knockdown using siRNA in the DR-GFP U2OS cells (siCtr is scramble).

## DISCUSSION

Genomic instability is known to be a major factor in cancer development, thus understanding the systems that influence genomic stability in the normal somatic tissue is key to understanding cancer predisposition. Homologous recombination is a key DNA repair pathway that is usually considered to have high fidelity. However, either too much or too little HR can be deleterious, resulting in genomic alterations that can promote cancer development. To examine HR through normal somatic development we use the $p^{un}$ assay system, often observing changes that are too infrequent to be observed by tissue culture systems and without confounding issues of multiple altered genetic backgrounds as are often present in established tissue culture systems. We have used this strategy successfully for a number of genetic models with differing results; ATM, p53, GADD45a, BLM, and PARP1 suppress $p^{un}$ events, while BRCA1 and BRCA2 are necessary for a subset of $p^{un}$ events [24], [25], [33], [37], [38]. Although the initiating lesion in the $p^{un}$ assay is unknown, data from our laboratory and others predict that spontaneous HR events can be initiated in response to DNA replication stress [22], [24], [33]. Additionally, Saleh-Gohari *et al.* demonstrated that endogenous damage (*e.g.* DNA single strand breaks) is most similar to damage-induced collapsed replication forks, and that these lesions are most likely substrates for spontaneous HR events [39].

The primary objective of this study was to assess the effect of ATR loss on spontaneous HR frequency. In order to bypass the embryonic lethality associated with complete loss of ATR [3] or *Atr*kinase dead [4] mouse models, we utilized the recently developed *in vivo* conditional mouse $p^{un}$ assay [24]. In the current study, we found that conditional loss of ATR led to a 2.5-fold reduction in the frequency of spontaneous HR, suggesting that ATR promotes this type of repair (Fig. 3bi). In response to site-directed DSBs, Wang *et al.* also showed a 2-fold reduction in ATR kinase dead cells [17], and we also observed a similar reduction of HR following ATR knockdown using a similar *in vitro* assay (Fig. 4). ATR is believed to be the primary signaling kinase responding to replication stress in order to prevent deleterious lesions (*e.g.* replication fork collapse or DSBs) that might be caused by replication fork stalling. Previous studies have found that ATR deficient cells have increased levels of the DSB marker H2AX following exposure to agents that induce replication stress [9], [18]. Therefore, we suggest that ATR reduces the incidence of DSBs that can occur as a result of replication fork collapse by promoting HR at a stalled replication fork (Fig. 5). In apparent contrast to this model, an increase in both spontaneous and replication damage-induced RAD51 foci in the absence of ATR has been reported [9], [18]. A potential explanation for this discrepancy is that accumulation of RAD51 foci measures HR initiation [40], which may not be inhibited by the absence of ATR, while the study by Wang *et al.* and the work presented here measured the completion of an HR event. Therefore, it is plausible to hypothesize that ATR function is not required for the initiation of HR, but rather a later step that results in the removal of RAD51, thus allowing for the completion of HR. Alternatively, but not necessarily mutually exclusively, the absence of ATR and a lack of S-phase arrest may not allow sufficient time for the completion of HR before a merging replication fork will stall at the same lesion. Such an event would result in replication fork collapse, DNA breakage and thus micronuclei formation. Such a "replication catastrophe" with ATR deficiency was recently described[41] and shown to result from depletion of RPA. RPA normally binds single stranded DNA during DNA replication and is known to interact with homologous recombination proteins to facilitate this DNA repair reaction [42].

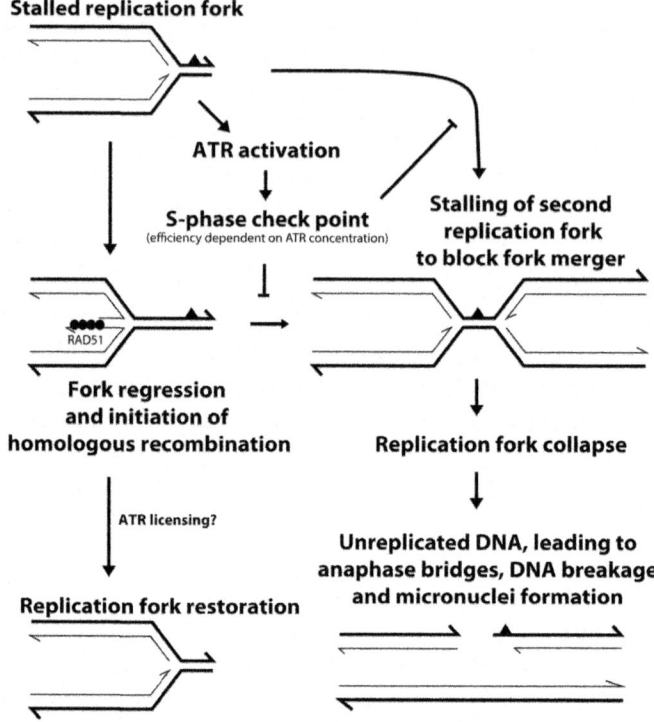

**Figure 5:** Model for the relationship between ATR, homologous recombination and chromosomal stability.

In the presence of replication stress from endogenous lesions, ATR is activated initiating an S-phase arrest to block aberrant merger of another replication fork at the same lesion. Some stalled replication forks will be substrates for HR (depicted here is the formation of a chicken foot structure that acts as a RAD51 substrate). If HR proceeds as normal, then the replication fork will be restored. However, we propose that this progression is dependent upon sufficient time to complete the HR reaction and possibly a more direct licensing of a later step in HR by ATR kinase activity (CHK1, BRCA1 and BLM, for example – not shown). If HR does not restore the stalled fork, then it may collapse, potentially leading to chromosomal breaks and the production of micronuclei. Thick lines are parental strand DNA, thin lines are daughter strand DNA, half arrowheads represent 3′ ends, the solid black triangle represents a DNA lesion and solid black circles represent RAD51 protein.

Of note with the current study, in comparison to prior studies using the $p^{un}$ assay, the conditional loss of ATR is the only genetic alteration that has resulted in a size reduction of the eye and RPE. Furthermore, fragmented nuclei as identified by nuclear localized β-galactosidase staining in the conditional $p^{un}$ system, have only been observed in the partial and complete loss of ATR. Both of these unique observations attest to the hypothesis that ATR is essential for mitigating endogenous-derived genomic instability, most likely resulting from replication stress.

Homologous recombination is an essential process as evident by the number of genes involved in HR that are essential for mouse embryonic development [43]. This is in part due to increased levels of DNA damage and subsequent genomic instability. As mentioned above, cells lacking ATR have increased levels of DNA damage, and we also observed the product of an increased level of DNA damage in ATR deficient RPE, the accumulation of micronuclei (Fig. 2a and c). Our data therefore further supports a model in which a reduction in HR from loss of ATR leads to an increase in DNA damage and chromosomal instability (Fig. 5). This is also evident from the loss of RPE cellular morphology. It is noteworthy to mention that a recent study investigated the affect of TrpCre-1 on RPE biology. In this paper, the authors showed that Cre alone caused a decrease in cell density with an increase in abnormal cell morphology [44]. We also saw this phenomenon in our *Trp1-Cre*$^{tg/o}$ RPEs, but loss of ATR greatly augmented these effects. Perhaps this is due to the cell's inability to properly respond to DNA damage induced from illegitimate Cre activity [45]. However, the ability of RPE cells to persist through to adulthood with this type of DNA damage having occurred during embryonic development, and the ensuing chromosomal instability, micronuclei and multiple nuclei, suggests something more. RPE cells frequently undergo azygotic mitosis towards the end of the development of the tissue, resulting in many RPE cells with two nuclei (but not more). This suggests that these cells may be able to retain viability when mitotic division is unsuccessful or smaller micronuclei are formed, despite being a primary tissue. This may explain the presence of multi nucleated and enlarged RPE cells when ATR is deficient. In a recent publication, Eykelenboom *et al* described a similar phenomenon following loss of ATR in the DT40 tissue culture system, however their cells died following the inappropriate progression through G2-M without completion of replication [46]. However, their results substantiate the requirement of ATR in preventing cell division in the absence of complete DNA replication and that without this checkpoint the production of micronuclei will ensue. Of note, the $p^{un}$ assay relies on the specific deletion of chromosomal DNA, generally considered a deleterious event due to the potential loss of genetic material.

Our results would suggest that such genetic loss is preferential to the more catastrophic consequence of continuing through cell cycle without completing DNA replication.

Although *Atr* conditional heterozygous RPE did not display a HR haploinsufficiency phenotype, we did observe that *Atr* heterozygosity impacted both accumulation of micronuclei and cell density (Fig. 2). It is interesting to compare this observation with Seckel syndrome, the human disease associated with *ATR* mutation. Seckel syndrome results from the hypomorphic effect of inheriting a single *ATR* mutation, which consequently reduces ATR function and is characterized by severe microcephaly and proportional primordial dwarfism and skeletal abnormalities [47]. One cellular phenotype of Seckel syndrome is increased spontaneous and replication stress-induced micronuclei [34]. This is in agreement with the present study where we observe that both *Atr* conditional null, as well as *Atr* heterozygous cells has increased micronuclei frequency (Fig. 2c). Of interest, we have not observed these micronuclei when we used the same system to conditionally delete BRCA1 or BRCA2, similarly impairing HR (*data not shown*). If the absence of ATR results in increased replication fork collapse, DSBs and an inability to utilize HR to repair these lesions, then one would expect chromosomal fragmentation and thus micronuclei. However, we also noted a significant increase in micronuclei when only one copy of ATR was present, suggesting increased spontaneous damage and ensuing chromosomal instability. This observation fits with the known decreased S-phase arrest associated with *ATR* heterozygosity [5].

A recent report by Ruzankina *et al.* also conditionally deleted ATR in adult mouse tissues. The authors concluded that loss of ATR caused the premature appearance of age-related phenotypes and increased the deterioration of tissue homeostasis [31]. Similar to that study, we also observed a mosaicism of *Atr* deletion in our RPE, particularly in the normal sized *Atr* conditional null RPE (Fig. 1b). This is most pronounced in the more distal region of RPE where proliferation is greatest (many tightly packed cells produced in a short period at the end of RPE development), and suggests that a number of cells escaped Cre activity thereby remaining heterozygous and were capable of proper organ development. Supporting this interpretation of our observation is that the petal length (Fig. 1c) and HR frequency of the outer region without staining (therefore lacking Cre activity) (Fig. 3biii) of the normal sized *Atr* conditional null eyes are not different from *Atr* conditional heterozygous and *Trp1-Cre^{tg/o}* RPE. These observations also belie the inherent issues of trying to measure HR in the context of modulating an essential gene such as *Atr*. In any such study it is necessary to control for the competition between the cells impaired in progression through cell division and proliferation because of the loss of an essential gene with cells that have a growth advantage due to retaining a functional copy of the essential gene. As such, our result that demonstrates the requirement of ATR in HR is probably a conservative measurement.

This study is another example of the ability of the conditional $p^{un}$ assay in measuring the effect of essential genes on HR frequency. More importantly, we have provided direct evidence that ATR is necessary for completion of HR events during somatic development in response to the normal levels of endogenous replication stress that occurs. The reduction of HR in the absence of ATR correlates with increased DNA damage and abnormal cell morphology.

## ACKNOWLEDGMENTS

We are grateful to members of the Bishop Lab for their comments on the manuscript and the UTHSCSA Department of Lab Animal Services for the care of animals.

## AUTHOR CONTRIBUTIONS

Conceived and designed the experiments: AJRB. Performed the experiments: ADB BWS AG SST. Analyzed the data:AJRB ADB BWS AG EJB SST. Contributed reagents/materials/analysis tools: EJB. Wrote the paper: ADB BWS AJRB.

## REFERENCES

1.    Myung K, Datta A, Kolodner RD (2001) Suppression of spontaneous chromosomal rearrangements by S phase checkpoint functions in Saccharomyces cerevisiae. Cell 104: 397–408. doi: 10.1016/s0092-8674(01)00227-6

2.    Budzowska M, Kanaar R (2009) Mechanisms of dealing with DNA damage-induced replication problems. Cell Biochem Biophys 53: 17–31. doi: 10.1007/s12013-008-9039-y

3.    Brown EJ, Baltimore D (2000) ATR disruption leads to chromosomal fragmentation and early embryonic lethality. Genes Dev 14: 397–402.

4.    de Klein A, Muijtjens M, van Os R, Verhoeven Y, Smit B, et al. (2000) Targeted disruption of the cell-cycle checkpoint gene ATR leads to early embryonic lethality in mice. Curr Biol 10: 479–482. doi: 10.1016/s0960-9822(00)00447-4

5.    Garg R, Callens S, Lim DS, Canman CE, Kastan MB, et al. (2004) Chromatin association of rad17 is required for an ataxia telangiectasia

and rad-related kinase-mediated S-phase checkpoint in response to low-dose ultraviolet radiation. Mol Cancer Res 2: 362–369.

6.  Tibbetts RS, Cortez D, Brumbaugh KM, Scully R, Livingston D, et al. (2000) Functional interactions between BRCA1 and the checkpoint kinase ATR during genotoxic stress. Genes Dev 14: 2989–3002. doi: 10.1101/gad.851000

7.  Lopes M, Cotta-Ramusino C, Pellicioli A, Liberi G, Plevani P, et al. (2001) The DNA replication checkpoint response stabilizes stalled replication forks. Nature 412: 557–561. doi: 10.1038/35087613

8.  Cha RS, Kleckner N (2002) ATR homolog Mec1 promotes fork progression, thus averting breaks in replication slow zones. Science 297: 602–606. doi: 10.1126/science.1071398

9.  Brown EJ, Baltimore D (2003) Essential and dispensable roles of ATR in cell cycle arrest and genome maintenance. Genes Dev 17: 615–628. doi: 10.1101/gad.1067403

10. Helleday T, Lo J, van Gent DC, Engelward BP (2007) DNA double-strand break repair: from mechanistic understanding to cancer treatment. DNA Repair (Amst) 6: 923–935. doi: 10.1016/j.dnarep.2007.02.006

11. Pichierri P, Rosselli F, Franchitto A (2003) Werner's syndrome protein is phosphorylated in an ATR/ATM-dependent manner following replication arrest and DNA damage induced during the S phase of the cell cycle. Oncogene 22: 1491–1500. doi: 10.1038/sj.onc.1206169

12. Davies SL, North PS, Dart A, Lakin ND, Hickson ID (2004) Phosphorylation of the Bloom's syndrome helicase and its role in recovery from S-phase arrest. Mol Cell Biol 24: 1279–1291. doi: 10.1128/mcb.24.3.1279-1291.2004

13. Li W, Kim SM, Lee J, Dunphy WG (2004) Absence of BLM leads to accumulation of chromosomal DNA breaks during both unperturbed and disrupted S phases. J Cell Biol 165: 801–812. doi: 10.1083/jcb.200402095

14. Hu B, Wang H, Wang X, Lu HR, Huang C, et al. (2005) Fhit and CHK1 have opposing effects on homologous recombination repair. Cancer Res 65: 8613–8616. doi: 10.1158/0008-5472.can-05-1966

15. Sorensen CS, Hansen LT, Dziegielewski J, Syljuasen RG, Lundin C, et al. (2005) The cell-cycle checkpoint kinase Chk1 is required for mammalian homologous recombination repair. Nat Cell Biol 7: 195–201. doi: 10.1038/ncb1212

16. McCabe N, Turner NC, Lord CJ, Kluzek K, Bialkowska A, et al. (2006) Deficiency in the repair of DNA damage by homologous recombination

and sensitivity to poly(ADP-ribose) polymerase inhibition. Cancer Res 66: 8109–8115. doi: 10.1158/0008-5472.can-06-0140

17. Wang H, Wang H, Powell SN, Iliakis G, Wang Y (2004) ATR affecting cell radiosensitivity is dependent on homologous recombination repair but independent of nonhomologous end joining. Cancer Res 64: 7139–7143. doi: 10.1158/0008-5472.can-04-1289

18. Chanoux RA, Yin B, Urtishak KA, Asare A, Bassing CH, et al. (2009) ATR and H2AX cooperate in maintaining genome stability under replication stress. J Biol Chem 284: 5994–6003. doi: 10.1074/jbc.m806739200

19. Brilliant MH (2001) The mouse p (pink-eyed dilution) and human P genes, oculocutaneous albinism type 2 (OCA2), and melanosomal pH. Pigment Cell Res 14: 86–93. doi: 10.1034/j.1600-0749.2001.140203.x

20. Brilliant MH, Gondo Y, Eicher EM (1991) Direct molecular identification of the mouse pink-eyed unstable mutation by genome scanning. Science 252: 566–569. doi: 10.1126/science.1673574

21. Oetting WS, King RA (1999) Molecular basis of albinism: mutations and polymorphisms of pigmentation genes associated with albinism. Hum Mutat 13: 99–115. doi: 10.1002/(sici)1098-1004(1999)13:2<99::aid-humu2>3.3.co;2-3

22. Galli A, Schiestl RH (1995) On the mechanism of UV and gamma-ray-induced intrachromosomal recombination in yeast cells synchronized in different stages of the cell cycle. Mol Gen Genet 248: 301–310. doi: 10.1007/bf02191597

23. Karia B, Martinez JA, Bishop AJ (2013) Induction of homologous recombination following in utero exposure to DNA-damaging agents. DNA Repair (Amst).

24. 24. Brown AD, Claybon AB, Bishop AJ (2011) A conditional mouse model for measuring the frequency of homologous recombination events in vivo in the absence of essential genes. Mol Cell Biol 31: 3593–3602. doi: 10.1128/mcb.00848-10

25. Claybon A, Karia B, Bruce C, Bishop AJ (2010) PARP1 suppresses homologous recombination events in mice in vivo. Nucleic Acids Res 38: 7538–7545. doi: 10.1093/nar/gkq624

26. Farago AF, Awatramani RB, Dymecki SM (2006) Assembly of the brainstem cochlear nuclear complex is revealed by intersectional and subtractive genetic fate maps. Neuron 50: 205–218. doi: 10.1016/j.neuron.2006.03.014

27. Claybon A, Bishop AJ (2011) Dissection of a mouse eye for a whole mount of the retinal pigment epithelium. J Vis Exp.

28.    28. Bishop AJ, Kosaras B, Carls N, Sidman RL, Schiestl RH (2001) Susceptibility of proliferating cells to benzo[a]pyrene-induced homologous recombination in mice. Carcinogenesis 22: 641–649. doi: 10.1093/carcin/22.4.641

29.    Gunn A, Bennardo N, Cheng A, Stark JM (2011) Correct end use during end joining of multiple chromosomal double strand breaks is influenced by repair protein RAD50, DNA-dependent protein kinase DNA-PKcs, and transcription context. J Biol Chem 286: 42470–42482. doi: 10.1074/jbc.m111.309252

30.    Pierce AJ, Johnson RD, Thompson LH, Jasin M (1999) XRCC3 promotes homology-directed repair of DNA damage in mammalian cells. Genes Dev 13: 2633–2638. doi: 10.1101/gad.13.20.2633

31.    Ruzankina Y, Pinzon-Guzman C, Asare A, Ong T, Pontano L, et al. (2007) Deletion of the developmentally essential gene ATR in adult mice leads to age-related phenotypes and stem cell loss. Cell Stem Cell 1: 113–126. doi: 10.1016/j.stem.2007.03.002

32.    Mori M, Metzger D, Garnier JM, Chambon P, Mark M (2002) Site-specific somatic mutagenesis in the retinal pigment epithelium. Invest Ophthalmol Vis Sci 43: 1384–1388.

33.    Bishop AJ, Hollander MC, Kosaras B, Sidman RL, Fornace AJ Jr, et al. (2003) Atm-, p53-, and Gadd45a-deficient mice show an increased frequency of homologous recombination at different stages during development. Cancer Res 63: 5335–5343.

34.    Alderton GK, Joenje H, Varon R, Borglum AD, Jeggo PA, et al. (2004) Seckel syndrome exhibits cellular features demonstrating defects in the ATR-signalling pathway. Hum Mol Genet 13: 3127–3138. doi: 10.1093/hmg/ddh335

35.    Xia B, Sheng Q, Nakanishi K, Ohashi A, Wu J, et al. (2006) Control of BRCA2 cellular and clinical functions by a nuclear partner, PALB2. Mol Cell 22: 719–729. doi: 10.1016/j.molcel.2006.05.022

36.    Polo SE, Blackford AN, Chapman JR, Baskcomb L, Gravel S, et al. (2012) Regulation of DNA-end resection by hnRNPU-like proteins promotes DNA double-strand break signaling and repair. Mol Cell 45: 505–516. doi: 10.1016/j.molcel.2011.12.035

37.    Bishop AJ, Barlow C, Wynshaw-Boris AJ, Schiestl RH (2000) Atm deficiency causes an increased frequency of intrachromosomal homologous recombination in mice. Cancer Res 60: 395–399.

38. Reliene R, Bishop AJ, Li G, Schiestl RH (2004) Ku86 deficiency leads to reduced intrachromosomal homologous recombination in vivo in mice. DNA Repair (Amst) 3: 103–111. doi: 10.1016/j.dnarep.2003.09.010

39. Saleh-Gohari N, Bryant HE, Schultz N, Parker KM, Cassel TN, et al. (2005) Spontaneous homologous recombination is induced by collapsed replication forks that are caused by endogenous DNA single-strand breaks. Mol Cell Biol 25: 7158–7169. doi: 10.1128/mcb.25.16.7158-7169.2005

40. Yeung PL, Denissova NG, Nasello C, Hakhverdyan Z, Chen JD, et al. (2012) Promyelocytic leukemia nuclear bodies support a late step in DNA double-strand break repair by homologous recombination. J Cell Biochem 113: 1787–1799. doi: 10.1002/jcb.24050

41. Toledo LI, Altmeyer M, Rask MB, Lukas C, Larsen DH, et al. (2013) ATR Prohibits Replication Catastrophe by Preventing Global Exhaustion of RPA. Cell 155: 1088–1103. doi: 10.1016/j.cell.2013.10.043

42. Park MS, Ludwig DL, Stigger E, Lee SH (1996) Physical interaction between human RAD52 and RPA is required for homologous recombination in mammalian cells. J Biol Chem 271: 18996–19000. doi: 10.1074/jbc.271.31.18996

43. Friedberg EC, Meira LB (2006) Database of mouse strains carrying targeted mutations in genes affecting biological responses to DNA damage Version 7. DNA Repair (Amst) 5: 189–209. doi: 10.1016/j.dnarep.2005.09.009

44. Thanos A, Morizane Y, Murakami Y, Giani A, Mantopoulos D, et al. (2012) Evidence for baseline retinal pigment epithelium pathology in the Trp1-Cre mouse. Am J Pathol 180: 1917–1927. doi: 10.1016/j.ajpath.2012.01.017

45. Loonstra A, Vooijs M, Beverloo HB, Allak BA, van Drunen E, et al. (2001) Growth inhibition and DNA damage induced by Cre recombinase in mammalian cells. Proc Natl Acad Sci U S A 98: 9209–9214. doi: 10.1073/pnas.161269798

46. Eykelenboom JK, Harte EC, Canavan L, Pastor-Peidro A, Calvo-Asensio I, et al. (2013) ATR activates the S-M checkpoint during unperturbed growth to ensure sufficient replication prior to mitotic onset. Cell Rep 5: 1095–1107. doi: 10.1016/j.celrep.2013.10.027

47. O'Driscoll M, Ruiz-Perez VL, Woods CG, Jeggo PA, Goodship JA (2003) A splicing mutation affecting expression of ataxia-telangiectasia and Rad3-related protein (ATR) results in Seckel syndrome. Nat Genet 33: 497–501. doi: 10.1038/ng1129

# Chapter 9

## DIVERSITY AND RECOMBINATION OF DISPERSED RIBOSOMAL DNA AND PROTEIN CODING GENES IN MICROSPORIDIA

Joseph Edward Ironside

Institute of Biological, Environmental and Rural Sciences, Aberystwyth University, Aberystwyth, United Kingdom

## ABSTRACT

Microsporidian strains are usually classified on the basis of their ribosomal DNA (rDNA) sequences. Although rDNA occurs as multiple copies, in most non-microsporidian species copies within a genome occur as tandem arrays and are homogenised by concerted evolution. In contrast, microsporidian rDNA units are dispersed throughout the genome in some species, and on this basis are predicted to undergo reduced concerted evolution. Furthermore many microsporidian species appear to be asexual and should therefore exhibit reduced genetic diversity due to a lack of recombination. Here, DNA sequences are compared between microsporidia with different life cycles in order to determine the effects of concerted evolution and sexual reproduction upon the diversity of rDNA and protein coding genes. Comparisons of cloned rDNA sequences between microsporidia of the genus Nosema with different life cycles provide evidence of intragenomic variability coupled with strong purifying selection. This suggests a birth and death process of evolution. However, some concerted evolution is suggested by clustering of rDNA sequences within species. Variability of protein-coding sequences indicates that considerable intergenomic variation also occurs between microsporidian cells within a single host. Patterns of variation in microsporidian DNA sequences indicate that additional diversity is generated by intragenomic and/or intergenomic recombination between sequence variants. The discovery of intragenomic variability coupled with strong purifying selection in microsporidian rRNA sequences supports the hypothesis that concerted evolution is reduced when copies of a gene are dispersed rather than repeated tandemly. The presence of intragenomic variability also renders the use of rDNA sequences for barcoding microsporidia questionable. Evidence of recombination in the single-copy genes of putatively asexual microsporidia suggests that these species may undergo cryptic sexual reproduction, a possibility with profound implications for the evolution of virulence, host range and drug resistance in these species.

## INTRODUCTION

Microsporidia are near-ubiquitous intracellular parasites of animals and protists. They are closely related to fungi, although their taxonomic status (as a clade within the fungi or a sister clade to the fungi) remains the subject of debate [1], [2]. Microsporidia possess the smallest genomes of any eukaryotic organisms and cause a variety of important medical, veterinary, agricultural and ecological impacts [3]. Many studies of microsporidia and other parasites have attempted to classify strains by amplifying and sequencing variable regions of the genome such as the ribosomal internal transcribed spacer region (ITS) [4], [5], [6], [7], [8]. The ITS is also proposed as a universal DNA barcode marker for fungi [9]. However, given the reduced and rearranged nature of microsporidian ribosomal DNA (described below), "universal" fungal primers are unlikely to amplify the microsporidian ITS reliably. The dispersed nature of rDNA repeats in some microsporidian species [10] also calls into question the assumption that repeats will be homogenised by concerted evolution.

Genes coding for ribosomal RNA subunits usually occur as multiple copies within eukaryotic genomes. They are typically organised into rDNA units, each consisting of 18S, 23S and 5.8S subunits, separated by two internal transcribed spacer regions (ITS1 and ITS2). Within most eukaryotic genomes, rRNA genes exist as tandem arrays, with each rDNA unit separated by an intergenic spacer (IGS). The units of these tandem arrays are subject to homogenization by unequal crossing over and gene conversion [11], a process known as concerted evolution. A result of concerted evolution is that RNA units at different positions within the genome of a species tend to be more similar than RNA units at equivalent positions in the genomes of different species.

However, in rare cases, rDNA units are dispersed throughout the genome rather than organised in tandem arrays. Because unequal crossing over and gene conversion occur less frequently between sequences on heterologous chromosomes than on homologous chromosomes [12], concerted evolution is predicted to act less strongly upon dispersed rDNA units than upon tandem repeats. This prediction is supported by the finding that sequences belonging to the dispersed rDNA units of Apicomplexa often cluster phylogenetically between species rather than within species [13]. This indicates that concerted evolution is insufficient to homogenise these dispersed units and suggests that their evolution results instead from a birth and death process in which new copies are repeatedly produced by gene duplication and removed by gene deletion [14]. A similar process appears to underlie the evolution of dispersed 5S rRNA units in fungi [15], plants [16] and animals [17]. Dispersed 5S rDNA units also appear to evolve through a combination of concerted and birth-and-death processes in fish [18] and in the mussel Mytilus [19]. Intragenomic variation of rDNA can cause errors when rDNA is used to discriminate strains or species, as in the case of the putative apicomplexan species Eimeria mitis and Eimeria mivati [20].

Within Nosema, a genus of microsporidian parasites, one species N. apis possesses tandemly repeated rDNA units [21] while in another species N. bombycis, rDNA units are dispersed over multiple chromosomes [10]. Furthermore, within isolates of Nosema bombycis from the hostBombyx mori, rDNA units are highly variable, differing with regard to nucleotide sequence, subunit organisation and the presence of transposable elements [22]. This suggests that, inNosema, the evolution of rDNA units may occur through a birth-and-death process rather than through concerted evolution.

Some, but not all, Nosema species also possess an unusual "reversed" arrangement of rDNA subunits [23], [24] in which the large 18S subunit occurs upstream of the smaller 16S subunit. The two subunits are separated by a spacer region, described by Huang et al. [23] as an internal transcribed spacer (ITS), although no evidence of transcription is presented. Downstream of the 16S subunit is a small, 5S subunit, separated from the 16S subunit by a second spacer region, described by Huang et al. [23] as an intergenic spacer (IGS). OtherNosema species possess an arrangement of the rDNA unit more common among microsporidia, with the 16S subunit positioned upstream of the 18S subunit and separated from it by a short ITS region. Intragenomic variation in the ITS of N. bombi [25], [26] indicates that these species may also experience relaxed convergent evolution, despite evidence that the rDNA unit is repeated tandemly [21]. In N. ceranae, a 5S subunit occurs upstream of the 16S subunit, separated from it by an IGS [8]. The presence of two different subunit arrangements in different species of the same genus suggests a birth and death process in which a mutation of the subunit arrangement in one rDNA unit spread to other locations in the genome concurrently with the extinction of the original arrangement [24].

Species within the genus Nosema also vary markedly in their transmission strategies and life cycles [24]. While some species are transmitted vertically from female hosts to their offspring, others are transmitted horizontally via infectious spores and a third group combine these two modes of transmission [24]. Most Nosema species produce diplokaryotic stages that develop in direct contact with the host cell's cytoplasm but some species also produce unikaryotic stages that develop within a sporophorous vesicle, usually in groups of eight

[27], [28]. Such species were fomerly allocated to the genus Vairimorpha but this is now acknowledged to be synonymous with Nosema [29]. The unikaryotic phase is thought to be associated with sexual reproduction and species lacking this phase are often assumed to be asexual [24]. By affecting the degree to which parasite populations mix, these differences in life cycle and transmission are predicted to have important implications for the degree of intragenomic and intergenomic diversity found in Nosema parasites within a single host.

For the purposes of this study, Nosema species were divided into three groups (Table 1), depending on their life cycles. Group 1 contains species which lack the putatively sexual unikaryotic phase and the ability to undergo horizontal transmission. Because all transmission between hosts is vertical, typically involving approximately 200 parasite cells, a bottleneck in population size occurs each host generation [30], [31]. It is therefore predicted that the genetic diversity of group 1 species should be eroded rapidly, resulting in intergenomic near-homogeneity within any given host. Most diversity observed in multicopy regions of DNA should therefore represent intragenomic variation between copies. Group 2 contains species which are capable of horizontal transmission (with or without supplementary vertical transmission) but lack a unikaryotic phase in their life cycle and are presumed to be asexual. Mixing of cell lineages through coinfection is likely to occur in Group 2 species and so intergenomic diversity within hosts is predicted to be higher than in species of Group 1. Group 3 contain species that are horizontally transmitted and have lifecycles containing a unikaryotic phase, suggesting that they undergo meiosis and sexual reproduction. As sexual, horizontally transmitted species Group 3 species are expected to show higher levels of recombination than species of groups 1 or 2.

**Table 1:** Assignment of Nosema species to life cycle groups based on transmission and sexuality

| Group | Parasite | Host(s) |
|---|---|---|
| 1: No unikaryotic phase (putatively asexual). No horizontal transmission. | Nosema granulosis | Gammarus duebeni |
| 2: No unikaryotic phase (putatively asexual). Horizontal transmission. | Nosema bombycis | Bombyx mori |
| | | Trichoplusia ni |
| | | Tyria jacobaeae |
| | Nosema apis | Apis mellifera |
| | Nosema lymantriae | Lymantria dispar |
| 3. Unikaryotic phase (putatively sexual). Horizontal transmission. | Vairimorpha cheracis | Cherax destructor |
| | Vairimorpha disparis | Lymantria dispar |
| | Vairimorpha necatrix | Lacanobia oleracea |

doi:10.1371/journal.pone.0055878.t001

In this study, the genetic diversity of Nosema species belonging to the three groups was examined by comparing the genetic diversity of the protein-coding genes RNA Polymerase II(RPB1), Elongation Factor-1 alpha (EF-1α) and Surface antigen protein 30.4 (SAP30.4) between these species. The process of evolution of rDNA units in various Nosema species was then investigated by analysing the pattern of diversity of rDNA sequences within and between isolates from different geographical locations. If concerted evolution occurs then the diversity of rDNA sequences should be lower in Group 1 species than in horizontally transmitted Group 2 or Group 3 species. Furthermore, rDNA diversity should be structured primarily between species and between populations within species rather than between copies within the genome. Additionally, if the species lacking a unikaryotic spore cycle are truly asexual, then Group 3 species should demonstrate evidence of meiotic recombination in single copy genes while species of groups 1 and 2 should not.

# RESULTS

## rDNA Sequences

The diversity of cloned ribosomal DNA sequences appears similar in N. granulosis (group 1), N. bombycis (group 2) and V. cheracis (group 3). In all three species, sliding window analysis of cloned rDNA sequences (Genbank JX213695–213745, JX213774–213781) indicates that the ITS (Figure 1) and IGS (Figure 2) spacer regions of the rDNA repeat unit have elevated nucleotide diversity when compared with the flanking 18S, 16S and 5S rRNA genes. This indicates that purifying selection eliminates mutations in the rRNA genes but is relaxed in the spacer regions. There is no significant difference in ITS diversity between isolates of N. bombycis (Figure 3) and no significant differences in IGS diversity between isolates of N. bombycis or N. granulosis (Figure 4). In each of these cases, the diversity of pooled sequences from all isolates is not significantly greater than that of individual isolates, indicating that most genetic diversity in rDNA sequences occurs within isolates rather than between isolates.

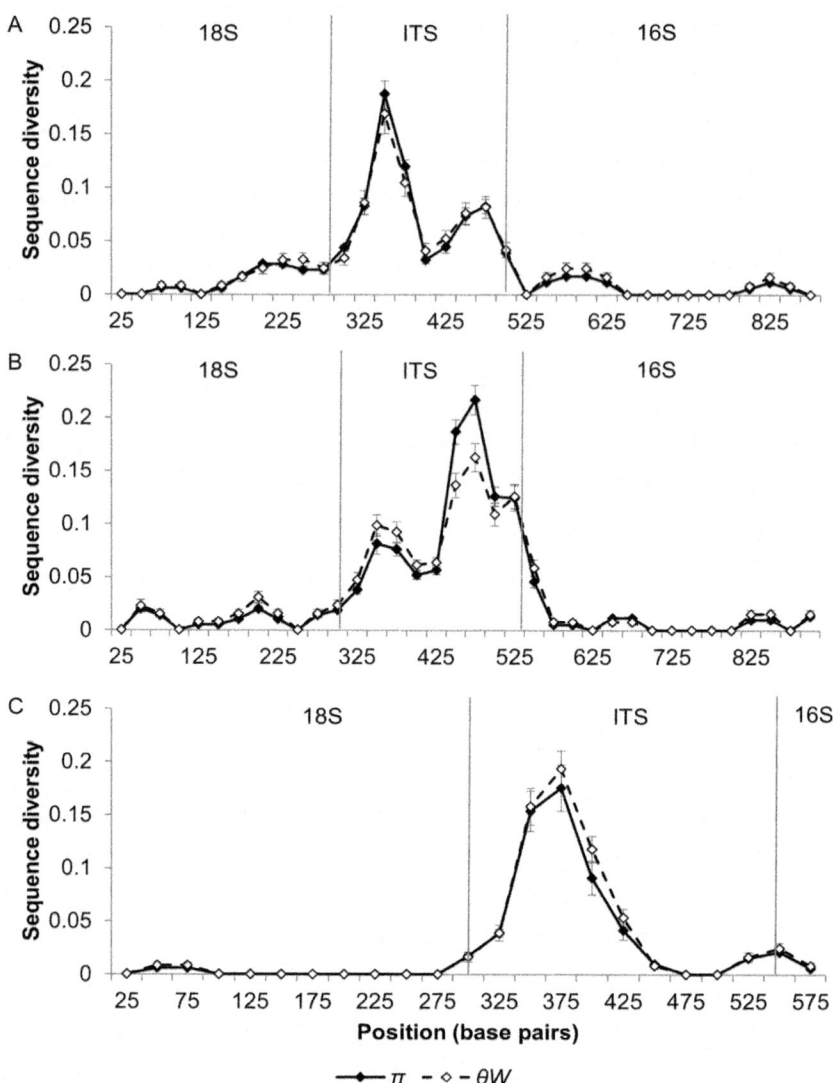

**Figure 1:** Sliding window analysis of nucleotide diversity in the ITS and flanking 18S and 16S rDNA regions of N. bombycis (Part A), N. granulosis (Part B) and V. cheracis (Part C). A sliding window of 50 base pairs is used, with an increment of 25 base pairs. Nucleotide diversity is calculated as the average heterozygosity per site ($\pi$) and the average number of nucleotide differences per site ($\theta_w$). Error bars show the standard error for each window.

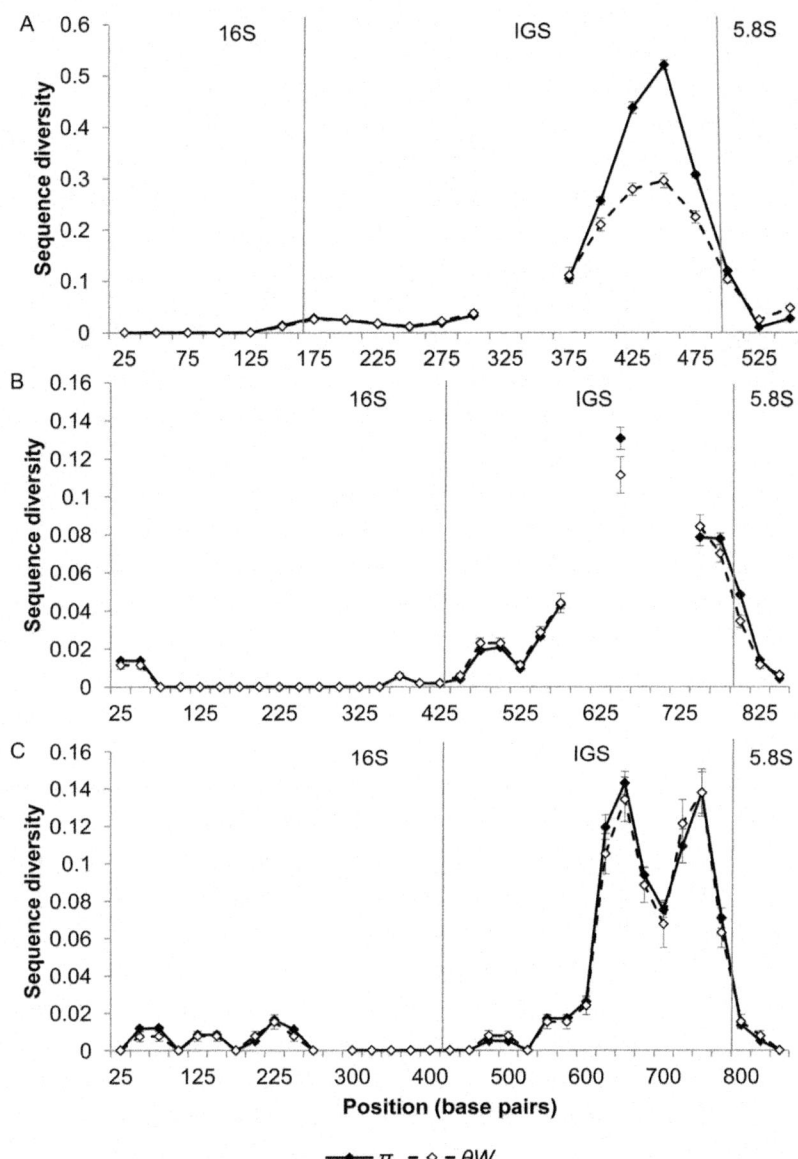

**Figure 2:** Sliding window analysis of nucleotide diversity in the IGS and flanking 16S and 5.8S rDNA regions of N. bombycis (Part A), N. granulosis (Part B) and V. cheracis (Part C). A sliding window of 50 base pairs is used, with an increment of 25 base pairs. Nucleotide diversity is calculated as the average heterozygosity per site ($\pi$) and the average number of nucleotide differences per site ($\theta_w$). Error bars show the standard error for each window. Regions in which sequences could not be aligned due to multiple insertions and deletions are indicated by missing data (breaks in the line).

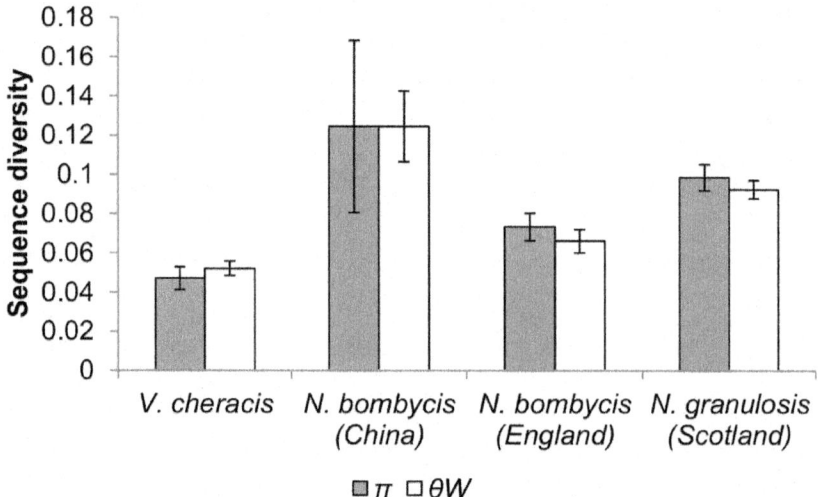

**Figure 3:** Nucleotide diversity of the ITS and flanking 18S and 16S rDNA regions of N. bombycis, N. granulosis and V. cheracis isolates. Nucleotide diversity is calculated as the average heterozygosity per site ($\pi$) and the average number of nucleotide differences per site ($\theta_w$). Error bars show the standard error for each isolate.

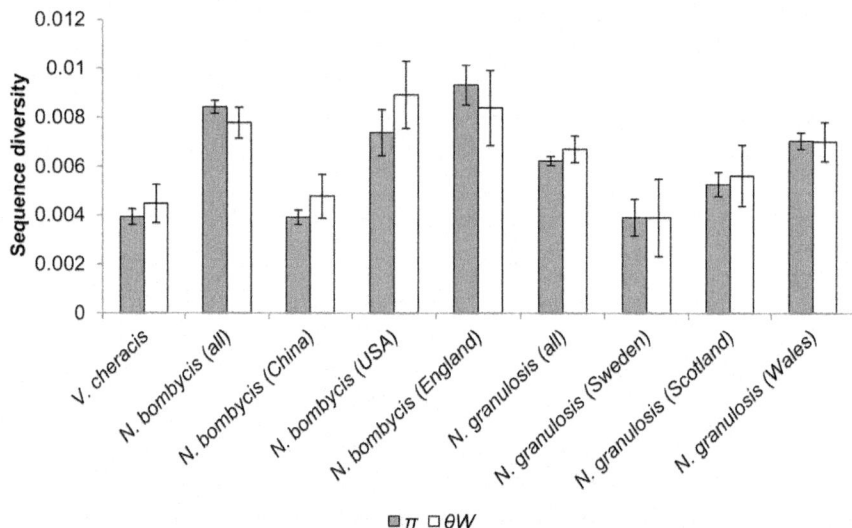

**Figure 4:** Nucleotide diversity of the IGS and flanking 16S and 5.8S rDNA regions of N. bombycis, N. granulosis and V. cheracis isolates. Nucleotide diversity is calculated as the average heterozygosity per site ($\pi$) and the average number of nucleotide differences per site ($\theta_w$). Error bars show the standard error for each isolate.

Haplotype networks indicate that rDNA haplotypes of N. bombycis, N. granulosis and V. cheracis fall into distinct clusters (Figure 5). While IGS sequences of the group 1 species N. granulosis do not cluster within isolates, the rDNA sequences of the group 2 species N. bombycis show clustering within isolates (Figure 5). These results are confirmed by permutation tests of $F_{ST}$ which indicate significant population structure between species and among isolates of N. bombycis but not among isolates of N. granulosis (Table 2). Recombination events were detected through analysis of cloned ITS and IGS sequences in all three species (Table 3).

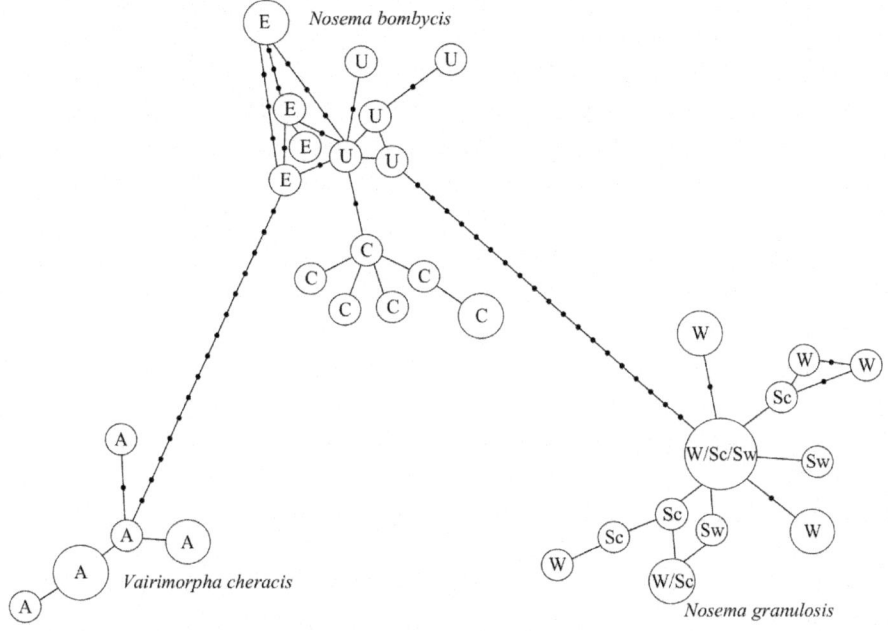

**Figure 5:** Haplotype network of cloned rDNA sequences containing the IGS and flanking 16S and 5.8S rDNA regions of N. bombycis, N. granulosis and V. cheracis isolates. Branch lengths are proportional to the number of mutations separating haplotypes. Areas of circles are proportional to the number of clones containing each haplotype. Isolates are labelled as follows: A = V. cheracis (Australia), E = N. bombycis (England), U = N. bombycis (USA), C = N. bombycis (China), W = N. granulosis (Wales), Sc = N. granulosis (Scotland), Sw = N. granulosis (Sweden).

**Table 2:** Population structure of ribosomal IGS sequences from N. bombycis,N.granulosis and V. cheracis isolates

| Isolate | Sc | W | Sw | C | U | E | A |
|---------|------|------|------|------|------|------|------|
| | | | $F_{ST}$ | | | | |
| Sc | | 0.02 | 0.10 | 0.92*** | 0.89*** | 0.88*** | 0.93*** |
| W | 0.00013 | | 0.05 | 0.90*** | 0.87*** | 0.87*** | 0.91*** |
| Sw | 0.00049 | 0.0003 | | 0.93*** | 0.90*** | 0.89*** | 0.94*** |
| C | 0.05296 | 0.05207 | 0.05266 | | 0.31** | 0.45*** | 0.91*** |
| U | 0.05049 | 0.04961 | 0.0502 | 0.00258 | | 0.17** | 0.87*** |
| E | 0.05382 | 0.05294 | 0.05353 | 0.00552 | 0.00176 | | 0.85*** |
| A | 0.05903 | 0.05938 | 0.0605 | 0.04216 | 0.03901 | 0.03771 | |
| | | | $D_a$ | | | | |

Comparisons were made using Wright's index of fixation ($F_{ST}$) and net nucleotide substitutions per isolate ($D_a$). Levels of significance: *0.05, **0.01, ***0.001. Isolates are labelled as follows: A = V. cheracis (Australia), E = N. bombycis (England), U = N. bombycis (USA), C = N. bombycis (China), W = N. granulosis (Wales), Sc = N. granulosis (Scotland), Sw = N. granulosis (Sweden).
doi:10.1371/journal.pone.0055878.t002

**Table 3:** Recombination events detected in alignments of cloned sequences using RDP4

| Gene | Species | Break points | Recombinant sequences | Parental sequences | Methods | p |
|------|---------|-------------|----------------------|-------------------|---------|---|
| IGS | N. granulosis | 561–673 | Wales_Frd9, Wales_Frd11 | Sweden_Nyd17, Wales_Frd4 | 3Seq | 0.004 |
| | | 487–630 | Sweden_Nyd6 | Wales_Frd10, Scotland_Jsb5 | SiScan | 0.048 |
| | | 604-? | Scotland_Jsb5 | Wales_Frd12, unknown | SiScan | 0.021 |
| | | ?-677 | Sweden_Nyd8 | Wales_Frd10, unknown | SiScan | 0.019 |
| | V. cheracis | 169–255 | Vch8 | Vch2, Vch7 | MaxChi, SiScan, 3Seq | 0.001 |
| | | ?-221 | Vch6 | Vch2, Vch1 | MaxChi | 0.009 |
| ITS | N. bombycis | 359–452 | China_Bm8 | England_Nty8, unknown | GeneConv, BootScan, MaxChi, Chimaera, 3Seq | 0.019 |
| Rpb1 | N. bombycis | 2421–2648 | China_Bm9, England_Tj9 | England_TJ5, USA_Tn5 | SiScan | 0.012 |
| | | 484-? | China_Bm9 | England_TJ5, unknown | MaxChi, 3Seq | 0.018 |
| | V. necatrix | 446-? | Vnec3 | Vnec1, Vnec2 | RDP, GeneConv, BootScan, 3Seq | 0.048 |

Only recombination detection methods providing statistically significant support for a given recombination event are listed. P-values are for Bonferroni-corrected multiple comparisons.
doi:10.1371/journal.pone.0055878.t003

Sequence diversity within the short ITS regions of N. apis (group 1) and V. necatrix (group 2) does not appear to be elevated compared to the flanking 18S and 16S regions while analysis of the cloned ribosomal sequences (Genbank JX213654–213662, JX213789-JX213795) provides some evidence of recombination (Table 3).

## Protein-coding Sequences

Direct sequencing of fragments of the largest subunit of RNA Polymerase II (RPB1) amplified from N. granulosis, V. cheracis and N. apis (Genbank JX213746–213748, DQ996235, DQ996230) revealed little genetic diversity within isolates (Table 4, Figure 6). However, fixed differences indicated genetic divergence between the two British isolates of N. granulosis (from Wales and Scotland) and the Swedish isolate (Table 5). RPB1 sequences from V. necatrix, V. disparis and N. lymantriae (Genbank DQ996236, JX213748, JX213749) exhibited levels of diversity slightly higher than those of N. granulosis (Table 4). In contrast, RPB1 sequences from N. bombycis (Genbank DQ996231) displayed extremely high levels of synonymous nucleotide diversity. Many polymorphisms are shared between N. bombycis isolates (Table 6) and there are relatively few fixed differences between isolates. Direct sequencing revealed no indels in RPB1 in any species and the vast majority of single nucleotide polymorphisms occurred at third codon positions (Table 4, Figure 6). This indicates strong purifying selection to maintain gene function, confirming that the amplified fragments belong to functional genes. Analysis of six cloned RPB1 sequences (Genbank JX213750–JX213755, JX213663–JX213670, JX213796–JX213798) indicated recombination events in N. bombycis and V. necatrix but provided no evidence of recombination in N. apis (Table 3).

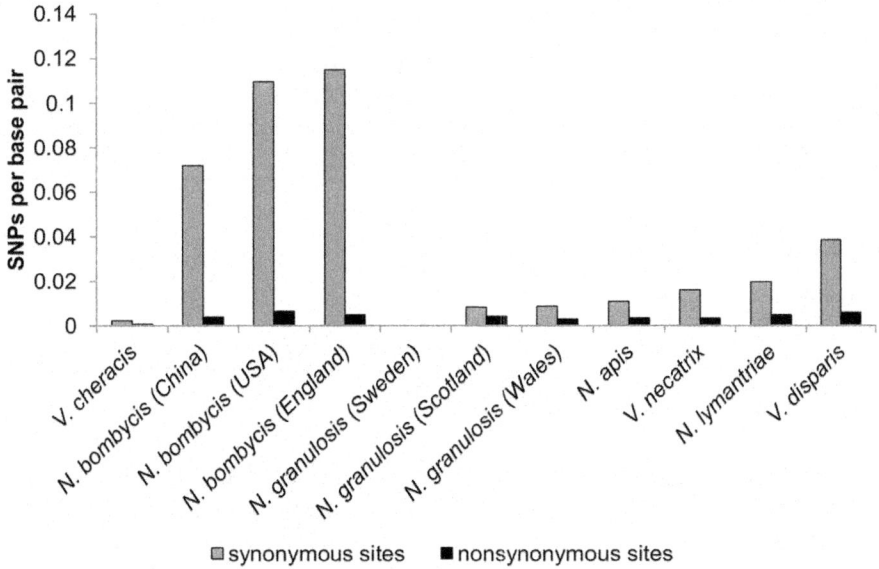

**Figure 6:** Numbers of synonymous and nonsynonymous single nucleotide polymorphisms (SNPs) detected by direct sequencing of the RPB1 gene from Nosema isolates.

**Table 4:** Synonymous and nonsynonymous substitutions detected by direct sequencing of protein-coding genes from Nosema isolates

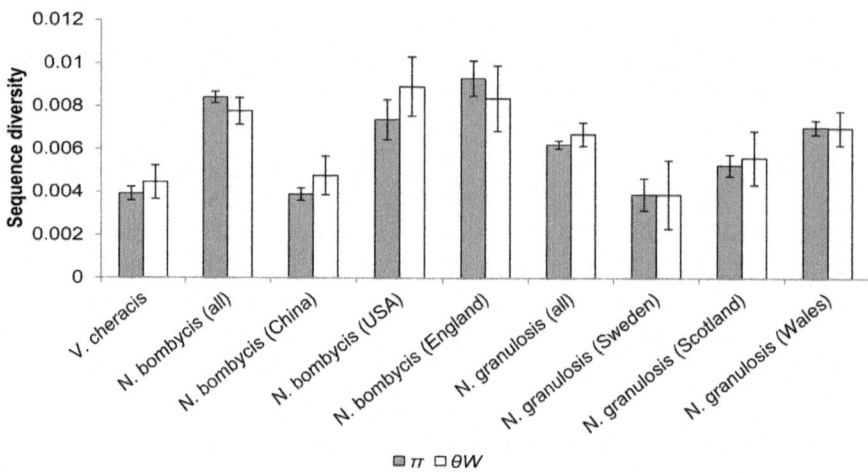

**Table 5:** Pairwise comparisons of RPB1 sequences between N. granulosisisolates, showing frequency of shared polymorphisms, unique polymorphisms and fixed differences

| | | Shared/unique polymorphisms | |
|---|---|---|---|
| | **Scotland** | **Wales** | **Sweden** |
| Scotland | | 4/3 | 0/0 |
| Wales | 0 | | 0/0 |
| Sweden | 12 | 12 | |
| | | Fixed differences | |

doi:10.1371/journal.pone.0055878.t005

**Table 6:** Pairwise comparisons of RPB1 sequences between N. bombycis isolates, showing frequency of shared polymorphisms, unique polymorphisms and fixed differences

| | China | Shared/unique polymorphisms USA | England |
|---|---|---|---|
| China | | 31/130 | 39/115 |
| USA | 3 | | 58/119 |
| England | 13 | 11 | |
| | | Fixed differences | |

doi:10.1371/journal.pone.0055878.t006

Fragments of EF-1α and SAP 30.4 amplified from N. bombycis (Genbank JX213671–JX213694, JX213799, JX213756–JX213773) display moderate levels of synonymous nucleotide diversity (Table 7). Single nucleotide polymorphisms occur in both genes within all three isolates, mainly at third codon positions (Table 4). Direct sequencing of EF-1α revealed a single, polymorphic indel within the isolate from the USA. This is 391 bp in length and is located within a spliceosomal intron. Haplotype networks indicate strong clustering of EF-1α by isolate, but only weak clustering of SAP 30.4 (Figure 7). However, permutation tests of $F_{ST}$ indicate restricted gene flow between isolates in both EF-1α ($F_{ST}$ = 0.38, P<0.0001) and SAP 30.4 ($F_{ST}$= 0.20, P<0.05). Analysis of cloned sequences provides no evidence for multiple recombination events in either of these genes (Table 3).

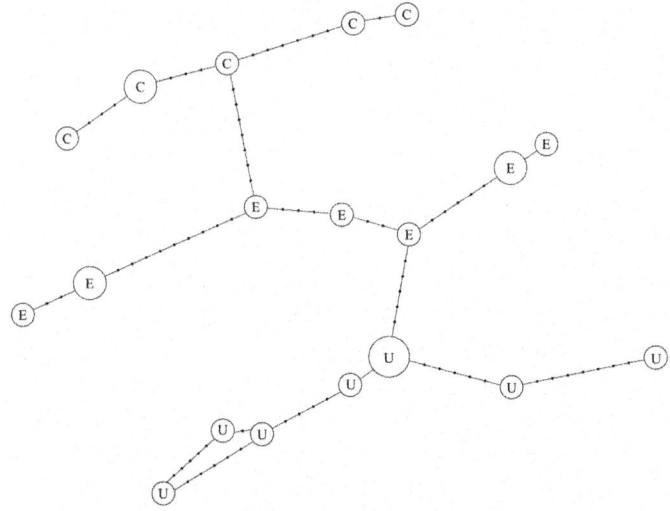

**Figure 7:** Haplotype network of cloned EF-1α sequences from N. bombycisisolates. Branch lengths are proportional to the number of mutations separating haplotypes.

Areas of circles are proportional to the number of clones containing each haplotype. Isolates are labelled as follows: E = England, U = USA, C = China.

**Table 7:** Genetic diversity in cloned sequences from Nosema samples

| Sequence | Species | N | Length (bp) | $\pi \pm SD$ | $\theta_W \pm SD$ |
|---|---|---|---|---|---|
| 18S-ITS-16S | N. bombycis (all) | 7 | 899 | 0.0230±0.01314 | 0.0235±0.00333 |
| | N. bombycis (China) | 2 | 899 | 0.0288±0.02650 | 0.0265±0.00553 |
| | N. bombycis (England) | 5 | 899 | 0.0224±0.01273 | 0.0210±0.00341 |
| | N. granulosis (Scotland) | 8 | 944 | 0.0320±0.01738 | 0.0318±0.00373 |
| | V. cheracis | 7 | 604 | 0.0204±0.01258 | 0.0225±0.00398 |
| 16S-ITS-18S | N. apis | 9 | 782 | 0.0053±0.00384 | 0.0072±0.00185 |
| | V. necatrix | 7 | 795 | 0.0029±0.00172 | 0.0031±0.00126 |
| 16S-IGS-5S | N. bombycis (all) | 17 | 617 | 0.0718±0.02588 | 0.0515±0.00583 |
| | N. bombycis (China) | 7 | 617 | 0.0557±0.02494 | 0.0446±0.00638 |
| | N. bombycis (USA) | 5 | 617 | 0.0741±0.03504 | 0.0579±0.00787 |
| | N. bombycis (England) | 5 | 617 | 0.0888±0.04282 | 0.0707±0.00870 |
| | N. granulosis (all) | 19 | 1022 | 0.0161±0.00740 | 0.0148±0.00241 |
| | N. granulosis (Scotland) | 5 | 1022 | 0.0180±0.01031 | 0.0170±0.00334 |
| | N. granulosis (Wales) | 11 | 1022 | 0.0167±0.00778 | 0.0149±0.00263 |
| | N. granulosis (Sweden) | 3 | 1022 | 0.0100±0.00746 | 0.0100±0.00302 |
| EF-1α | N. bombycis (all) | 24 | 923 | 0.01302±0.00069 | 0.01134±0.00182 |
| | N. bombycis (China) | 6 | 923 | 0.00803±0.00173 | 0.00713±0.00184 |
| | N. bombycis (USA) | 9 | 923 | 0.00838±0.00174 | 0.00798±0.00178 |
| | N. bombycis (England) | 9 | 923 | 0.01095±0.00131 | 0.00837±0.00183 |
| Sap 30.4 | N. bombycis (all) | 18 | 736 | 0.01076±0.00076 | 0.00751±0.00172 |
| | N. bombycis (China) | 5 | 736 | 0.01005±0.00239 | 0.00913±0.00244 |
| | N. bombycis (USA) | 5 | 736 | 0.00842±0.00247 | 0.00783±0.00226 |
| | N. bombycis (England) | 8 | 736 | 0.00907±0.0011 | 0.00629±0.00182 |
| RPB1 | N. bombycis (all) | 6 | 2646 | 0.03132±0.00322 | 0.03161±0.00229 |
| | N. apis | 7 | 1571 | 0.00029±0.0002 | 0.00041±0.00041 |
| | V. necatrix | 3 | 1056 | 0.00631±0.00202 | 0.00631±0.002 |

Nucleotide diversity is calculated as the average heterozygosity per site ($\pi$) and the average number of nucleotide differences per site ($\theta_W$).
doi:10.1371/journal.pone.0055878.t007

# DISCUSSION

The findings of this study provide evidence for intergenomic and intragenomic diversity of rDNA and protein coding sequences within isolates of microsporidia acquired from single hosts. Very high levels of diversity are present in the rDNA intergenic spacer regions and internal transcribed spacer regions of V. cheracis, N. bombycis and N. granulosis, including evidence of multiple insertion, deletion and substitution events. In contrast, the functional rDNA subunits are relatively conserved. This suggests that the similarity of microsporidian rRNA gene products is maintained by purifying selection rather than by concerted evolution [14]. High levels of diversity also occur within the sequences of the protein-coding genes RPB1, EF-1α andSap30.4. Again, the occurrence of the vast majority of polymorphic sites at synonymous positions suggests that the high rates of DNA sequence evolution known to occur in microsporidia [32], [33] are countered by strong purifying selection.

Sequence diversity of the rDNA IGS and ITS of the Group 1 species N. granulosis is similar to those of N. bombycis (Group 2) and V. cheracis (Group 3). Given that N. granulosis lacks horizontal transmission, its intergenomic diversity should be reduced due to repeated bottlenecks during vertical transmission. The similarity of rDNA diversity between species from groups 1, 2 and 3 therefore suggests that this diversity is primarily intragenomic.

Although most rDNA diversity occurs within, rather than between isolates, rDNA haplotypes form clusters within each of the three microsporidian species. This suggests that the length of evolutionary time since these species diverged has been sufficient to obscure shared histories of rDNA copies between species through slow concerted evolution and/or cumulative birth and death events. rDNA haplotypes also cluster within isolates of N. bombycis. However, rDNA haplotypes do not cluster within isolates of N. granulosis. This may reflect the fact that the N. bombycis isolates were collected from three different host species on three different continents while the N. granulosis isolates were collected from a single host species in north-west Europe.

The presence of more than two distinct haplotypes of the genes EF-1$\alpha$ and Sap30.4 within N. bombycis isolates indicates that this diversity is not simply due to heterozygosity within diploid cells, but is likely to be due to intergenomic diversity. Synonymous diversity of RPB1 within isolates of V. cheracis, V. necatrix, N. apis, N. lymantriae and V. disparis is much lower than that of N. bombycis and is more similar to that of N. granulosis despite the fact that the former five species are horizontally transmitted. In contrast, the diversity of RPB1 at non-synonymous sites in N. bombycis is similar to those of the other Nosema species, producing values of Ka/Ksthat are an order of magnitude lower in N. bombycis than in the other species.

The unusually high diversity of RPB1 suggests that it may occur as multiple copies in N. bombycis. However, the genes EF-1$\alpha$ and Sap30.4 also have high synonymous nucleotide diversity in N. bombycis, suggesting that a whole genome duplication in N. bombycis followed by divergence of paralogous copies may have increased the intragenomic diversity of all three genes. In support of this hypothesis, the estimated genome size of N. bombycis (15.33 Mbp) is greater than those published for other Nosema species [34], [35], [36], [37], all of which fall in the range 9.25–10.56 Mbp. However, in this case alleles would be expected to cluster into two divergent groups, present in all isolates and with little evidence of recombination between them. This does not appear to be the case for RPB1, EF-1$\alpha$ or Sap30.4. It therefore seems more likely that the higher RPB1 diversity observed in N. bombycis is intergenomic, resulting from a more diverse population of parasites within each host than is found in the other Nosema species.

Recombination was detected between rDNA sequences in N. granulosis, N. bombycis and V. cheracis. Recombination was also detected in the protein-coding gene RPB1 in N. bombycis and V. necatrix. This is surprising given the lack of unikaryotic stages in the life cycles of N. granulosis and N. bombycis, previously taken to indicate a lack of meiosis. Recombination between homologous and/or non-homologous copies of the rDNA unit within each genome may have occurred at mitosis. This possibility is suggested by studies of other apomictic species, which have revealed highly homogeneous rDNA arrays, indicating that mitotic recombination is sufficient for concerted evolution to occur in the absence of sex [38], [39]. However, it does not explain the evidence for multiple recombination events in RPB1 in N. bombycis unless this gene occurs as multiple copies.

Alternatively, N. bombycis and N. granulosis may possess cryptic meiotic stages, making them capable of sexual reproduction. Unikaryotic cells have occasionally been observed in N. bombycis [40] while recombination between rDNA units has been detected in the honeybee pathogen Nosema ceranae [41] and is also suggested as an explanation for incongruous 16S and 18S rDNA phylogenies in a Nosema species from the host Pieris rapae [42]. The potential for Nosema strains to exchange genes through sexual reproduction has important implications for the evolution of virulence and drug resistance in these damaging pathogens.

In Nosema species possessing the rearranged version of the rDNA repeat unit (V. cheracis, N. bombycis and N. granulosis in this study), evidence for intragenomic diversity suggests that rDNA spacer regions are unsuitable for typing of strains. Evidence of intergenomic diversity also suggests that coinfection with multiple strains and recombination between strains is common. However, the high level of diversity found at synonymous sites in protein-coding genes such as RPB1 suggest that these may be of use in strain typing. In the case of N. granulosis, for example, British and Swedish isolates could not be distinguished through analysis of the ribosomal IGS but could be distinguished clearly by comparing RPB1 sequences.

In Nosema species possessing the non-rearranged version of the rDNA repeat unit (N. apis, V. necatrix, N. lymantriae and V. disparis in this study). The ITS region is very short (33–34 bp) and does not appear to be substantially more variable within isolates than the flanking rRNA subunit genes. This may be taken as evidence that the arrangement of rDNA units in tandem repeats has led to stronger concerted evolution in these species than in species possessing the rearranged rDNA unit. However, diversity within single hosts has been detected in the ITS of N. bombi [25], [43] and in intergenic regions flanking the rDNA units of N. apis [21] and N. ceranae[41].

In conclusion, the findings of this study provide some evidence for concerted evolution in the dispersed rDNA units of microsporidia. This is sufficient to produce clustering of rDNA copies within species, and even within populations, but is insufficient to prevent diversification of rDNA copies within microsporidian genomes. Additional genetic variation is provided by intergenomic diversity of microsporidia within single hosts and by recombination between the resulting heterogeneous strains. However, microsporidia appear to be subject to strong purifying selection, with the result that most genetic diversity occurs within rDNA spacer regions or at synonymous sites in protein-coding genes.

# MATERIALS AND METHODS

## Samples

Three isolates of N. bombycis spores were obtained from the hosts Bombyx morii (China),Trichoplusia ni (USA) and Tyria jacobaeae (England). The Chinese isolate originated in cultured silk moths, the American isolate was purchased from the American Type Culture Collection (strain ATCC 30702) and the English isolate was obtained from a wild cinnabar moth population near Plymouth. The American isolate was originally described as N. trichoplusiae[44]. However, N. bombycis and N. trichoplusiae have since been shown to be synonymous[45]. Three isolates of N. granulosis were obtained from wild populations of the amphipod crustacean Gammarus duebeni from Anglesey (Wales, UK), Isle of Cumbrae (Scotland, UK) and Nynashamn (Sweden). Field studies at Plymouth, Nynashamn, Anglesey and Isle of Cumbrae did not involve protected species and were undertaken on public land that was not subject to any form of protection. No specific permits were required for the described field studies.

A single isolate of the putatively sexual, octosporous species Vairimorpha cheracis, originally obtained from a population of the Australian yabby Cherax destructor was donated by the University of New England, Australia. An isolate of N. apis was obtrained from honey bees (Apis mellifera) in Ireland and an isolate of V. necatrix was obtained from a laboratory population of the moth Lacanobia oleracea maintained at Central Science Laboratories, England. Isolates of Vairimorpha disparis and Nosema lymantriae from a laboratory population of Gypsy Moth Lymantria dispar, were donated by Illinois Natural History Survey, USA and used with their permission.

Each isolate was obtained from a single host individual and dissected using flame-sterilised forceps to avoid cross-contamination. The only exception to this was the case of N. apis in which spores were harvested from several honey bees from the same colony. Spores of N. bombycis (China and USA), Vairimopha disparis and Nosema lymantriae were purified using a percol gradient. Other isolates were not purified. Total DNA (containing host and microsporidian DNA) was extracted from infected tissues. Genomic DNA was extracted from all isolates using Qiagen's DNeasy® DNA purification kit, according to the manufacturer's instructions.

## PCR, Cloning and Sequencing

Sequences for all PCR and sequencing primers are provided as supporting information. All of these primers were designed to be microsporidian-specific in order to avoid amplification of host DNA. A fragment of rDNA was amplified from each N. bombycis, N. granulosis and V. cheracis isolate using the primers HG4F and 5SR. The rDNA fragment was approximately 550 bp in length and contained the intergenic spacer (IGS) separating the 16S and 5S ribosomal RNA subunits. A second rDNA fragment, approximately 830 bp in length and containing the internal transcribed spacer (ITS) separating the 16S and 18S subunits was amplified from the single V. cheracis isolate, the Scottish N. granulosis isolate and the N. bombycis isolates from China and England using the primers ILSUF and 530R. Fragments of rDNA were also amplified from N. apis and V. necatrix isolates using the primers HG4F and HG4R. The resulting rDNA fragments were approximately 800 bp in length and contained the internal transcribed spacer (ITS) separating the 16S and 18S ribosomal RNA genes.

A fragment of the largest subunit of the housekeeping gene RNA polymerase II (RPB1) was amplified from each N. granulosis and N. bombycis isolate and from the isolates of N. apis, V. cheracis, V. necatrix, V. disparis and N. lymantriae. In each case, the general microsporidian primers RPB1F, RPB1R, AF1, AF3 and GR1 [46] were used to obtain preliminary sequence data. This was then used as a template to design species-specific primers. A fragment of the housekeeping gene elongation factor-1 alpha (EF-1$\alpha$) was also amplified from each N. bombycis isolate with the primers EF1$\alpha$F and EF1$\alpha$R while the gene for spore surface antigen protein 30.4 (Sap 30.4) was amplified from the three N. bombycis isolates using the primers SAPF and SAPR. Southern blot analysis indicates that RPB1 occurs as a single copy inVairimorpha necatrix [47]. Sequence similarity searches of the genomes of N. ceranae (the onlyNosema genome currently available) and E. cuniculi (the only fully assembled microsporidian genome) were performed for RPB1,

EF1-α and Sap 30.4 using BLASTn and tBLASTx, implemented on the NCBI website with an alignment score cut-off of 200. These detected a single copy of RPB1 and EF1-α in each genome. No sequences similar to Sap 30.4 were detected in either genome, suggesting that this gene has evolved recently in the lineage containing N. bombycis.

All PCR products were sequenced directly in both directions using BigDye® terminators on an ABI 3100 high throughput sequencer. Chromatograms of RPB1, EF-1α and Sap 30.4sequences were studied carefully by eye and double-peaks indicating single nucleotide polymorphisms within a isolate were noted. This could not be accomplished for rDNA sequences because multiple indels resulted in overlapping sequences. PCR products were cloned using TOPO TA according to the manufacturer's instructions. Plasmid DNA was purified using Qiagen's QIAprep® Spin DNA purification kit and sequenced with primers T7 and T3. For products greater than 1 kb in length, internal primers were designed to allow full sequencing. Extensive overlap between fragments amplified by general and species-specific primer pairs ensured that most regions of DNA were amplified and sequenced at least twice, independently. This increased the likelihood of detecting PCR and sequencing artefacts.

## Analysis

Sequences were aligned using Clustal W and adjusted manually. Fasta files of sequence alignments are provided as supporting information. In the cases of the rDNA spacer regions (ITS and IGS), multiple, overlapping indels rendered it impossible to align sequences with confidence in highly variable regions. Such hypervariable regions were therefore excluded from the alignments. In order to avoid cloning artifacts, differences in sequence between cloned DNA fragments were accepted as genuine polymorphisms only if they corresponded to double peaks obtained through direct sequencing. The numbers of cloned rDNA and protein-coding sequences used in the analyses are provided in Table 7.

Changes in nucleotide diversity along each rDNA sequence alignment were calculated as the average heterozygosity per site ($\pi$) and the average number of nucleotide differences per site ($\theta_w$) within a sliding window of 50 base pairs with an increment of 25 base pairs, implemented in Proseq 3.2 [48]. Haplotype networks were constructed for the cloned rDNA sequences obtained from N. bombycis, N. granulosis and V. cheracis, and for the EF-1α sequences obtained from N. bombycis. Connection distances between haplotypes were calculated using Arlequin [49] and the network was visualised using a force-directed algorithm, implemented in Hapstar [50].

Where several different sequences were cloned from the same gene within the same host species, recombination events were detected using RDP4 [51]. This program allows a sequence alignment to be analysed simultaneously with the RDP [52], BootScan [53], GeneConv [54], MaxChi [55], Chimaera [56], SiScan [57], 3Seq [58] and LARD [59] methods to provide a single multiple comparison (MC) Bonferroni-corrected p-value for each recombination event.

## ACKNOWLEDGMENTS

I would like to thank Leellen Solter for providing isolates of Nosema lymantriae and Vairimorpha disparis.

## AUTHOR CONTRIBUTIONS

Conceived and designed the experiments: JEI. Performed the experiments: JEI. Analyzed the data: JEI. Contributed reagents/materials/analysis tools: JEI. Wrote the paper: JEI.

## REFERENCES

1.  Voigt K, Kirk PM (2011) Recent developments in the taxonomic affiliation and phylogenetic positioning of fungi: impact in applied microbiology and environmental biotechnology. Applied Microbiology and Biotechnology 90: 41–57. doi: 10.1007/s00253-011-3143-4

2.  Lee SC, Corradi N, Doan S, Dietrich FS, Keeling PJ, et al.. (2010) Evolution of the sex-Related Locus and Genomic Features Shared in Microsporidia and Fungi. Plos One 5.

3.  Keeling PJ, Fast NM (2002) Microsporidia: Biology and evolution of highly reduced intracellular parasites. Annual Review of Microbiology 56: 93–116. doi: 10.1146/annurev.micro.56.012302.160854

4.  Sak B, Kvac M, Petrzelkova K, Kvetonova D, Pomajbikova K, et al. (2011) Diversity of microsporidia (Fungi: Microsporidia) among captive great apes in European zoos and African sanctuaries: evidence for zoonotic transmission? Folia Parasitologica 58: 81–86. doi: 10.14411/fp.2011.008

5.  Santin M, Fayer R (2009) Enterocytozoon bieneusi Genotype Nomenclature Based on the Internal Transcribed Spacer Sequence: A Consensus. Journal of Eukaryotic Microbiology 56: 34–38. doi: 10.1111/j.1550-7408.2008.00380.x

6.  Wilkinson TJ, Rock J, Whiteley NM, Ovcharenko MO, Ironside JE (2011) Genetic diversity of the feminising microsporidian parasite Dictyocoela:

New insights into host-specificity, sex and phylogeography. International Journal for Parasitology 41: 959–966. doi: 10.1016/j.ijpara.2011.04.002

7.    Li JL, Chen WF, Wu J, Peng WJ, An JD, et al. (2012) Diversity of Nosema associated with bumblebees (Bombus spp.) from China. International Journal for Parasitology 42: 49–61. doi: 10.1016/j.ijpara.2011.10.005

8.    Huang WF, Bocquet M, Lee KC, Sung IH, Jiang JH, et al. (2008) The comparison of rDNA spacer regions of Nosema ceranae isolates from different hosts and locations. Journal of Invertebrate Pathology 97: 9–13. doi: 10.1016/j.jip.2007.07.001

9.    Schoch CL, Seifert KA, Huhndorf S, Robert V, Spouge JL, et al.. (2012) Nuclear ribosomal internal transcribed spacer (ITS) region as a universal DNA barcode marker for Fungi. In Press.

10.   Liu HD, Pan GQ, Song SH, Xu JS, Li T, et al. (2008) Multiple rDNA units distributed on all chromosomes of Nosema bombycis. Journal of Invertebrate Pathology 99: 235–238. doi: 10.1016/j.jip.2008.06.012

11.   Dover G, Coen E (1981) Spring-Cleaning Ribosomal DNA - a Model for Multigene Evolution. Nature 290: 731–732. doi: 10.1038/290731a0

12.   Goldman ASH, Lichten M (1996) The efficiency of meiotic recombination between dispersed sequences in Saccharomyces cerevisiae depends upon their chromosomal location. Genetics 144: 43–55.

13.   Rooney AP (2004) Mechanisms underlying the evolution and maintenance of functionally heterogeneous 18S rRNA genes in apicomplexans. Molecular Biology and Evolution 21: 1704–1711. doi: 10.1093/molbev/msh178

14.   Nei M, Rooney AP (2005) Concerted and birth-and-death evolution of multigene families. Annual Review of Genetics 39: 121–152. doi: 10.1146/annurev.genet.39.073003.112240

15.   Rooney AP, Ward TJ (2005) Evolution of a large ribosomal RNA multigene family in filamentous fungi: Birth and death of a concerted evolution paradigm. Proceedings of the National Academy of Sciences of the United States of America 102: 5084–5089. doi: 10.1073/pnas.0409689102

16.   Negi MS, Rajagopal J, Chauhan N, Cronn R, Lakshmikumaran M (2002) Length and sequence heterogeneity in 5S rDNA of Populus deltoides. Genome 45: 1181–1188. doi: 10.1139/g02-094

17.   Vierna J, Gonzalez-Tizon AM, Martinez-Lage A (2009) Long-Term Evolution of 5S Ribosomal DNA Seems to Be Driven by Birth-and-Death Processes and Selection inEnsis Razor Shells (Mollusca: Bivalvia). Biochemical Genetics 47: 635–644. doi: 10.1007/s10528-009-9255-1

18. Pinhal D, Yoshimura T, Araki C, Martins C (2011) The 5S rDNA family evolves through concerted and birth-and-death evolution in fish genomes: an example from freshwater stingrays. Bmc Evolutionary Biology 11: 151. doi: 10.1186/1471-2148-11-151

19. Freire R, Arias A, Insua AM, Mendez J, Eirin-Lopez JM (2010) Evolutionary Dynamics of the 5S rDNA Gene Family in the Mussel Mytilus: Mixed Effects of Birth-and-Death and Concerted Evolution. Journal of Molecular Evolution 70: 413–426. doi: 10.1007/s00239-010-9341-3

20. Vrba V, Poplstein M, Pakandl M (2011) The discovery of the two types of small subunit ribosomal RNA gene in Eimeria mitis contests the existence of E. mivati as an independent species. Veterinary Parasitology 183: 47–53. doi: 10.1016/j.vetpar.2011.06.020

21. Gatehouse HS, Malone LA (1998) The ribosomal RNA gene region of Nosema apis(microspora): DNA sequence for small and large subunit rRNA genes and evidence of a large tandem repeat unit size. Journal of Invertebrate Pathology 71: 97–105. doi: 10.1006/jipa.1997.4737

22. Iiyama K, Chieda Y, Yasunaga-Aoki C, Hayasaka S, Shimizu S (2004) Analyses of the ribosomal DNA region in Nosema bombycis NIS 001. Journal of Eukaryotic Microbiology 51: 598–604. doi: 10.1111/j.1550-7408.2004.tb00592.x

23. Huang WF, Tsai SJ, Lo CF, Soichi Y, Wang CH (2004) The novel organization and complete sequence of the ribosomal RNA gene of Nosema bombycis. Fungal Genetics and Biology 41: 473–481. doi: 10.1016/j.fgb.2003.12.005

24. Ironside JE (2007) Multiple losses of sex within a single genus of microsporidia. Bmc Evolutionary Biology 7: 48. doi: 10.1186/1471-2148-7-48

25. O'Mahony EM, Tay WT, Paxton RJ (2007) Multiple rRNA variants in a single spore of the microsporidian Nosema bombi. Journal of Eukaryotic Microbiology 54: 103–109. doi: 10.1111/j.1550-7408.2006.00232.x

26. Tay WT, O'Mahony EM, Paxton RJ (2005) Complete rRNA gene sequences reveal that the microsporidium Nosema bombi infects diverse bumblebee (Bombus spp.) hosts and contains multiple polymorphic sites. Journal of Eukaryotic Microbiology 52: 505–513. doi: 10.1111/j.1550-7408.2005.00057.x

27. Canning EU, Curry A, Cheney S, Lafranchi-Tristem NJ, Haque MA (1999) Vairimorpha imperfecta n.sp., a microsporidian exhibiting an abortive octosporous sporogony inPlutella xylostella L. (Lepidoptera

: Yponomeutidae). Parasitology 119: 273–286. doi: 10.1017/s0031182099004734

28. Pilley BM (1976) A new genus, Vairimorpha (Protozoa: Microsporida), for Nosema necatrix Kramer 1965: Pathogenicity and life cycle in Spodoptera exempta(Lepidoptera: Noctuidae). Journal of Invertebrate Pathology 28: 177–183. doi: 10.1016/0022-2011(76)90119-1

29. Baker MD, Vossbrinck CR, Maddox JV, Undeen AH (1994) Phylogenetic-Relationships among Vairimorpha and Nosema Species (Microspora) Based on Ribosomal-Rna Sequence Data. Journal of Invertebrate Pathology 64: 100–106. doi: 10.1006/jipa.1994.1077

30. Weedall RT, Robinson M, Smith JE, Dunn AM (2006) Targeting of host cell lineages by vertically transmitted, feminising microsporidia. International Journal for Parasitology 36: 749–756. doi: 10.1016/j.ijpara.2006.02.020

31. Dunn AM, Terry RS, Taneyhill DE (1998) Within-host transmission strategies of transovarial, feminizing parasites of Gammarus duebeni. Parasitology 117: 21–30. doi: 10.1017/s0031182098002753

32. Corradi N, Akiyoshi DE, Morrison HG, Feng XC, Weiss LM, et al.. (2007) Patterns of Genome Evolution among the Microsporidian Parasites Encephalitozoon cuniculi,Antonospora locustae and Enterocytozoon bieneusi. Plos One 2.

33. Nassonova E, Cornillot E, Metenier G, Agafonova N, Kudryavtsev B, et al. (2005) Chromosomal composition of the genome in the monomorphic diplokaryotic microsporidium Paranosema grylli: analysis by two-dimensional pulsed-field gel electrophoresis. Folia Parasitologica 52: 145–157. doi: 10.14411/fp.2005.019

34. Cornman RS, Chen YP, Schatz MC, Street C, Zhao Y, et al. (2009) Genomic Analyses of the Microsporidian Nosema ceranae, an Emergent Pathogen of Honey Bees. Plos Pathogens 5: 14. doi: 10.1371/journal.ppat.1000466

35. Kawakami Y, Inoue T, Ito K, Kitamizu K, Hanawa C, et al. (1994) Comparison of chromosomal DNA from four microsporidia pathogenic to the silkworm Bombyx mori. Applied Entomology and Zoology 29: 120–123.

36. Malone LA, McIvor CA (1993) Pulsed-field gel-electrophoresis of DNA from four microsporidian isolates. Journal of Invertebrate Pathology 61: 203–205. doi: 10.1006/jipa.1993.1036

37. Munderloh UG, Kurtti TJ, Ross SE (1990) Electrophoretic characterization of chromosomal DNA from two microsporidia. Journal of Invertebrate Pathology 56: 243–248. doi: 10.1016/0022-2011(90)90107-h

38. Crease TJ, Lynch M (1991) Ribosomal DNA variation in Daphnia pulex. Molecular Biology and Evolution 8: 620–640.

39. McTaggart S, Dudycha JL, Omilian A, Crease TJ (2007) Rates of recombination in the ribosomal DNA of apomictically propagated Daphnia obtusa lines. Genetics 175: 311–320. doi: 10.1534/genetics.105.050229

40. Streett DA, Lynn DE (1984) Nosema bombycis Replication in a Manduca sexta Cell-Line. Journal of Parasitology 70: 452–454. doi: 10.2307/3281586

41. Sagastume S, del Aguila C, Martin-Hernandez R, Higes M, Henriques-Gil N (2011) Polymorphism and recombination for rDNA in the putatively asexual microsporidianNosema ceranae, a pathogen of honeybees. Environmental Microbiology 13: 84–95. doi: 10.1111/j.1462-2920.2010.02311.x

42. Choi Y, Lee Y, Cho KS, Lee S, Russell J, et al. (2011) Chimerical nature of the ribosomal RNA gene of a Nosema species. Journal of Invertebrate Pathology 107: 86–89. doi: 10.1016/j.jip.2011.02.005

43. Cordes N, Huang WF, Strange JP, Cameron SA, Griswold TL, et al. (2012) Interspecific geographic distribution and variation of the pathogens Nosema bombi andCrithidia species in United States bumble bee populations. Journal of Invertebrate Pathology 109: 209–216. doi: 10.1016/j.jip.2011.11.005

44. Tanabe AM, Tamashir M (1967) The biology and pathogenicity of a microsporidian (Nosema trichoplusiae sp. n.) of the cabbage looper, Trichoplusia ni (Hubner) (Lepidoptera: Noctuidae). Journal of Invertebrate Pathology 9: 188–&.

45. Pieniazek NJ, daSilva AJ, Slemenda SB, Visvesvara GS, Kurtti TJ, et al. (1996) Nosema trichoplusiae is a synonym of Nosema bombycis based on the sequence of the small subunit ribosomal RNA coding region. Journal of Invertebrate Pathology 67: 316–317. doi: 10.1006/jipa.1996.0049

46. Cheney SA, Lafranchi-Tristem NJ, Bourges D, Canning EU (2001) Relationships of microsporidian genera, with emphasis on the polysporous genera, revealed by sequences of the largest subunit of RNA polymerase II (RPB1). Journal of Eukaryotic Microbiology 48: 111–117. doi: 10.1111/j.1550-7408.2001.tb00422.x

47. Hirt RP, Logsdon JM, Healy B, Dorey MW, Doolittle WF, et al. (1999) Microsporidia are related to Fungi: Evidence from the largest subunit

of RNA polymerase II and other proteins. Proceedings of the National Academy of Sciences of the United States of America 96: 580–585. doi: 10.1073/pnas.96.2.580

48. Filatov DA (2002) PROSEQ: A software for preparation and evolutionary analysis of DNA sequence data sets. Molecular Ecology Notes 2: 621–624. doi: 10.1046/j.1471-8286.2002.00313.x

49. Excoffier L, Laval G, Schneider S (2005) Arlequin (version 3.0): An integrated software package for population genetics data analysis. Evol Bioinform 1: 47–50.

50. Teacher AGF, Griffiths DJ (2010) HapStar: automated haplotype network layout and visualization. Mol Ecol Res 11: 151–153. doi: 10.1111/j.1755-0998.2010.02890.x

51. Martin DP, Lemey P, Lott M, Moulton V, Posada D, et al. (2010) RDP3: a flexible and fast computer program for analyzing recombination. Bioinformatics 26: 2462–2463. doi: 10.1093/bioinformatics/btq467

52. Martin D, Rybicki E (2000) RDP: detection of recombination amongst aligned sequences. Bioinformatics 16: 562–563. doi: 10.1093/bioinformatics/16.6.562

53. Martin DP, Posada D, Crandall KA, Williamson C (2005) A modified bootscan algorithm for automated identification of recombinant sequences and recombination breakpoints. Aids Research and Human Retroviruses 21: 98–102. doi: 10.1089/aid.2005.21.98

54. Padidam M, Sawyer S, Fauquet CM (1999) Possible emergence of new geminiviruses by frequent recombination. Virology 265: 218–225. doi: 10.1006/viro.1999.0056

55. Smith JM (1992) Analyzing the mosaic structure of genes. Journal of Molecular Evolution 34: 126–129. doi: 10.1007/bf00182389

56. Posada D, Crandall KA (2001) Evaluation of methods for detecting recombination from DNA sequences: Computer simulations. Proceedings of the National Academy of Sciences of the United States of America 98: 13757–13762. doi: 10.1073/pnas.241370698

57. Gibbs MJ, Armstrong JS, Gibbs AJ (2000) Sister-Scanning: a Monte Carlo procedure for assessing signals in recombinant sequences. Bioinformatics 16: 573–582. doi: 10.1093/bioinformatics/16.7.573

58. Boni MF, Posada D, Feldman MW (2007) An exact nonparametric method for inferring mosaic structure in sequence triplets. Genetics 176: 1035–1047. doi: 10.1534/genetics.106.068874

59. Holmes EC, Worobey M, Rambaut A (1999) Phylogenetic evidence for recombination in dengue virus. Molecular Biology and Evolution 16: 405–409. doi: 10.1093/oxfordjournals.molbev.a026121

# CITATION

## CHAPTER 1

Vitali Alexeev, Alyson Pidich, Daria Marley Kemp and Olga Igoucheva (2013). Recombinant DNA Technology in Emerging Modalities for Melanoma Immunotherapy, Melanoma - From Early Detection to Treatment, Dr. Ht Duc (Ed.), ISBN: 978-953-51-0961-7, InTech, DOI: 10.5772/55357.

## CHAPTER 2

Jorge Angel Ascacio-Martínez and Hugo Alberto Barrera-Saldaña (2012). Genetic Engineering and Biotechnology of Growth Hormones, Genetic Engineering - Basics, New Applications and Responsibilities, Prof. Hugo A. Barrera-Saldaña (Ed.), ISBN: 978-953-307-790-1, InTech, DOI: 10.5772/38978.

## CHAPTER 3

Kei Adachi and Hiroyuki Nakai (2011). The Role of DNA Repair Pathways in Adeno-Associated Virus Infection and Viral Genome Replication/Recombination / Integration, DNA Repair and Human Health, Dr. Sonya Vengrova (Ed.), ISBN: 978-953-307-612-6, InTech, DOI: 10.5772/24265.

## CHAPTER 4

Carolina Quayle, Carlos Frederico Martins Menck and Keronninn Moreno Lima-Bessa (2011). Recombinant Viral Vectors for Investigating DNA Damage Responses and Gene Therapy of Xeroderma Pigmentosum, DNA Repair and Human Health, Dr. Sonya Vengrova (Ed.), ISBN: 978-953-307-612-6, InTech, DOI: 10.5772/22946.

# CHAPTER 5

Bhupal Ban and Diane A. Blake (2012). Recombinant Antibodies and Non-Antibody Scaffolds for Immunoassays, Advances in Immunoassay Technology, Dr. Norman H. L. Chiu (Ed.), ISBN: 978-953-51-0440-7, InTech, DOI: 10.5772/35983.

# CHAPTER 6

Shi Z, Wedd AG, Gras SL (2013) Parallel In Vivo DNA Assembly by Recombination: Experimental Demonstration and Theoretical Approaches. PLoS ONE 8(2): e56854. doi:10.1371/journal.pone.0056854.

# CHAPTER 7

Lee C, Hong S, Lee MH, Koo H-S (2015) A PHF8 Homolog in C. elegansPromotes DNA Repair via Homologous Recombination. PLoS ONE 10(4): e0123865. doi:10.1371/journal.pone.0123865.

# CHAPTER 8

Mathiasen L, Bruckmann C, Pasqualato S, Blasi F (2015) Purification and Characterization of a DNA-Binding Recombinant PREP1:PBX1 Complex. PLoS ONE 10(4): e0125789. doi:10.1371/journal.pone.0125789.

# CHAPTER 9

Brown AD, Sager BW, Gorthi A, Tonapi SS, Brown EJ, Bishop AJR (2014) ATR Suppresses Endogenous DNA Damage and Allows Completion of Homologous Recombination Repair. PLoS ONE 9(3): e91222. doi:10.1371/journal.pone.0091222.

# CHAPTER 10

Ironside JE (2013) Diversity and Recombination of Dispersed Ribosomal DNA and Protein Coding Genes in Microsporidia. PLoS ONE 8(2): e55878. doi:10.1371/journal.pone.0055878.

# INDEX